TRIADIC AVENUES OF INDIA'S CULTURAL PROSPECTS

TRIADIC AVENUES OF INDIA'S CULTURAL PROSPECTS

PHILOSOPHY, PHYSICS AND POLITICS

ANIL KUMAR SARKAR
M.A., Ph.D., D. Litt.
California State University, Hayward

South Asian Publishers * New Delhi

SOUTH ASIAN PUBLISHERS PVT. LTD.
50 Sidharth Enclave, P.O. Jangpura, New Delhi 110014

Distributed in the United States by

SARKAR PUBLICATIONS & INTERNATIONAL
RESEARCH INSTITUTE
818 Webster Street, Hayward, California 94544, USA

ISBN 81–7003–177–X

Published by South Asian Publishers Pvt. Ltd.
50 Sidharth Enclave, P.O. Jangpura, New Delhi 110014
and printed at Rangmahal Daryaganj
New Delhi 110002. Printed in India.

DEDICATED TO

Those Great Personalities who Brought in India New Democracy Non-violently Eliminating the British Rule of India, They Placed the Future of India to a Novel Process with a Radical Change Where Philosophy No More Soars in the Domain of an Abstract Ultimate – Physics no more Depends on a Restricted Final Theory – and Politics No More Loses Itself in Lifeless Ideology.

Contents

Preface

Now that India is independent it has to examine its cultural prospects from an intercultural perspective, so that philosophy will no more be looked upon as restricted to its discourses – subjective or objective – it has to transcend the restricted spheres of some bygone deliberations resting on logic, epistemology or psychology. One has to internationalize one's perspective by a progressive common sense and insight, to touch the transcendentally advanced currents of not merely philosophical or religious deliberations in relation and contrast.

This will generate a common interest towards a path of peace in which all countries or persons will be commonly interested.

Similarly, while specializing in various scientific thinking from physics et al., the aim should be to step into a similar transcendent order of detached thinking in an inter-scientific enterprise in which whole world will be interested. That subject is fostered in relation to the different scientific spheres – to master the technologies in different areas, for the world today is oriented towards a crossing of boundaries from one country to another, as also from one to another subject – macrocosmic to the microcosmic, or from the terrestrial to the galactic.

Hence one is for a progressive stretching of one's sphere of knowledge to outer spaces, or to the sphere of the unknown as detached from the already known universe. It is a march from the visible to the spheres beyond it. Man wants to gain progressive powers through science and technology, of a new kind, to stand on par with the advanced nations of the world, for, without progressive developments of many new types of hybrid advanced sciences and technologies – a country cannot be self-sufficient, nor can it be able to maintain the standard of its defence against any foreign invasion or aggression.

If again, a country's interest lies in progressive business prospects, one should be deeply awakened to some new ways of international trade. In the sphere of politics, too, one has to frame its own constitutions through a democratic frame of reference, taking necessary advice from specialists in this area of international politics, as all experiential processes are interrelated and cannot be exclusively cultivated as isolated from all other experiential fields or countries. Humanity finally is operative as participants of an 'United Nations', no longer subject to one supreme 'superpower' but to a number of them, in relational pacts, bound by international laws. Though experience is primarily a process for a country's internal welfare and prosperity, it has to maintain its self-existence in international contexts, also, with some sense of international morality or adherence to some international laws.

The 'Preamble' following the preface of this work, therefore, goes at some lengths with some ideas in international and intercultural contexts, and tries to free itself from the trammels of mere abstract thinking as prevailed in the earlier periods of one's cultural history. Every self-reflective person or country has to develop a thinking freed from abstract ideas, transcending the past cultural processes of other countries, and developing a scheme of deliberation pertaining to other countries. The day is no longer for mere exclusive theorizing – from philosophy and sciences, to politics. It steps into a new emergent history.

All these different types of broad-based interrelated disciplines have been clarified in progressive details of this new fabric, through its chapter II, and through its later parts by a reference to some select and intensive studies. This book suggests all-through that we no longer live in the world of absolutes, but are confronting every moment a world of infinite probabilities. The confronted world as presented, is not any more a stable phenomenon, but a possible oscillating symbol – something emergent – a 'may be' or with a consequent 'may not be'.

February 1995
Hayward, California

ANIL KUMAR SARKAR

Preamble

After the completion of my four volumes of *Dynamic Facets of Indian Thought*, taking note of the specialists of Indian thought, both Indian and non-Indian – I indicated my impartial attitude towards a review of Indian thought, from the Vedic period to the present-day *movements*.[1]

Now my pathway is beyond the *movements*, which were operative socio-politically before Indian independence. After independence, India has its independent government to be operative in international situation.

With this as the objective of the Indian minds – the Indian Government will change progressively. In an aspect it will be progressively developing a contrasting outlook, retrospectively examining what was during the British rule of India, restricted in many directions by foreign rule, and now prospectively what is the future possibility of India, given the fruits of independence.

The Indian Government will progressively devise possible improved governmental mechanisms with innovative plans. There is indeed a great possibility for the future of India.

All-through the development of Indian thought, from period to period, in continuity and development – I pointed out in my writings that Indian thought, with an insightful modification, had been found progressively effective in facing the growing complexity of the different periods, within the country, or in relation to other immigrant or invading countries.

The problems all-through were broadly cultural, for, the Vedic thought from its inception, was not a unitary linear process, only focussing on to its own complex modes of many-sided developments, but it was a double-consciousness: in its transcendent aspect, it confronted the cosmic and social processes, viewing the presented consciousness by a transcending consciousness, and at the same time it faced other cultural processes – both transcendent and non-transcendent – in varied senses, present as other dominant currents, prevalent within India.

The reference is here to Jainism, fostering an overall transcendental attitude, from a different perspective as an alternative process, coexisting with the visions of the seers of the Vedic sages.

The Jaina *tirthankaras* or the liberate personalities – claim to have transcended the entangling passions that subject one to the presented panoramic universe, with many sided confused experiences, and they also have a challenging interpretation of the confronted universe – running parallel with the Vedic visions of the confronted universe, presupposing a transcendent background of experience controlling the confronted.

Unlike the Vedic seers – the Jaina Tirthankaras – had a distinct view of the emergence of the confronted universe, and also a distinct view of the roles of the liberate personalities, as helping the ignorant masses towards an elevated liberate condition. To the liberate Jaina the cosmic process is dual – 'upward' and 'downward.' The role of the Jaina liberate persons (*tirthankaras*) – is to emerge in critical or crisis periods to redirect the universe to the 'upward process,' resolving the problems of a 'crisis situation.'

Both were operative in different 'symbolical meditative directions' – none of them were 'intellectualizing' in 'abstractive repetitive senses.' Both enjoyed 'freedom' in a post-intellectual sphere.

Besides these two dominant modes – there was another great civilization, located to the North-West India, facing progressively the vast masses of the original Indian inhabitants, and tribal peoples of the different pockets of India – stretching to the East and also to the South – of Indian subcontinent. But the civilization, it referred to was an established form of civilized processes – a remnant of the civilized processes of Mesopotamia, Babylonia (modern Iraq) and Egypt and of other adjacent places, which was stretched to India. These stretches were along the river Indus from the North-West to South-West, flowing into the Arabian sea.

It was a civilization of progressive possibilities of a city-state with grandeur and dignity, but in a process of decay or destruction.

Then possibly there was a descent of that civilization perhaps to some pockets of the Indian subcontinent, as evident from the progressive archaeological findings of the present century.

From the beginning, it must be pointed out, that all these civilized processes, in different locations of undivided India were many-sided modes of *contrasting* deliberative processes – they had a keen sense of the *presence* of the *confronted universe*, as 'objectively and socially present,' with the towering mountains and flowing rivers, subject to elemental forces, and also dependent on surrounding other countries bordering and confronting them. But they were in a kind of *perishing occasion.*

They cherished distinct views of cosmology, from their respective philosophic attitudes, but mostly restricted to Indian geography and culture.

If any one of the present century thinkers, like a French thinker Bergson, tries to interpret the concrete confronted sensuous-intellectual universe – as a vital vibrating process – he will be sure to be critical about the past abstract modes of western thought from Plato to Hegel – a retrospective patchwork – a post-mortem examination of facts. None of the structures exist for ever, even when constructed with a *vision*.

While Bergsons's 'vision' of the prospective universe was of a living vibrating confrontation towards a 'post-intellectual' *intuitive* possibility – and a *vital urge* – it was not just an 'intuitional claim' of Husserl, in post-Hegelian thought. It was not – of a confrontation from a transcendent de-conceptualized *abstraction* – one's ego, and the objective universe of objects. Bergson's vision was of a living possibility, by a *contrast.*

Bergson's position was not also a claim of Kierkegardian 'personality' in 'existence,' prior to one's conceptual-empirical process – reversing Descartes' position of 'I think, therefore I exist.'

In contrast to these two 'abstract modes' of early western thought Bergson's position was of a confrontation of a living universe – with a post-intellectual intuitional possibility – understood from a concrete biological possibility.

Bergson's was a reinterpretation of 'intellect' from a biological perspective, evolving into some form of transcendent order of new experiential processes.

But this position of Bergson is typically subjective, even when the basis of it was a claim from a biological basis. It totally ignored the confronted objective-social universe, and stayed as an 'argument merely.'

The afore-mentioned cultural processes of Vedic India, in development – developed its dual consciousness of the transcendent as a continuous self-shining psychic process and possibility – enlightening the confronted psychic-cosmic objective universe of 'differences' – with no 'absolute split.' The Vedic or the Jainist vision of the confronted – point to both 'ends' – with a 'luminosity' of 'infinity,' without restricting it to a 'thought' or intuition.' The Vedic-Jainist process was replete with the floodlights of symbolic process and possibility.

The presented universe as experienced by a living psyche – both 'subjectively' and 'objectively' – were/are confronted 'processes and possibilities.' They, as presented were/are 'reversible' orders-placed with the 'transcendent milieu' – with an ever-fresh possibility, for, the Indian mind-was/is very alert to maintain its overall transcendent attitude with progressive modification through its depth-insightful processes.

It is for ever also very alert in scrutinizing the characters of the confronted situations. Hence, to an insightful Indian – there will be always a 'relativity' between the transcendent and the confronted.

Further, if any reader has patiently followed the general trends of Indian thought, as developed by me through its different volumes, and also by my other books on Indian philosophy he will find that the later post-Vedic Upanişadic thought was challenged by a developing process of Buddhist thought. I have developed these processes by a 'contrast' through the centuries, with a title, *Changing Phases of Buddhist Thought*, in several improved and advanced editions.[2]

I also have developed Indian thought from century to century to the present times, in relation to the advanced currents of Western thought, with some stretches to advanced currents of American process-philosophers, who claim some specialized knowledge of the advanced currents of Buddhist thought.

This again, I have indicated by my recently published book, with the title, *Buddhism and Whitehead's Process Philosophy*.[3] If one pursues this book one will, surely, hesitate to persist with a long-refuted notion that the Indian thought, is after all 'subjective.' Indian thought from the beginning as analysed never stood for an 'abstract analysis' but it was entertained always as a concrete phenomenon confronting a man, and is a posterior process, after one is sensibly aware through different sense-organs that there is an 'objective universe' co-present with the individual organism, having a 'thought' of a sort.

Hence a person not having this background of Indian thought, will wrongly separate the 'subjective' aspect of experiential aspect from the objective process, on which the 'subjective ideas' are based.

His will be a failure to understand India's realization of the importance of the objective universe and its changing geography and a spread of culture.

The notion of the subjective or the objective has to be viewed always, in conformity with the distinct cultural process, that governs in one's thinking, and in togetherness. The 'subjective' and the 'objective' have a common prospect, they are constantly replaced by a posterior transcendence. One should always note that the Indian thought was never in the stage of mere subjective or objective bifurcatory process, it advocated always a continuous integral experiential process from a two dimensional perspective: transcendental and confronted, considering the total experiential process in continuity, including the subjective and the objective, and also the transcendent processes beyond these orders to a fresh order with similar or yet not similar possibilities. In that transcendent order, there are distinct subject matters for study, but never ignoring the over all transcendent cultural process, preceding or following any existing order.

But, in this somewhat monotonous process, one should not think that India was ever exclusively 'subjective,' ignoring the 'objective.' For the same reason, it is never 'objective' exclusively.

Here I am drawing attention to a transcendent process, which is beyond any specific 'subjective' or 'objective' process. India's different alternative movements of the nineteenth and twentieth centuries were definitely not merely 'subjective' or 'objective' but transcending both.[4] This is demonstrated by India's achievement of independence from the undoubtedly the then superior powers of the world by the comprehensive realistic steps that India somehow could devise to adopt. This was a success, and not a imaginary affair.

In the sphere of experiential process, or in the understanding of the different geographical areas, India had no separative spirit.

Like the Vedic and the Jainist transcendental symbolic process the posterior post-Vedic and post-Jainist thought was followed by Upanisadic-Buddhist process-in a different order of symbolic process moving towards a posterior process of Vedantic movements in association with Islamic mystic and symbolic processes-of post-intellectual yogic and trantric processes.

Avoiding the personalist statements, I will humbly suggest, that the Indian thought in cherishing its distinct transcendent attitudes, admits both the confronted subjective and the objective orders, their continuous relationships and transcendence with novel prospects. It was not operative in an abstract aspect. It was a concrete project or confrontation of objective situations, in relation to varieties of other experiential processes. The Indian mind does not give importance to either, in their exclusive 'abstract contexts,' for, they are transient, perishable or even transforming processes of nature.

Observing the two orders of the presented universe the Indian mind, was operative through some disciplined practices or ways of life. Hence, finally the Indian mind was ready to abandon the presence of any personality as such, to deal with the possibility of an experiential process in continuity.

If one finds in this pursuit, somewhat of a *subjective* element, still lingering, then, in that stage of its operation of that process, there are a series of alternative projects of transcendent higher orders, according to one's distinct acquisitions, with progressive disciplines, found among the liberate persons from the Vedic to the Upanisadic periods and through the Buddhist and Vedantic periods to the present century to the 'Movements' as referred to.

This discipline has been progressively improved upon. Jainism has a different cultural background and hence it has a different interpretation of the confronted universe, with different cosmological attitudes, but like the Vedic attitude, it views that the experiential situation is two-dimensional—transcendent and confronted.

Both these cultural processes have to be distinguished from the Indus Valley Harappan and Mahenjodaro civilization, which is a remnant of city civilization, found in a ruined condition, hence it is very difficult for one to trace back to its original presuppositions. It might serve as only a contrast in relation to the parallel Vedic and Jainist civilizations. The Vedic civilization thrives in a forest environment in contrast to city life and its constructive socio-cultural process of the Indus Valley civilization, but inspite of its concrete expressions of an advanced city life and civilization, it was not without some mystic elements as found in archaeological discoveries and of the Saiva cult demonstrating Man-Woman unity.

The Jaina mode of life and thought, fosters a final ascetic goal, but it is also a prescription of a social life by way of concession in different periods of reigning kings of India and their subjects. So the Jaina movements throughout the centuries were a concrete tackling of the crisis situations, in different periods of ruling kings and in the present century both Gandhi and Sri Aurobindo, two different types of political leaders – though both Hindus – were influenced by Jainism, in two different ways, each advocating some aspects of non-violent principles. But both these leaders of Indian 'movements' were interested in a 'liberate form of life' in contrast to ordinary social life, where the two processes were socially interlinked, one respecting the other.

Be it noted, that the Indian thought from the Vedic period itself is 'symbolic' and 'hypothetical' and not a categorical or theoretical assertion, but undoubtedly concerned with the shaping of life in a concrete liveable way. Hence their attitudes are tolerant, fostering a process of coexistence and mutual understanding.

When this current was further improved upon by the advanced Buddhist touches of the final character of the transcendent process, as a complete 'Void', not that it is 'Nothing', but that it only flouted any intellectual or empirical formalist or abstract determination, in a restrictive sense, as developed in their Monastic forms of life. It comes as a challenge to Western thought, which is restrictively operative through its conceptual and empirical modes, stretching to its progressive transcendental phenomenological or existentialist modes.

At this stage, the western mind, discovered that they have wasted their different historical periods with conceptual-empirical disputes and ended up with some form of 'dogmatism.'

This end-process of the Western world led them to innumerable wars, ending up with two World Wars and with a future possibility of *Star Wars* or *Nuclear Wars*, because the western mind cannot transcend its 'intellectual forms' or 'abstractions' or 'ideologies.'

Observing all these anti-humanist developments the Buddhist thinkers, like other Indian insightful thinkers are conscious and also tolerant of the western power aspirants as they are busy in controlling other powers by diplomacy and threats.

The Indian mind as a whole, knows that none of them are completely free from some intellectualist or empiricist proneness, or 'ideological' dispositions and conflicts. So these restrictive modes, or attitudes are not powerful enough, but they sow the seed of their own destruction by conflict, tension and war.

But the Indian thinkers from the Vedas are aware of the fact that the confronted 'power' that a passionate man exercises is nothing but an attachment to an 'abstract intellectual form' which is already dead and hence is 'powerless.' Hence those who remain confined to some kind of reflecting through abstract intellectual forms, are never free from their self-created 'intellectual castles.'

The Western World is trying its best to break through all cultural barriers, and so-called geographical distinctions and progressively darting towards a universe of possibilities. But its orientation is now imbedded in ideological proneness. They have to break so many 'German Walls' between several countries and cultures.

By a contrast the Vedic sages, as Swami Prabhavananda pointed out, fostered the insightful notion of a confronted universe, as apparently evolving, as a possible presentation, but transcendentally based, hence has to be understood symbolically, as a possibility. The confronted world-process, as accepted is not with a categorical assertion of some restricted sort. Hence the Vedic sages stand for a beginningless universe, evolving and manifesting on a transcendent foundation, which in its core is a habitat of infinite possibilities. It is cyclic and meditative. The reference is to the first volume of *Dynamic Facets . . .*

In this volume, in different symbolical senses, the consequent processes of Vedic cultural processes were reshaped in the Upanisadic and later Indian thought of Bhagavatgita and subsequent auxiliary processes. They have been developed in sequence in that volume, as vast spheres of confronted possibilities, understood as concrete emergences. They could be transcended not merely 'subjectively' but 'objectively' as well. None of the assertions are categorical, they are all hypothetical and symbolical and not restrictively 'subjective' or 'objective', in the Western conceptual or empirical or any other senses. They are dual, but finally transcendentally oriented, in both ways, hence they are not 'subjective,' as such, they are progressively disciplined and transforming, 'objective' or 'super-objective possibilities,' modifying or replacing wholly the 'subjective' and 'objective processes' of any restrictive sort.

This philosophic attitude as an over all evaluating process of confronted possibilities, is different from the varied confronted processes, if categorically understood as dealing with concrete matters of interests. The Jaina transcendental-confronted mode also is only tentatively admitting the possible confronted universe and the mind in continuous modification or transformation to an end-process towards an integral transcendent vision.

Needless to mention, that the second volume of the same work, develops the Jainist process as also the Vedic process, stretched to the Buddhist process and other processes, as progressive alternative spheres of confronted possibilities, in a fully concrete sense – a relatively dominant current in a post-Upanisadic and post-Bhagavadgita process, as an advanced process, in India and in the Asian countries in the same symbolical concrete ways.

The contention of the third volume is post-Buddhist Vedantic process, in an ideal sense, but a concrete revolutionary process and activity, as a revival of the past in progressive symbolically oriented pathways, transcendentally based, giving a new advanced interpretation of the earlier activity processes of the Upanisads or the Bhagavatgita, assimilating or refuting some of the contentions of Buddhist thought in general, but nor totally ignoring it, as the impact of Buddhist current remains always sustaining the core of Indian thought continuously, governing both philosophic and social life.

The culminating process is of the 'Movements' of various alternative sorts, of the present century, after the British impact of the West in all areas of experiential process through its English method of educational system, heading towards a novel political phase of parliamentary form of Government of a new variety after the replacement of the British rule of India, through a process of transfer of rule as indicated in the fourth volume of this work. In this connection one should take note of Patel's political activity process, which by a single stroke dispersed the several rules of the Maharajas of India.

The fourth volume makes it clear that the over-all philosophic attitude is important, but with it, it has to pay due attention to the varied confronted subjects of varied interests, specially the scientific and the technological and their principles et al., so that by a depth study of those areas a further improvement of the philosophic consciousness can be effectively realized in concrete objective directions, besides adhering to the over-all cultural prestigious basis. But everywhere the story is one of the extensions of the fields of alternative 'possibilities,' with patience and tolerance, as clearly indicated by K.C. Bhattacharyya in his very insightful deliberations, suggesting that the finally transcendent possible modes, are not touched by one's 'thought,' they have their existence, sue generis, in the order of 'freedom,' as nonrestricted 'subjective' or 'objective' processes or possibilities. India always prepares a ground for tolerance and understanding of mutual contributions – give and take.

Hence the post-independence period of India today is no longer merely cultural, in a restrictive sense, it is so in 'supra-subjective' and 'supra-objective' sense. A superb process of 'give and take' and 'let live.' Indian thought – at this period of its history – is operative with a new nationalist version confronting India's social and economic needs, in the contexts of international political and scientific situations, of the world as a whole, beyond all geographical barriers, of India, with social changes, changing the shape of India and the world, as a whole, in which India has an important role to play. There is no gain in wars, is its basic slogan from the days of Emperor Asoka.

This is the 'felt need' for this new volume with the title, *Three Avenues of India's Cultural Prospects – Philosophy, Physics and Politics*. I feel a new spirit to articulate this process, to generate a new spirit to the young minds of this century, which are generally swayed by the general currents of the world, without looking back to one's own heritage, which can be utilized from a new perspective absorbing all best of the foreign currents, from an intercultural perspective.

Now India has to face both of the rapidly changing processes of cultural and social changes within the country, with a broad defensive attitude for maintaining in fact Indian independence and not alone its culture, but also of its scientific and technological efficiencies, playing its independent role and influencing the world at large with its uniqueness, 'tolerance,' 'patience' and 'coexistence' and with its superior orders of scientific and technological researches.

The pre-independent India always elevated its transcendental cultural background and assets and now there is a reasonable departure and progress beyond abstract adherence to a traditional culture, in many-sided scientific and technological, environmental aspects, crossing geographical boundaries. The initial plan in the defence of India needs a rapid programme for updating its nuclear assets for peaceful purposes, but cleaning the cities of India, as any clean city of the United States or any of the Asian countries, as Singapore or Japan, by educational changes of Indian minds and Governmental enforcements.

These steps will indicate India's choice not only for environmental cleanliness, but also some expression of general peoples' developing aesthetic preferences.

It is a suggestion for change of one's tastes and choices or preferences. But India is taking time, almost to the impatience of a sober personality, who has become habituated to stay in a foreign country, for some palpable gain or genuine 'self-improvement' availing of the opportunities, but also choosing sufferings, as deliberate decisions.

It should be mentioned that India throughout its history, not only prescribed an ultimate aim or objective of life, as based on 'cleanliness,' in both confronted and transcendent spheres. India was all-through aware of day to day socio-political life, of mankind in general, with progressive interests in the pursuits to efficient socio-cultural process, with a progressive scientific and technological perfection, for modernizing India.

Not only a cleanliness of the temples, or the Ashramas et al. are primary requisites, following modern sanitary systems with maintenance provisions through well-paid officers of custodian services but also by launching 'educational curricula' from scientific perspectives to present India in a decent form.

India is also known to the whole world for its subtle artistic taste in art, archaeology and music, besides its varieties of literary expression from the myths to the intricate sciences –mathematical and non-mathematical.

This at once will change the apparent look of poverty of India, to a new look of prosperity.

Every Indian should learn to feel that it is no longer a dependent country, it is only dependent upon one's own self-perfection helping the process of social, cultural and scientific-technological prospects and possibilities towards a fair and improved dynamic politics, which will be clarified in the final part of this work.

It should be made really an exhibit for the whole world beginning with the improved facilities for the tourists. The obstinate adherence to one's confused and clumsy culture, or habits is no good. Every Indian has to learn to face life, as it is, from a different perspective, taking poverty or disease as a concrete human social phenomenon, in all parts of the world, including the United States, but its apparent look has to be changed.

This can be solved by concrete methods of planning and policies, from one place to another, by changing one's habit of life, by judicious preference 'sacrificing' voracious eating, by 'refined ways of eating,' where science, engineering and technologies in interdependence have important roles to play.

But now is the time for giving every aspect of experiential process, not only equal but sometimes prior or posterior attention, for improving the status and life-processes in those complicated areas, where public opinion has an increasing value.

The present day international community situation is not merely competitive, but it is a process towards a progressive claim towards superiority or self-improvement, by critical judgment through learning by travelling, both within one's own country, and choosing to travel foreign countries according to one's means and judicious savings, now that the world has become shorter than before.

Whatever complexity, there might have been within India in the past, in an aspect, was comparatively simple, in contrast to the present day confronted universe of all emergent countries from their respective pasts, which also like India had a past cultural history with a complexity, rising progressively to the stage of developing countries, with scientific and technological efficiency, but one has to construe them as comparatively 'simple,' in contrast to the complexities that are emerging in the present world of 'emerging nations' or competitive 'military efficiency' or 'superiority.' Here my attention is towards a definitely improved mode of 'human psychology' of progressive disciplines, by slow steps or practice.

This writer, finally indicates that this complex international world-situation-with growing complexities, supplementing it not merely by aspiration as by controlled habits of disciplines. He also demonstrates simultaneously a multifaceted mode of life and existence, as a broad symbolical process, philosophically, psychologically, mathematically and finally scientifically and technologically, through post-Einsteinian physics and astrophysics, proceeding towards many-sided aesthetic processes, through the concrete researches, as suggested by eminent Indian scientists, among others of the present century, preparing a ground for 'peaceful coexistence.'

India's three great scientists to be mentioned are: C.V. Raman, S. Chandrasekhar and George Sudarshan. The first and the second have already become nobel laureates and third one is in the way of getting it, for his discovery of 'Tachyon,' as having a speed more than light.

If Raman during his lifetime had a worldwide reputation, he also influenced many scientists of his time and later scientists of India and other countries, by Raman's new interpretation of 'radiation' theory in 'non-linear' direction to 'broaden aesthetic processes,' was not possible through British Rayleigh's 'linear scattering radiation process.'

This scientific achievement is also an achievement in an 'aesthetic aspect,' which is scientifically approved. So a basis for a new revolutionary attitude can be formulated wherein both the scientist and an efficient aesthete can be operative together joining hands with a physicist or an astrophysicist, to open up new avenues for progressive scientifically oriented cultural processes.

Raman's discovery of non-linear many-dimensional scattering ray's known as 'Raman's Rays' was a revolutionary scientific discovery. His 'inelastic scattering process, in contrast to Rayleigh's 'elastic scattering' has given a new avenue for energetic component with the atomic and photonic process to penetrate into deepest recesses, facilitating surgical operations through 'laser rays' in multifarious ways with accuracy and to the focal points.[5]

The present day scientists – Indians and non-Indians – have become so alert and keen in the appreciation and understanding of the progressive aspects of research in the direction of laser' theory of light, they apply the laser rays to medical science for surgical operations, through computer science and technology, in all spheres of experiential processes with different skills.

The computer science can stretch itself to 'surgery,' mathematics, physics and astrophysics. This undoubtedly signalized a possible great scientific future for India and for the world as a whole.

So, if any one still persists in holding the view that the Indian thought is after all 'subjective' and negligent of objective or sociological and scientific processes is wrong.

It is an attachment to a restrictive theorizing tendency, a kind of deliberate 'ignorance' still persisting, in some non-Indian countries.

This writer's, book, with the title, *Zero – Its Role and Prospects in Indian Thought* – is developed with a further suggestion, by an expression: 'Its Impact on Post-Einsteinian Astrophysics.' It is post-theorist's stand in the real spirit of 'freedom' from restricted 'ideology.'[6]

The advance of knowledge through this symbolic way from the perspective of zero will spell progress in all fields of experiential processes for which India's credit will remain for all times to come.

This extended process is addressed to this trend to indicate that India was alert to scientific and technological modes of deliberation all-through the centuries, through a progressive mathematical symbolism, crossing all confronted experiential boundaries, to the transcendent heights of progressive 'meditative possibilities.'

India's technological history can be found in ship-building from the mythical history of the great merchant like 'Chand-sadagar' to Narottam Morarjee of the present century. In order to preserve India's past history, Morarjee refused to sell to the British, his 'Scindia Steam Navigation Company Ltd.', Bombay, inspite of a tempting offer.[7]

Everywhere there is a reference to a transcendent philosophic basis. This aspect of India's cultural, scientific and technological history with a farsight will revolutionize the world, from another perspective.

After Indian independence this process could be judiciously extended and developed according to country's economic solvancy.

That one should not be attached to any kind of restrictive theorizing tendency, in any area of experiential process, neglecting other aspects. This is a new message of India, to other countries of the world, all-through its history and is well indicated in this writer's another published book, with the title, *Experience in Change and Prospect Pathways from War to Peace.* Here the lesson is given that the theories have only a relative place, in cultural history of mankind, but they are not 'ultimates.'[8]

India is operative through its transcendent mathematical symbolism, de-theorizing all processes. These are indicated by my above two processes. They are again associated with my third work.

This writer's third important work is published, with the title, *The Mysteries of Vajrayana Buddhism from Atisha to Dalai Lama.*[9]

This book indicates its further advanced thinking as crossing the boundaries of this life by flouting that 'death' is a terminating process. This is an attempt to 'reach out' beyond this 'life-process' intercepted by a confronted 'death-process.' The attempt of the 'liberate persons' of India's different exponents of 'transcendent orders' of experiential process is to probe beyond the barrier of 'death-process.'

This is not a universal truth, but at least true of some disciplined or liberate persons, who do not lose their 'continuity of experiential process,' for, they are not subject to their confronted restricted knowledge of reality only. They have acquired the skill of going beyond the confronted 'subjective and objective processes' of 'death.' It is learning to control the vital and mental processes, and set them to a transcendent level of 'continuity.'

It is difficult to cross the boundaries of this life, also, if not adequately disciplined, but if so disciplined, something can be discerned of the future life beyond death, one must pay attention to that as a new avenue for reflection. The importance of this uncertain life-process, can be stabilized further by opening up the possibility of progressive knowledge of the future birth-processes, in progressive disciplined aspects and as well stretching the theoretical interests to the practical and the aesthetic spheres, as an experiential continuum.

But this process is a disciplined transcendental psychological project, and not either restricted empirical psychology or some form of spiritualism or

Shamanism.

Thus one will find, in the present century of complexity, a clue to 'one's inner life.' It has to be examined with a complexity, as of the outer confronted processes. A human individual, therefore, is facing a world of complexity within and without – he has to live helplessly in this vast universe, suffering in each one's particular ways, both within and outside.

But there is a prospective civilized perspective which mankind can share if by a disciplined life, one can discover the possibility of crossing the boundaries between lives, making the present life as a basis or platform for experiment and research. India is stepping to such possibilities. My researches are directed to this direction.

If the scientists raise the question of any 'scientific verification' one must be competent to challenge the scientists by demonstrating to them that some Vajrayana specialists, both Indian and Tibetan, could demonstrate their powers of controlling the 'elements' of nature as to transform the 'confronted nature' to a new 'middle' situation between this life, and life after death, which is an intermediate process, a 'new middle', to be projected to a fresh birth-cycle.

This is an excellence which one has to cultivate as a scientist cultivates through technology like the processes of 'ship-building,' 'aeroplanes,' 'Satellites,' or 'Voyagers.'

Here in this work, it has been shown that the Indian and Asian thinkers in general, who were influenced by India's Buddhist thought in development, never subjected themselves to restrictive subjective or objective processes, confining oneself to this present life process alone.

They wanted as well to probe into the mysterics of the interrelated life-process linking the present with future possible lives. They were aware of them and their presence, with a development, but not in their ultimacy. They thought of vital-mental controlling of those processes through Yogic-Tantric disciplines of the body, breath and the mental processes through progressive 'concentration' and transcendent 'insightful processes.'

The varied many dimensional experiential processes were understood insightfully and hypothetically with an 'If..then' process, symbolically, not merely logically, but mathematically and scientifically and insightfully, without any restrictive qualification.

Hence, this writer, after a broad survey of the development of Indian thought, tries to suggest that when the Indian thought is a disciplined insightful process, it is not a superficial subjective or objective process, conforming to a restrictive 'ism' hence vulnerable to any criticism.

It formulates an advanced view of philosophical perspective with a depth-psychology and discipline, so that it runs quite competently with the advanced processes of modern developments of scientific thinking, specially in the aspects of physics, astrophysics and cosmology, no longer fostering any dogmatic or restrictive attitude, alone.

The interest of Indian thinkers, was always with a universalist spirit from the Vedic period. This attitude of maintaining a continuous possibility of experiential process has been intensified further, with the increasing consciousness of one's own country in relation to all other countries from an intercultural perspective.

The main concern of this present volume will be examined only from this broad perspective as far as possible without running into any superstitious ideas or processes. It is a book of progressive development of 'ideas' in 'contrast' and so possible processes are not excluded.

It is heading towards a new kind of transcendent disciplined psychology, not realized so far, hence mankind in general, was troubled by the growing processes of criminal behaviours. A man posing that he or she, being abnormal has no control of his/her instinctive dispositions, adoring Freud as ultimate. This wrong psychology is a phase of one's egoistic tendency. It has a place and persistence in the cultural history of mankind, but it can be gradually removed with considerable patience and discipline.

The writer is deeply aware of the results of the First and Second Wars, and various emergence of abnormal psychologies and criminal behavioural processes, followed by the long cold war period, and the rivalry of the superpowers, with trials and errors of the dominance of the one or the other hemispheres of the world, and creating the problems further by participating in the Vietnam war, with changes in the European situation, and bringing in the varied problems of Middle East, now after so many other events, the Middle East is presently heading towards a very complicated international situation in the Gulf area by Saddam Hussein's invasion of Kuwait, resulting in the Gulf Crisis, affecting the entire South Asian countries.

All nations of the world are interested in the availability of the oil in the Middle East. While with the new political move of Russia of a reduction of arms, and demolishing of the German wall between East and West Germany, and helping in the process of their unification – a new peace-movement has been set in process, by the unique activity of Gorbachev.

His foresight of a new socio-political change through a scheme 'perestroika,' is an innovative suggestion, and now with a credit of a nobel prize he is trying his best to bring down some form of peace in international situation.

But there are complexities of world-situation, after the Gulf War, and there is no hope for solution of all problems so easily and within a short time. The possible peace in the Middle East, is now threatened with a possibility of the third world war.

This writer, therefore, has given the full importance to Dale Riepe's very valuable and enlightening work, with the title, *Indian Philosophy Since Independence*,[10] but not his contention that Indian thought is 'subjectivist,' giving no importance to objective process.

India's move from its inception is culturally oriented in the sense of a transcendent as operative with the confronted situation, as both 'subjective' and 'objective', with a possibility of 'transcending' both, insightfully finding out their restrictiveness, if any.

An insightful Indian is still defining his philosophic ideas, to some specific concepts, as characteristic of some abstract thinkers, of the West.

This general trend is commonly present in India, so this writer suggests, that there is a way out by a different sort of 'disciplined training'; there is a possibility of treating any confronted situation, 'subjective' or 'objective' insightfully with an inter-cultural outlook in supra-subjective and supra-objective sphere and possibility.

This writer, with all his experience of the past history of India, as already referred to stressing its dominant cultural stress specially Buddhist and Vedantist, with a continuity, now posits that there is a march beyond such exclusive cultural stress. It is not losing its past but it is an admission of some other improved ways through insightful 'scientific developments' – taking hints from Whitehead's symbolic mathematical thinking, and philosophic pro-Buddhist thinking, crossing the boundaries of 'restrictive intellectual thinking,' by a transcendent march towards post-intellectual – 'wisdom mode' process and possibilities towards 'Creativity' going beyond all forms to the formless continuity.[11]

If Dale Riepe, chooses to modify his self-oriented contentions, he will get sufficient materials to understand that Indian thought is in no sense 'only subjectivist.' An Indian is also prepared for transcending the restricted 'objective outlook.'

India is for a 'progressive world-process, transcending both confronted 'subjective' and 'objective' processes, in a particular situation. Its effort also is in a further direction as pointed out already: it is towards even transcending the confronted 'death-experience' – whether any disciplined person has access into the next life to continue his 'unfinished activity-process' as claimed by the adepts of Vajrayana Buddhism.

This transcendence is possible from a transcendent disciplined perspective. But an exercise in this difficult aspect might be the interest for a group of such persons or may be of a single person.

In India such a pursuit is found among the Buddhists, Jains, Vedantists, or the Yogins and Tantrics, in different major religions of India, including Islamic Sufism, as variously expressed from Kabir, Nanak, Dadu et al., with a further movement towards Baul and Nathist movements in Bengal, or Swaminarayan movements in Gujarat. These processes are still continuous with intercultural effects – from Africa and Europe to the Americas.[12]

But 'transcendence' in a general philosophical or scientific area is found if a person, with some discipline can think of transcending one's specific thought, for another kind of experiential process. It is an exercise in 'freedom' from thought itself for a rise to a 'placid condition' to further orders of 'insightful processes.' This transcendence is both 'subjective' and 'objective', as both of these processes are 'confronted experiences' with a possibility of a replacement in a different order of emergence.

One has to understand the depths of these varied modes of India's progressively insightful notions, and their suggestions for a process of a final emptiness of all thought, or of the intellectualist propensity – if one minutely studies the disciplined modes as expressed from a widely comparative perspective of a very illuminating paper, entitled, "The Experience of Nothingness in Buddhism and Existentialism" by Ramkant Sinari, in *Contemporary Indian Philosophy*, Series Two, Edited by Margaret Chatterjee.[13]

This paper, if taken in relation to my mode of deliberation has its undeniable value.

In this paper, Sinari, points out referring to the transcendent background of all confronted processes—subjective or objective—that it is a uniquely 'ego-exploring attitude of philosophy, probably the most penetrating in its concept of 'Nothingness.'

This 'ego-exploring' technique or 'atmology' is to direct consciousness towards itself, to intensify self-transcending process, and to put it in a position to observe its own 'roots,' since the time of the Upanisads.

Several schools of philosophy – non-Vedic as Jainism, and Vedic as Samkhya, Yoga, Mimamsa, and the Vedanta – demonstrate varied trends in general and remain totally committed to some form of 'atmology' as a way of attaining other-worldly salvation.

But soon they find some kind of insufficiency. Some Buddhist thinkers of India, following the Buddha were quick enough to shuffle their position with a skill and insight. They discovered that in an aspect of transcendent depths, Atman fades into Nothingness, when it grants a profound silence as its background, to determine a further possibility, avoiding any restricted dogmatism – including Atman.

The Buddha and the Buddhists, in detaching themselves from all proneness towards ratiocinative tendency, and from different levels of personality-process, daringly suggest that there is 'no personality' finally—to indicate an understanding of a process towards a complete Nothingness – declaring 'No Buddha.'

Asvaghosa provided a 'potential background-statement for Nothingness,' from the 'intellectualist perspective when he practiced continuous meditative process and possibility towards, *tathata* or 'suchness', where there is no more any place for restrictive intellectual mode—it is stepping to a condition of 'bodhi' or enlightenment or 'Silence.'

This disciplined tendency in India's advanced thought is a special feature of Indian mind, offering it, a field of progressive insightful approach to face the problems of a situation, not in terms of any specific thought, but in going beyond its clutches, to a 'transcendent insightful experience,' spontaneously or suddenly' as a flash, in contrast to the continuous nourishing of 'intellectual forms.' Real meditation is an insightful merge (*Nirvana*).

If this meditative transcendence is understood from the perspectives of Western thought it is a complete separation from the intellectualist mode, from Aristotle to Hegel, and also from the post-Hegelian Phenomenological and Existential movements of Husserl and Kierkegaard and their later followers, for all of these thinkers were not completely free from some specific modes of conceptualism or empiricism – as they were all in some sense 'reactionaries' – hence ego-centric in different-senses.

But the Buddhist total detachment as stretched from Asvaghosa, Nagarjuna and others through the centuries was beyond all forms of conceptual thinking, in contrast to the then philosophical thinking of different types—by placing oneself to a transcendent liberate condition. They were all referring to a profound transcendent situation of stability as 'peace' or a 'placid condition' beyond all intellectual restlessness and restriction – A continuous swing from *Nirvana* to *Shanti*.

But this ascent to an exalted condition was not one of 'darkness' or 'blankness but it was one beyond any restricted 'personality-process' disciplines, fostering varied interests of life.

Nagarjuna, after Asvaghosa, in India, by progressive elimination of all 'intellectual forms' dialectically eliminated all intellectual choices by 'neither nor,' and placed one to a condition of transcendent 'silence or freedom' from intellectual forms or sufferings. His was a rise to a trans-intellectual process or *Prajna* or 'wisdom order' of 'insightful experience.'

In a sense, it is a foresight of a possibility. If applied to scientific sphere of experience it is beyond post-Einsteinian, special and general theory of relativity and a passage to a transcendent possibility of a vast openness or 'field' beyond the gulf between the 'particles' and 'waves' of Heisenberg, in post-Einsteinian, aspect of scientific possibility, in stretch to the present world of progressive researches.

Coming back to the philosophical sphere of post-Hegelian thought, Sartre's 'transcendent Nothingness' in phenomenological-existentialist process is not a condition of complete dissolution of intellectuality, it is only a 'vacuity' or a 'lack,' of a craving for re-establishment of some form of existentialism by a contrast.

This trans-empirical Nothingness, of Sartre, is different from the 'completely negative field' of Nagarjuna in Later Buddhist thought.

Nagarjuna's post-intellectual 'insightful pursuit' by a detachment from all 'intellectual forms' by a 'dialectic of neither-nor' process is a realization of the 'experiential process,' not intellectuality but 'insightfully' without any 'intellectual forms.'

It is understanding without any medium, it is 'only understanding' in a transcendent order, with a further possibility. The insightful pursuit has no station but an open signal.

Even if one tries to understand 'symbolically', as if a 'Being' concealing itself in Nothingness, is allowing a possibility to pop up as one finds in Heidegger in the 'form' of relativity between 'Being' and 'Nothingness' where there is no real ascent to a transcendent possibility of insightful experience as in Buddhist meditative exercise.

Assuming that Heidegger's 'presented universe' emerges and vanishes in the transcendent habitat of 'Nothingness' is still an insistence on an 'objectivity' or 'Beingness.' It is a return to Heidegger's existentialist position, an objectively oriented functioning of a restricted science or logic.

Sartre's metaphorical statement is: 'Nothingness lies coiled in the heart of Being . . . like a worm.' It is only a contrasting outlook with no improvement if viewed from the Buddhist post-intellectual transcendent experiential possibility.

But in this insightful experiential process, I find a return again to a suppressed 'intellectual form, which is not in the line of transcendent freedom from intellectuality, as in Buddhism.

In the words of Sinari, again, 'Just as Nagarjuna sees no difference between *tathata* and *sunyata* (or Nothingness) so, Heidegger and Sartre look upon the transcendental vastness of Being as void,' or 'Void' as 'Being.' Man undoubtedly is a 'self-surpassing being.'

But I cannot agree with this equation as suggested between Western Existentialism, however advanced, and Buddhism in development.

I hold to my thinking that in these Existentialists, Man cannot cross the boundaries of 'relativity.' He is stationed in the world, yet the World cannot own him.

The human consciousness is a ceaseless search for something 'beyond' what is given, as present, beyond – the positive and the other-worldly.

If so, my experiential process soars above the restricted transcendent possibility of Sartre and Heidegger. We pass over every 'actualized possibility' to a further transcendent level of balance, and freedom, which is an 'indefinable possibility,' only 'insightfully understood' with no 'forms.' (Cf. 'Mathematical and the Physical World' by Morris Kline Thomas Y., Crowell Co., N.Y. Appollo ed., 1969).

But one must know that 'transcendence' is not a return to any 'past' *intellectual form*.

If Sinari's statement is taken as 'insightful' without any intellectual form then it is enough to touch my interpretation of Buddhism, and also my comments on advanced scientific thinking in the aspect of post-Einsteinian thinking. But

I would choose a further lift.

This ever-fresh further transcendent possibility can be found in 'Whitehead', if he is interpreted with a stretch but with a suggestion of the return to the past.

According to this writer, Nagarjuna's dialectic of 'neither-nor' is a useful instrument to perform the final scavenger's business. In Nagarjuna's thinking there is no 'Bradley's something or somehow.' For this insistence, Bradley remained in the sphere of 'intellectuality' – or "Appearance." But Nagarjuna soared beyond.

Both Buddhism and Atheistic Existentialism of Heidegger and Sartre are committed to such a study but they fail to reach its further possibility by retaining still an element of intellectual proneness, by relating 'nothingness' with 'Being.' It is a continuity of a 'past intellectuality.'

Dale Riepe's demand of a stable, fixed or restricted objective world, does not carry any substantial value from the advanced perspectives of either Nagarjuna and his later followers in Indian and Tibetan Vajrayana Buddhism. Both these trends foster a higher possibility for life. Riepe merely returns to an intellectual form.

Dale Riepe's grasp of the conceptually or empirically oriented experiential process of the philosophical deliberations of the world, as a whole, is only a common interest of mankind in general, including several of the Russian Indologists, as Stcherbatsky and Anikeev, as frequently referred to by Dale Riepe.

But there is still a process beyond in a post-intellectual and post-empirical sphere – one may still be occupied with a process of updating confronted conceptual-empirical process of knowledge by the advanced studies, as suggested, with no return to any intellectual past.

Such an individual may, as well plunge himself to the process of progressive transcendence, and totally free himself from Western conceptual-empirical touches—if he tries to understand the core of Buddhist transcendent 'wisdom' or 'insightful experience,' without any intellectual form – with no return but continuous advance.

Their's was a proneness for the 'past' intellectuality, which needs a prospective 'insightful' a process of a further transcendence.

This writer's attitude is a progressive departure or transcendence, by a thorough study of the depth-processes of the early Indian thinkers as Radhakrishnan, Dasgupta, Hiriyanana, Ranade, and Datta, among others.

They all prescribe a dual-way, but with a further possibility of 'transcendence.'

From the depth-processes of these early Indian thinkers, one has to scrutinizingly study the positions of Debiprasad Chattopadhya and others, to bring out the exact 'core processes' of Indian thought, why for example, from the very beginning, the Indian Vedic thought was so 'indifferent to' theistic or non-theistic modes.

They were not the real exalted transcendent philosophic ways. Even the term 'atheistic mode' has to be variously interpreted not only from the Vedic perspective, but also from the Jainist and Buddhist perspectives in general, not trying to interpret the disciplined mode of life from any restricted theistic perspective.

Hence Chattopadhya's interpretation of 'atheistic mode' in Indian thought needs considerable modification or qualifications. It is a general discipline that can be shared by any follower of any aspect of experience. But all in different cultural contexts, and different historical situations and perspectives.

Every pursuit needs a kind of discipline appropriate to it, hence the expressions 'theistic' and 'atheistic' have to be carefully used, without relating them to any kind of Western thought, with a past string or pull.

The Indian transcendent disciplines suggest something higher than the theistic or atheistic modes of deliberation, when, in Buddhist thought, there is a total detachment from the 'intellectual forms' or intellectuality.

Man has to distinguish between 'past' and the 'present' experiential processes by ways of 'progressive contrasts' as also Whitehead understands.

His interpretation of 'Western cultural processes'—has to be completely freed from any attachment to restricted conceptualism or empiricism.

He did not exclude the Indian Buddhist thought, if there was any specific transcendental mode, but my study of Whitehead in depth, tells me that his was not a detailed depth-study of Buddhism, he had a general knowledge with sufficient depths – hence he hated to return to the past.

This writer cannot, therefore, agree with Riepe's insistence on a specific importance given to the 'objective' universe, as such.

Indian advanced processes of philosophical and scientific thinking are non-different to both the 'subjectivist' and 'objectivist thought,' as both are confronted processes, are visions, not emerging out of a strenuous pursuit of discipline in the triadic path of Buddhist—Sila, Samadhi and Prajna, where there is a call for a discipline by a total abandonment of 'intellectual propensity,' and choosing the pathway of following 'non-intellectual' or 'post-intellectual' possibilities. The post-intellectual experience – as ever-present presence – becomes patent or manifest without the perished past. It is fresh and novel.

In this transcendent exalted possibility there is progressive concentration, in a disciplined line in continuity. Any specific 'subjective' or 'objective' situation is transcended every moment with perishing and emergence of experiential possibilities. They rest or depend upon a transcendent background, any position – occupies a 'middle,' which is continuously present, as a possible background, of any intellectual analysis.

A transcendent disciplined process is a 'remnant process' – as presupposition of all possibilities of a progressive development of or improvement upon the deliberations of any aspect of experience in interrelated contexts. This kind of experiential possibility will be evident in this work, in the advanced Buddhist thought and as well in the Vedantic developments till the present times, if Datta and others are referred to.

Even the advanced science is moving fast towards an assumption of transcendent basis confronting the galactic universe, which is only a 'middle' of a prospective new indeterminable 'remnant order' through 'Supernovic' efflorescences. It is a perished or burnt bird – Phoenix, with an emergence of new lease or cycle of a fresh life.

Hence this new volume, has chosen to deal with three important areas of experiential processes, in relation and possibility, from a continuously-remnant 'feeling' or 'presence' or present transcendent background, with a sub-title, (i) 'philosophy, (ii) physics, including astrophysics and cosmology, and (iii) politics,' interpreting them, as confronted processes, or as 'middles,' 'cyclic movements,' in their respective cycles in 'interaction,' with infinite possibilities from advanced international and interrelated processes – with a novel freshness – with no past proneness.

All these dynamic movements or pathways will be demonstrated in the final chapters on 'dynamic politics' in Part III of this work.

One has to study the depth-aspects of these three principal centres with all their novelties, with utmost scrutiny, respect, and impartiality, without falling into clutches of any specific past theory. The presented universe, if egoistically charged with a specific vision, is an 'illusory appearance' (in some form of Buddhist-Sankarite sense) from all order of experiential processes from the background of myths to the foreground of sciences in continuous developments.

From the Vedic period there was a distinction made between the transcendent and the confronted orders of life, so that liberate condition of life has to be fully lived with relative satisfaction in all the stages of life, understood in terms of the different stages to fulfil the four-fold objectives of life. These four-fold stages of life as prescribed in the Vedic social life—began with the life of a *student* with an aspiration of progressive knowledge, in a disciplined way, living the life of a disciplined student (Brahmacharya). It prepares one of the next stage of disciplined life of a *married person*, man and woman. Their objective lies in fulfilling the requirements of this period, by earning money, enjoying the pleasures of marriage life having children, and performing social duties (a life of Grihastha or householder).

Beyond this order there are two other prospective stages of *disciplined life*: a life separate from any family attachment in a *solitary place* (Vanaprastha), followed by a *total separation* – the life of a 'homeless' or an 'wanderer' (Sannyasin or ascetic).

This aspect of the life of *solitude* for each individual is in some of creative directions, hence it differs from person to person.

Each stage, varies from person to person, depending on how he or she would choose to shape each other's stages. I for myself will not choose to live the last two stages of life, in the milieu of ancient frame of reference, but I will shape it differently suiting my nature, environment, capacity and choice, as what have 'become' during the periods with a self-creative growth, for a life either in 'solitude,' or with a 'total detachment,' stationing myself to a convenient middle, between the perishing and emerging environment.

My final aim, of life as manifest in my creative activities – is for continuous 'self-improvement,' confronting the present situation, by progressive detachment, in other alternative transcendent ways of Buddhism, Jainism et al., in relativity and transcendence, visualizing my capacity by stationing myself to a self-created position through progressive exertion and insight, and through years facing many crises and sufferings and still existing with some relative self-confidence and alertness, in these days of scientific and technological possibilities, with a peaceful attitude.

Besides this Vedic form of life, there were other equally important forms of life prevalent in India, with a disciplined mode of life as of Jainism and Buddhism.

In Jainism, there are two orders, viz. (i) the order prescribed for the ascetics or monks, and (ii) the order for the ordinary householders, with relative or progressive disciplines, in conformity with Jainism.

This kind of two-way mode of life in a different sense is also practised by the Buddhists. The Buddhists have a relatively moderate or a softer way of life, in the context of the Buddhist Monks and the disciplined 'lay persons' – in contrast to the extreme attitudes of the Jaina ascetic mode of life practised by the householders and the Jaina monks of two types – Svetambaras (wearing white clothes) and Digambaras (remaining naked as indicating complete detachment).

I had associated myself with the Hindus, Jains, Buddhists, Jews, Christians, and Islamic, Zoroastrian communities, and varied other types of Indian people, including the tribal peoples, and peoples of other countries of the world, Polynesians of Hawaii or with the American Indians of New Mexico, with love and appreciation.

During my teaching for fifty years, I had taught some Buddhist Monks, in Ceylon and at the United States, and students of different parts of the world, while in my early teaching periods in Ceylon University, for over twenty years, and for over twenty years during my stay at the different periods since 1964, in Hayward, spending most of my time attending many international conferences at the United States, India and Canada. These are opportunities for forming and reforming myself with the ideas in interaction with my students going for doctorate degrees, or my learned colleagues of all types, interested in subjects from Philosophy to the Sciences and various aspects of Humanities' studies.

The term 'democracy' used in the context of Politics, in each country, has to be understood from their different types of political structures or constitutions and in their distinct historical and cultural backgrounds. But this also has lots of deviations and deviants. To talk in general ways is wrong as indulging in ambiguities or extremes, they are mere journalistic flares. But in each case, there are no fixed beliefs or interpretations, there is always a kind of 'decision,' as suitable for a country or culture. Each individual formulates in one's own way, with some ratiocinative arguments, as possible modes of choice. For each country, too, there is a distinct history of constitutional developments. Hence here also there are arbitrary decisions suiting to the country.

This work, therefore, is a work from a democratic perspective, with qualifications, in three parts, interconnecting them as possible processes, exhibiting human participations, of varied levels and tastes. The suggestion is for a revised study of these areas from the special Indian advanced cultural perspective, developing it from the Vedic period to this country, with a full knowledge of the developments from period to period.

In these areas, an intercultural perspective, has been pursued, for, India never isolated itself from all other advanced countries of the world long before the Christian era. It tried to maintain its cultural ideals, without subjecting itself to any specific theory, culture or country, as its stress was/is neither exclusively in subjective nor in exclusive objective process, but in a balanced and integral context with progressive insightful attitude.

India remained alert to confronting the experiential processes, with changing visions and aspirations as far as was possible to realize, without disconnecting oneself from the social or one's own psychic life.

Individual's psychic life is a series of 'lapses' from a basic deep-sleep to the emergent dreaming and waking processes, as it is, finally terminated by one's 'death-process.'

Hence Indian confrontation of experiential process is not an acceptance of the 'waking experience' as confronted, and 'as such,' as it takes varied turns and twists with the progressive disciplined modes of life one tends to realize.

One's confronted visions change. For an insightful Indian thinker, visions confronted, are interlinked psychic phenomena. These visions 'emerge' and 'lapse' but they are shunted back into the transcendental sleep, or into swoon or unconsciousness, during one's life-period, and after death, they pass into various unknowable processes, needing a disciplined mode of life or observation, for which one has to acquire special qualifications.

My varied types of books are attempts at revelations of such unknowable mysterious aspects or depths of life and death processes.

But there is a 'remnant' transcendent 'Nothingness' operative 'before' and 'after' all confronted experiential processes'—life and death processes.

I have given some glimpses of my reflections in my book on *Zero* . . . .

If one minutely studies India's cultural processes, in various historical periods – from the Vedic to the present century, which forms the rubrics of Hinduism in general – one will find that besides the recognition of the final philosophic-depth-process of a liberate condition, there was always a consideration of commonly shared socio-political and scientific-technological life.

Life is learning to live with others within a social milieu, peacefully exchanging one's own ideas with the ideas of others with tolerance, and patience, and to find out some prospective solutions, in an occasion of conflicting experience. But due to the presence of all possible types, problems of all sorts will persist for ever, and aim of success has to be understood always with a relative success, avoiding absolutist expectations or abrupt solution.

But, in India, often some peaceful discussion on various trends of Indian and Western thought, in togetherness is, a possible mode – which is very rare in England and the United States. I do not exclude this possibility in the case of some of my close friends, colleagues and devoted students.

India's main objective was how best one can be operative in a given 'confronted situation,' with a 'relative success,' tolerance and non-violence. There is no expectation of a final decision or evaluation. Alternative pathways are there.

Since Indian independence, India is confronting varied crisis-situations to tackle which, not merely an acquaintance with the basic or advanced cultural process of India, is sufficient, but what is more important is how the efficient philosophical, scientific or political ideas, can be applicable or implemented in Indian situation of growing or emergent problems, within India, or created by external international state of affairs, or sometimes by foreign 'interventions.'

Only through a successful implementation by India's policy makers, or by the continuous five-year plannings, some kind of social changes, can be brought about, with a relative rapidity, and with an eye to progressive provisions for defence from possible foreign attacks. These alone can tackle the critical crisis-situations both within and without the country. An efficient scheme of planning is necessary from these three broad aspects of experiential processes in interrelation, and applying the end-products to meet efficiently the multifarious emergent problems.

Without much argument, one must try to see that 'cleanliness' and 'basic honesty' in every sense are the prior concerns. Before thinking of 'eating', in a 'quantitative sense,' a man must learn to eat less than more. One has to form a qualitative judgement through some knowledge of health principles as traditionally prevalent, and also in terms of changing scientific knowledge and medical experience.

There is always a simple choice between extreme demands. My suggestions will be for increasing the good types of sanitary facilities, with scientific and technological ways for wayside resturants, gas stations, inns or motels, and public places, with similar facilities, all over the country, with proper scheme for maintaining the standard through high-salaried engineers or medical officers, in charge, and not merely depending on public discretion or giving individual freedom. This is high priority from my judgment, which will prevent opening many expensive hospitals, for, primary care is less expensive than actual facing of problems of diseases or sudden eruptions of epidemics, due to general insanitary conditions as prevalent in most parts of India.

In the first part of this new work, my project will be to shape Indian thought by pursuing the ideas and insights of some of the major thinkers, before and after Indian independence, and the present day Indian philosophers after Indian independence, in selective ways, pursuing two volumes on contemporary Indian philosophy. There are also other collected volumes on Indian contemporary philosophers. They are to be examined, not merely in terms of the number of 'papers' contributed, in a 'collective sense,' but from a select 'qualitative perspective' of evaluation from a progressive disciplined detached attitude. Here the suggestion is from a depth-study of their works. There is no need merely to refer to them.

They are to be studied in relation to some Asian countries wherever Indian philosophy influenced their dominant philosophical attitude and their trends. This study in interrelation is possible specially in the context of Buddhism and also in the context of other contemporary philosophical volumes, British, American and Australian countries, whenever felt necessary – to make Indian philosophy, comprehensive enough – as an international concern. But each mind whether of India, or of any other country – is differently oriented, hence there is always a confrontation of different phenomenological experiential processes, but one has to tolerantly accept the other, and make life judiciously livable, without trying to agree with others totally.

In all countries of the world, there was an earlier civilized period, before the broad Western impact.

This is true of American continent also, though a new hemisphere. No one can ignore its ancient tribal civilizations, as also surrouneing the processes of Island civilizations. In the New Hemisphere of North and South America, there were several civilizations as of the Aztecs, Incas et al., before the Spanish voyagers came and destroyed their civilizations.

The development of philosophical thinking is always possible alongside of other advanced philosophic currents. The editors' business therefore will be a qualitative selection from such advanced perspectives, as now the facilities for publications have increased. Moreover the computer sciences have magnified this process with galactic speed.

If the stress is given to the different collections on Indian philosophy of this country – then, also, one may as well try to find out some kind of differences among these chosen Indian scholars, and their interpretation of other major cultural processes, within the country, not recognized so far. This writer, therefore, will re-examine Indian thought with scrutiny as far as possible, bringing out the core of India's philosophic consciousness in process and possibility. The final aim will be to determine whether such philosophical attitudes fit in with the advanced scientific and political processes, as predominantly confronting one's attention for developing India's future with well-defined planning.

India will stand always for the development of 'transcendental psychology' in contrast to the confronted restricted 'conceptual and empirical' psychology of most of the other cultural processes of the world. This work is thus so selective and also eliminative at once.

Euro-American thought, or those who follow them are not for that reason carriers of advanced thoughts, if they do not develop the subtle processes of transcendental modes of deliberations in all aspects of confronted experiential processes. This work therefore does not give importance to a past, after a detachment with a discipline.

In part two, India's scientific philosophy or programme has to be formulated by India's 'transcendent Nothingness,' (Sunya), or 'Zero', when mathematically expressed. The zero is a symbolical representation of the 'transcendent Nothingness,' as confronting the positive numbers both at the beginning and at the end.

This will be briefly dealt with, repeating some of the main ideas as have been brought out by this writer's work, entitled, *Zero – Its Role and Prospects in Indian Thought* as mentioned.

Here the investigation will be an indication of the operative sphere of the human mind, which is always placed in the 'middle,' and is something 'positive' between the two extreme ends. 'Nothingness', as a transcendent process, is at the beginning and at the end of the 'Middle' or the 'Positive' situations. It is a symbolical representation of the transcendent process, which is a 'possibility' for all knowing.

This type of symbolical knowing from a mathematical perspective will be stretched to different sciences, and finally to physics and astrophysics.

From this broad insightful philosophic-scientific perspective one must know that if there is no final experience – then there is no final personality to adhere to, no science that can be categorically theorized, and, for that matter, no politically patterned form is 'absolute.'

One has to learn to be completely free from any specific conceptual or empirical mode of deliberation. The confronted universe, is phenomenologically present, but not in a specifically categorical existential sense.

It is not simply philosophical, or psychological, scientific or technological, but a mixture of all and none. Mankind's confronted visions are penetrated through and through with 'Nothingness', from both ends of a positive number from '1 to 9', if symbolically indicated.

The present work also takes note of the growing interests after India's independence, of the works of the specialists on India, as reflected in *Science and the Human condition in India and Pakistan*, edited by Ward Morehouse.[14]

There are other scientific works as *Science Policy and National Development*, by Vikram Sarabhai[15] and *History of Philosophy: Eastern and Western*, Vol.I.

In the case of *History of Philosophy: Eastern and Western*, reference is specially to Part III, Ch.XVI, Scientific Thought in Ancient India.[16]

Here also it will be shown that India pursues the advanced scientific trends with similar interests as of the other advancing countries not for any specific theorizing, but now with some knowledge, it can be operative harnessing its scientific resources to direct itself efficiently towards the processes of engineering and technology, to bring about social and economic changes as rapidly as possible. It is a scheme to remove speedily the world-wide problem of poverty, disease and political turmoil, by a 'psychic revolution,' to which a nation gets involved.

Here there is a reference to my advanced works, and also to the works of many Indians and non-Indians, if they address themselves to the development of the advanced trends of the world, bearing in mind that all these maladies of human socio-cultural situation can be tackled with a progressive transcendental insight and not merely by the facts as presented. India's programme for peace is not by the dangerous projects through nuclear arms, exhibiting the might of so-called superior nations.

Hence a comparative study with an 'integral insight' of India, will be an additional incentive for progressive enlightened insight for the scientific developments. My suggestion will be to be operative in this direction by small institutes managed by some salaried experts, whose services also will be valued by repeated public recognitions, and rewards. In the same critical and evaluative spirit, as the scientific discoveries and inventions, are evaluated from progressive integral insightful perspectives.

This means that scientific pursuit has to be made to discern the contribution of the Indian scientists, in shaping the scientific prospects of India, in particular, and of the world as a whole, in general, to bring out the unique speciality of Indian scientists.

This has to be recognized as a world-power, in a relative sense, for every Indian of the present generation has an obligation as an 'Indian,' in its true spirit. It is nothing but suggesting the Indian mind not to be forgetful of India, but by thinking primarily about India, even when the Indians become interested in the studies of scientific works of other advancing countries.

India's people living in any part of the world courageously should draw attention to the specialities of India's thinking in various experiential aspects by way of contrast in relation to the other countries' contribution, with mutual appreciation.

Unless one's own people speak about one's own country, the people of other countries will not learn nor will they be disposed to think about India.

This is evidenced in the innovative activities, in all aspects of life from agricultural interests to varied scientific and technological directions, and also to business enterprises of the immigrant Indians in America, their expert roles have changed the American ways of life, by various types of India's impact.

Part III will deal with a critical study of the essentials of Indian politics, its constitution, and its special operative schemes in contrast to the British and American constitution and parliamentary form of Government from each one's respective historical background as evolving and developing processes.

Here one has to find out the speciality of Indian Constitution, how India inherited from the interacting British parliament operative in England and India, tempting the Indian people with a possibility of a self-government, in course of time, but deferring it, till the India's insightful intellectuals created such critical situations as became difficult for the British liberals to defer further.

The transition from the British to India's democratically operative process, with some changes through the President and Prime Minister, with well-defined functional capabilities, and limitations, engaged the attention of the British and the Indian minds to be expressive in fruitful ways, so that there will be no conflict between the two heads for tackling or conducting all possible emergent 'crisis-situations,' as it happened between the Prime Minister Nehru and President Rajendra Prasad, with some difference of opinion in President Rajendra Prasad's attending the funeral of Patel, not following the formal procedures of the Constitution. This opened up discussion and clarification of other matters of the Indian Constitution.

Here one must know specifically that India does not address itself to mere conceptually or empirically oriented political principles or structures. India's speciality, even in the sphere of politics, is transcendentally oriented which takes due note of the confronted conceptual and empirical processes, with a possibility of their transcendence.

In the long run when there is a possibility of the collapse of all 'ideologies,' there should be a reasonable policy of a slow passage towards 'people's rule, when the Indian masses will be fully enlightened by real forms of education, which is a refined attitude, and not obstinately staying with some abstract unbreakable 'ideas.'

In the body of this work such special possibilities have been dealt with some details, in relevant contexts.

India's special political scheme is grounded on non-violence and nonalignment, both within and outside the country in international context. This process will be displayed at places, but India is not remaining indifferent to varied internal affairs, and the world-affairs.

Some select relevant works in this line, will have to be utilized, and understood as suiting the Indian frame of mind, and its international relation and policy, have to be improvised with insight and farsight, in relation to UN.

UN has to be recognized as India's ballast, in times of distress and in crisis when all powers are indifferent to India, and other developing and also super-powers that emerged after the end of Second World War. They displayed their strengths, without paying any heed to India's suggestions, but ultimately they came to their relative pragmatic senses after bitter international experiences, in course of time, though belatedly.

This broad attitude within India may be reflected through an organization as 'United States of India' (USI), which needs to be created. It should be used as an operative principle with UN, for India's development of an integral consciousness, as also to ventilate its concern for the confronted 'world-affairs.'

The present world is a dual life from national and international perspectives. This attitudinal development through proper education in schools, colleges, universities and institutions – will also be considered, by select representatives to USI, specially in the context of its initiation of a policy of 'nonalignment.'

'Nonalignment' – implies an awareness of the complexity of the political situation of India at a time, but remaining isolated, sticking to its policy of not-involving within and outside India, with any of the contending parties, but suggesting a judicious right attitude, through, its own organization, and sending representatives to the United Nations, to be operative with the developing emergent or non-emergent nations of the world.

Within the country also, this attitude has to be progressively developed, not allowing the country to remain as it is – separated as it is – by its various organizations, the special objective of USI will be to be operative with a progressive unifying principle, and with a detachment.

Hence the formation of this new organization as to be operative independently of the different conflicting parties. This writer, feels its special need.

This means more and more participation between India's USI, and international organization – UN. It is a suggestion for a new political insight.

This insightful operational scheme could be found in the first prime minister of India after independence. It remained an implied interest for later prime ministers of India, and then still later this scheme was responsibly followed by the ministers for external affairs, in different ways, according to the changing international situations.

I have developed this process at some length in the fourth volume of my *Dynamic Facets of Indian Thought*. Here also, with this new title, the contributions of the Indian politicians will have to be evaluated, by arranging for their education through different seminars, in the context of the international political situation and prospect. Some such ideas will be provided in the body of this work (USI) at places.

But in this statement I will not dogmatize, saying that the understanding of the principle of 'nonalignment' is confined to a particular family, man, or organization, but its understanding is a general affair and can be learnt.

India is a country of varieties and many dimensional experiential types. India changes slowly with insight, so that even when I am at Varanasi (Banaras), I still find the 'well fed' cows in front of a house, gracing a family, suggesting that India respects its animals like, cows, giving them proper respects, and indebtedness.

If Indian thought is considered from these triune contexts even in a general way, from a broad philosophic attitude, without claiming the full and comprehensive requirements of the specialized spheres of sciences and politics – a proper perspective of the advancing Indian philosophy and civilization – can be worked out, outmoding the defensive and protective propensity of Indian thought and life, as prevalent in the earlier works on India.

Lastly the specialized Indian minds, with a comprehensive knowledge, of this sphere and with the insights of the other spheres of experiential processes— have to be adequately recognized, with a comprehensive Philosophic-Scientific attitude – should one learn to face the counter-attacks of mis-representation of both Indian and non-Indian experts in general.

Too much allowance should not be given to the outsiders who do not have some real knowledge of India, specially in the contexts of transcendental spheres of experiential process as considered in this and other volumes by this writer.

This new work, with its title, will reflect India's highly complex triadic frames of reference, and their operational ways, with no adherence to any static ideology, for what is still worth, is the basic transcendent insightful mode of reflective activity, and not just discrete opinions or views as held by mankind in general.

This new volume may have an 'idealist outlook,' but in a sense, it fosters not any abstract or obstinate 'ism,' it flouts all such trends by facing the confronted universe, with a complete detachment or freedom from one's restricted intellectual disposition. India's eye is purely symbolical, with a detachment of freedom from intellectuality.

The author's earlier works will be shaped in these comprehensive directions, spreading its umbrella with a new flare in these relatively concrete spheres, as to create an international forum for study for the learned, or lovers of learning, of Indian thought, in three of the important areas of the civilized or 'civilizing' processes of the world as a whole.

These three spheres of life have been mentioned earlier, but these are to be observed as concrete objective phenomena changing their shapes as perishing and emerging characters not remaining static for ever in any sense.

A senior person of this century has three visions of this changing Universe, from the early years of the first world War to the period of Second World War, and the Indian independence in 1947, and also the later experiences to the nineties.

It is a vision of India after partition and dissection of its limbs, geographically, socially, linguistically and religiously. He has the shocking experience of varied changes of the value of rupees, and the difficulties of day to day life of one whose incomes are absolutely fixed, without any increase. For him, Indian independence was a shocking experience, yet he tolerates with a patience out of a spirit of helplessness. A sober Indian tries to maintain a composure, for, observations are varied.

The country apparently looks infected by poverty, but also there are a large number of persons occupied with various occupations, with business insight, and with new business enterprises, in imagination, so that if one travels from Bombay to Bangalore, one will find, that from a fairly stable and well-established place – one is confronting a place, whose previous look has changed from a calm more or less solitary place of healthy resort – to a complex centre of rapid constructions – it has become, a great centre of business, with immigrant peoples from outside India, and from peoples within India.

Besides all these changes, India is a great scientific centre with research possibilities in various advanced aspects.

If one travels to Benaras (Varanasi), from Bangalore, one will have still a look of its 'past' with buzzing people, and trespassing cows, dogs and other animals, on its streets, but not losing its ancient sacredness.

Who knows within a period of ten years, Benaras will become similarly re-constructed with modern types of buildings with varied business centres yet retaining the complexity of its past?

This time as I went to India, I travelled through Gaya through Palamau to Ranchi, a place sacred for my early life, I painfully marked Ranchi's turn to a crowded business centre, with a suffocating surrounding, losing its former solitude and healthy look.

India has changed both in its educational schemes, and in man's choice for professions. They no longer study for mere theoretical curiosity, but with a business insight.

Hence the story of the masses and their activities is not the consideration of the village peasants alone, as Jayprakas Narayan suggested, but also those created poor and deprived by changing circumstances – changed in terms of business possibilities, and in learning about international economic and political impacts, by some extracurricular vocational training.

I am referring here to the school teachers, college professors, and government servants of various departments of low and fixed incomes, changing their life from a prestigious position to the venomous situation of chill poverty. But they hope to change for the better through the younger generations – New children with mature ideas.

This book will not touch them, leaving this aspect to sensitive novelists to bring out their pictures of life.

This book is narrowly centred to the general learning aspects or facets of Indian thought, with gaps from the Vedas to the present times. But these 'gaps' are so many, and so complex, that no one however intellectual, can keep pace with them and their infinite possibilities. But Indian mind is moving towards business and other vocational enterprises, with a placidity and acceptance.

The general facets as mentioned are considered as dynamic and evolutionary in different historical periods, this writer, only, touches them in crest-aspects of Indian thought. They open up varied processes and facets, of India's complex activity-processes, in international contexts, with some kind of novelty and speciality.

This writer is extremely thankful to Mr. Nityananda Datta, the well-known co-director who worked for thirty years with late Mr. Satyajit Ray, for his criticisms and appreciations.

Mr. Datta also directed some of the films of public interests, and as well as some documentary films for the government of India, like the *Life and Activities of Swami Vivekananda.* Swami Vivekananda was an epoch-making personality of this early twentieth century.

Datta has also become very much interested in this writer's book on *Zero* and in talking about that book, he has become very much interested in the politics of India after the Gulf War, after his study of my works. This has given this writer some direct knowledge of India's political situation of this critical period.

With Satyajit Ray's recognition by international bodies by an Oscar Award – India's film industry has discovered a new Avenue for a future prospect, in a competitive 'business world' with a new prestige.

Mrs. Bharati Datta, wife of Nityananda Datta, with her Master's degree in philosophy, and specialist services in several shipping companies of India, took some courses on child development, and on the study of the process of aging and of old peoples (Gerentology), for some years when she came to America.

Now as an Activity Director of a Convalescent Home, she has become interested in the study of the old and handicapped people in general, from her specialized perspectives. She is taking care of the seniors and also children, as a new experience.

Her conversations with me of her experience of the convalescent seniors and handicapped children have opened for me a new insight for understanding the varied psychological problems in America.

These have led me to ponder over the day to day problems of this so-called affluent society of America.

I now begin to reflect that the consideration of the different varieties of transcendental psychological principles of India, discussed in many ofmy past works can be recommended to a society for self-development but those who are not so inclined, some kind of 'suffering' is likely to persist, if such 'empirically oriented' people multiply, their impulsive tendencies, without accepting some disciplines suggested in India's transcendent psychology.

In most of the chapters of this book, a reference to my publications, has been given, to emphasize the need for understanding the operative processes of the transcendent orders of experiential processes symbolically in confronting the apparent orders of conceptual and empirical orders of experiential processes.

REFERENCES

1. A.K. Sarkar, *Dynamic Facets of Indian Thought, in four Volumes*, Vol.I, Manohar Publications, New Delhi, 1980, Vols.II-IV, South Asian Publishers Pvt. Ltd., New Delhi, 1987-88.
2. A.K. Sarkar, *Changing Phases of Buddhist Thought*, in three editions, Bharati Bhawan, first edition, 1968, second edition, 1975, South Asian Publishers, third edition, 1983.
3. A.K. Sarkar, *Buddhism and Whitehead's Process-Philosophy*, South Asian Publishers, New Delhi, 1991.
4. A.K. Sarkar, *Dynamic Facets of Indian Thought*, Vol.IV, passim.
5. Rodney Loudon, *Quantum Theory of Light*, Clarendon Press, Oxford, 1983, Chs. 7-9, relevant places. In this book, I have pursued C.V. Raman's scientific researches in the area of physics but I have pointed out that he does not separate physics from the broad aesthetic aspects of experiential processes. Radiation determined in physical science runs into the vibrations of musical instruments, indicating that there is no gulf between these experiential processes, one has to learn the ways to connect the two.
6. A.K. Sarkar, *Zero: Its Role and Prospects in Indian Thought*, South Asian Publishers, New Delhi, 1992.
7. N.G. Jog (ed.), Narottam Morarjee, *Architect of Modern Indian Shipping*, The Scinthia Steam Navigation Company Ltd., 1977.
8. A.K. Sarkar, *Experience in Change and Prospect Pathways from War to Peace*, South Asian Publishers, 1989.
9. A.K. Sarkar, *The Mysteries of Vajrayana Buddhism from Atisha to Dalai Lama*, South Asian Publishers, 1993.
10. Dale Riepe, Research India Publications, Calcutta, 1979. Dale Riepe wrongly holds the view that Indian thought is 'subjective.' There is a difference between a 'transcendent attitude' in confronting the subjective inner experience and outer objective processes, where both the subjective and objective processes are confronted as other than the transcendent, and then these confronted processes are transcended to prepare the ground (open

space) for further emergent possibilities, to be insightfully examined again.

11. A.K. Sarkar, *Whitehead's Four Principles from West-East Perspectives*, 2nd ed., South Asian Publishers, New Delhi, 1994.
12. This writer here refers to his paper, 'Sree Swaminarayana and Taoism', published in Part II of a volume, entitled, *New Dimensions in Vedanta Philosophy*. This volume was published by 'Bochasanwasi Shri Akshar-purushottam Sanstha, Shahibag, Ahmedabad, 1981, forming part of 'Bhagawan Swaminarayan Bicentenary Commemoration Volume.' Swaminarayan was a post-Ramanujite thinker, he gave a new patnway beyond Ramanuja and Sankara—clarifying further that Indian thought remained continuously dynamic and progressive.
13. Series two of *Contemporary Indian Philosophy* by Margaret Chatterjee: London: George Allen & Unwin Ltd., 1974.
14. The Rockfeller University Press, 1968, first Indian edition, 1972, Popular Prakashan, Bombay.
15. Vikram Sarabhai, edited by Kamala Chowdhry, Foreward by Srimati Indira Gandhi, The Nehru Foundation, Ahmedabad, 1974, Macmillan Co. of India Ltd., 1974, Delhi, Calcutta, Bombay, Madras
16. George Allen & Unwin Ltd., London, first published, 1952, third impression, 1967.

CHAPTER I

Introduction

This book will perform its functions in a triadic manner in its three interrelated parts: (i) philosophy, (ii) physics and astrophysics, and (iii) politics – indicating India's prospective attitude in international contexts. It is a fitting addition to my long plan to this volume outlined earlier.

Any country, needing an over-all advancement in international context, should be operative with an advanced philosophical position, expressing an outlook for many-sided scientific and technological developments with relative restraints and limitations, acording to the country's resources and economic position, and cultural compatibility.

Its development of political consciousness, should be befitting to that country's international interest, position and policy, with a due respect and concern for other countries. India strictly follows these principles, to preserve the sanctity of independence in the milieu of international society, perhaps trying to display an idealist outlook by a general review of its own historical process along with a general vision of the other countries of the world.

India, through its history has observed the fights among the different heads of this fairly big subcontinent, and had also tolerated many invasions and foreign rules, but now after more than a thousand years of the Muslim rule ending with the rule of the great Moghul emperors and one hundred and fifty years of British rule of India – India's intelligentia and the masses in general – have chosen to be ruled by her own people from a unified constitutionally oriented democratic process.[1]

The constitutional pattern is a British-American mixture, to meet some needs suiting Indian frame of reference, its complexity, cultural variety, and continuity of civilization that no place in the world has ever experienced.

By way of contrast India wants to follow a path of non-violence, with a reasonable understanding of its own self-defence.

With a newly acquired independence in 1947, after the British left India, transferring to India, a selfgoverning power, India obtained a parliamentary form of democratc government befitting Indian people and its cultural background with a peaceful gesture.[2]

It was a gesture of recognition of India's three thousand or more years of civilization and culture, before the Christian era, and its competence to maintain and face independently, other nations of the world, with a potentiality for achieving the necessary efficiency, in science and technology, in international contexts.[3]

Now India given the freedom can tap the different sources of power through progressive scientific and technological developments, discoveries and inventions through its own people and contacting the foreign intelligentia.

This new movement, according to this writer, examining the trends of the world as a whole indicates a process that can be tackled efficiently in a triadic way, as suggested. This will remove the arbitrary and restricted operations of each of the pursuits in the direction of philosophy, physics and politics.

This goal was already achieved to a large extent, by India's own cultural, scientific, technological and aesthetic history with cooperative activities keeping the country always conscious of the developments in several of the non-Indian countries, and insightful persons, dedicated to revive the prospects of a great country like India.[4]

During the British period, since the nineteenth century a number of reformers arose in different parts of India, preceding Gandhi, who began a novel constitutional method of protest against the British rule, drawing attention to the discriminative attitude of the British rulers towards the Indians.

It was also an exploitation of India, and its people. But in the course of the British rule of India, some sober and liberal British administrators and personalities, both in England and India arose. They indicated pro-Indian concerns, and wished for a gradual transfer of power to the Indian people, by giving them some form of educational facilities, for a slow independence of the type of a self-rule or Dominian Status.[5]

For the details of this constitutional movement, and the other organizational processes, supporting or not supporting Gandhi and other of his group, some further materials are provided as grounds for the detailed histories of Indian independence, which are mentioned to some extent in the fourth volume of my *Dynamic Facets of Indian Thought.*[6]

The main purpose in that complex discourse in that volume, was to understand that the 'independence movement' was an interrelated phenomenon, in which Gandhi had an important role. But besides Gandhi, there were other great persons operative in different ways, making the British conscious of India's powers of possible leadership.[7]

This new volume will not however deal with the historical development of the philosophical deliberations of that aspect. Its immediate interest is in India's post-independent operative processes of not mere vapid philosophical deliberations, but concrete operative schemes of five-year plannings soon after Indian independence of quickly shaping India's future in triadic attitudinal ways.

India's is no longer a mere theoretical scheme and planning, but a practical scheme of a programme for a quickened process only when a country is fully conscious of its independence.

But there was also another phenomena of some major world powers, which became operative without really recognizing the legitimate role of minor powers that emerged simultaneously along with Indian independence.

To balance the operations of the different international powers, an impartial power was given to the authorities of the United Nations. But this amounted to an idealist programme, like similar organizations of the past. It was not efficient enough.

In the context of India for progressive transformation of India, in various spheres, scientific and technological training of the mind in such an international context was needed.

Hence through proper educational schemes, the provision for such education in theoretical and technological aspects was also a need. India had to make a programme of a dual process: scientific and technological. Without the cooperative operations of these processes, 'the psychological change' of India's mass mind cannot be effected.

Without an attitudinal change philosophically, scientifically and technologically, with a sober insightful political capability no quick and real socio-cultural change, within the country could be possible in the progressive aspects of self-improvement after independence.

Gandhi was here again a harbinger of this innovative process, by his psychological orientation, of different types of his people of his time – both before and after the British rule. His was a change of India's 'mass mind'.

Indian 'intelligentsia' and 'Indian masses' learnt of their novel operative schemes of non-violence and non-cooperation, boycott, going to jail, during the British rule – but these will not have much significance after the expiry of the Biritish rule.[8]

During the time of the British rule creation of a concern in the minds of the British and the Indians was a need to be operative psychologically and effectively, and this operational scheme was then inspired by an element of an appeal to a higher moral principle, and psychology, with a freedom to accept and follow.

But after the British rule, independent India has to prepare for a different practical scheme of action.

Gandhi's operative scheme also had affected the socio-political process of the world in general of his time, passing from the end of the first world-war and stepping into the processes of the Second World War, which changed the British mind, as also the refined minds of the world as a whole.

The British had to quit India, with a friendly gesture. At that historical period, there was some kind of higher morality still persistent. But time was changing fast with new visions and experiences, and so the so-called assumed moral ideas were changing, with new political visions, and growing scientific strengths of the different Superpowers, and their roles in international situation.

With an eye to equip India, with necessary perspective of changing the world by a sort of psychological impact on the other parts of the world – India, after independence, started its operational process, with great insight and diplomacy.[9]

This book provides a new philosophic position and attitude of India to tackle the process of India's internal prospects, and execution of those processes.

The progress in these aspects helped India to improve its internal strength of progress and defence. These improvements prepared the grounds for India's strong position abroad, not merely harping on its past rich heritage of the pre-independence or post-independent periods, but indicating how it can be operative very efficiently by the advanced philosophic and scientific attitude of India's intelligentsia who had some important roles, besides that of Gandhi, in other aspects of experiential processes.

India's philosophers and scientists expedited the socio-political processes, in bringing about widely comprehensive and advanced philosophical consciousness with a psychological attitude, reshaping its past heritage with an insight, as to bring out something very special which is not present in the philosophical traditions of the other countries.

Scientists, too, played their important roles through their discoveries and inventions not only in theoretical directions, but also in practical directions by understanding the great possibilities through its mathematical insights of zero, which is theoretically grounded in the transcendental notions of a final emptiness (*Sunya*) of all conceptual and empirical situations, for everything is in change with an unforeseen or unimaginable possibility, not in a 'categorical but in a many dimensional sense.'[10]

Here my suggestion will be for independent India, changing totally the monotonous abstract philosophic attitude, but examining the philosophical deliberations of the major European countries, and its transmission to the new Hemispheres of Americas, with some difference, for America, inspite of its Western heritage, claims to be an independent country with its history of independence. It will be an insightful creative activity through some books, to educate the minds of the Indians from the child to the grown ups, through schools, colleges and universities.

Hence for shaping the unique philosophic assets of India, before and after independence, I will try to review the dominant ideas by way of a contrast in relation to the advanced modes of other contemporary philosophic, scientific and political trends of some major parts of the world, through another programme of publications and educational curricula.[11]

The objective is to bring out some speciality in India's thinking, not just following the thoughts of other nations of the world, who have different cultural backgrounds.

Keeping these basic ideas in mind, one should examine very critically two of the volumes on contemporary Indian philosophy: (i) *Contemporary Indian Philosophy*, its second and enlarged edition, since 1952, and (ii) *Contemporary Indian Philosophy*, Series Two, 1974.

I will also refer to the work in two comprehensive volumes, with the title; (i) *History of Philosophy – Eastern and Western*. It deals with the Indian thought, from its pre-Vedic and Vedic stages, through later Vedic and non-Vedic Indian thought to gain some general ideas of Indian thought of religious and philosophic expressions, in relation to the contemporary Indian movements. The latest expressions of Indian thought in intercultural contexts both during and after the end of the British rule of India, were not an exclusive philosophical thinking, they were with some general references to Islamic and Sikh religions which flourished in India through the centuries, and developed both philosophic and religious consciousness from an advanced intercultural perspective.

Its advanced tremours in the early stage of Buddhism touched the post-cultural Chinese processes beyond its early cultural bases of Taoism and Confucianism, and also the Japanese Shintoism preparing the grounds for Chan and Neo-Chan processes in China and Neo-Shinto and Neo-Zen processes in Japan.

The Buddhist impact has already influenced Whitehead's process philosophy, and post-Einsteinian thinking of the contemporary west towards continuous field processes, closing the gulfs between special and general theories of relativity, and *particle and wave theories* of Heisenberg and Schroedinger.[12]

The entire cosmological energy-continuum is no more categorical or linearly based but interpersed through infinite possibilities in multi-dimensional non-linear directions.

Indian Buddhist thought has affected the American process-philosophers from a Buddhist perspective, of the present century, admitting at last that thought is self-transcending and hence opening up a path beyond all restricted theories conceptual or empirical transcending the multi-dimensional spheres of restrictive thinkings which prepare the grounds for sufferings or concerns of mankind in general.[13]

This advanced process as a possibility is found in the varied pursuits of Indian thought of mathematics and other sciences of ancient India, and aesthetics. Now these revolutionary processes have been continued in international contexts, by some Indian scientists from Raman and Chandrasekhar to Sarabhai and Sudarshan.[14]

In this narration, I find that all cultural processes, discover that the thought transcends itself, but the cultural processes, in turn, are confined to their restrictive thought without being capable of really transcending the restrictions of thought. Indian thinkers found out a possibility of transcending the sphere of thought, and discovered a transcendent insightful order of experiential processes.[15]

This is also true of Whitehead's 'wisdom order', not discerned in the past Western cultural processes, from Plato to Hegel. Whitehead, in an aspect, could touch the transcendent order of India's Buddhist life of discipline.

In appreciation of Whitehead's pro-Buddhist development of thinking, in the Naturalist socio-cultural situation, and admitting the prospect of thought beyond its sphere to a transcendent 'Wisdom order', Bernhard Mollenhauer says that Whitehead has performed the function of pursuing philosophical adventure as an eternal quest for a divine flame. Here the reference is to the short paper, entitled, 'Whitehead and the journey of the flame' by B. Mollenhauer, as mentioned, in *Darshana International.*[16]

Mollenhauer also in further appreciation of Whitehead, refers to Whitehead's career in the philosophical world like that of a Beethhoven, in the sphere of Music. Whitehead's career is indicated in three stages and periods: The first stage refers to his Mathematical Research, the second his pursuit and knowledge of Nature, and the third, his American period, began at an age when most people retire, he then presents himself as one of Harvard's most eminent philosopher.[17]

The vol.II of *History of Philosophy – Eastern and Western* from a historical perspective is a general survey of entire East Asian, Persian and Zoroastrian thought as developed in India, and western thought in development from the pre-Socratic stages in Greece through the medieval and modern periods, leading to the developments of contemporary trends of Euro-American thought in some aspects, and ending up with a general survey of science and philosophy.

But here also the Indian cultural process is not restricted to mere 'narration', but has the efficiency of transcending the sphere of thought, and realizing the order of experience in an insightful wisdom order of possibility. Indian philosophy – in progressive developments – continuously rejects the 'categorical assumptions' of the confronted universe.

This is a realization of the advanced processes of both philosophy and the sciences. It is a transition to a transcendent possibility of alternative modes, abandoning merely the confronted categoreal possibility. In dealing with this aspect only some essential currents of some thinkers will be brought out for the development of a general philosophic outlook and attitude, as an efficient operative scheme.

But here also I shall refer to a select group of Indian philosophers who exhibit an innovative scheme of a special sort in their studies of both Western and Indian thought.

The speciality of India's *thinking* was for *transcending*, *mere thinking*. It was a dealing with thought not conforming to its *forms* of thinking, but considering all its analysable possible *forms* of assertion, negation, their combination, and then an abandonment, or wiping out of all intellectual forms by a dialectic negation (by 'neither-nor') to rise to the transcendent 'insightful' or 'wisdom modes' or 'freedom' from 'intellectualism.'

This scheme of detachment from restricted intellectual forms rescues one from the intellectual trammels or sufferings.[18] I will not mention about the other eminent thinkers in the two volumes of Contemporary Indian Philosophy as I have dealt with most of them through the four volumes of my *Dynamic Facets of Indian Thought.*

No longer my desire or plan is for an insistence for thinking however important. My plan is how quickly one can free oneself from mere habit of thinking that is arguing for the sake of arguing, and not abandoning it when it is no more necessary. There is a process of before and after, in taking into consideration the intellectual mode, as a means to achieve a goal. It has to be understood symbolically.

What is most explicitly indicated in this volume is not merely the reflective activity of a thinker, but the basic life process as experienced by each of these participants, with some kind of sufferings, bereavements, or deprivations, due to the natural processes, or political forces, over which an individual has no control or very little control.[19]

Hence my thought process, in development of this fresh volume, in this connection will be an approach to philosophical thinking from a novel perspective. In these writings India's comprehensive attitude with a faithful representation of the major trends of the world are demonstrated.

Indian thinkers are not restrictively oriented to the developments of their own exclusive or philosophy, and not of others also with equal interest and realization of values, but the Indian thinkers in general are deeply interested in knowing and mastering the developments of thoughts, philosophic, scientific or other types as far as possible, as prevalent in other cultural centres of the world. I am here referring to those Indian thinkers who develop these special trends I am referring to, I am excluding other Indian thinkers who do not develop this main current.

India's approach to confront the world at large is with an awareness of its cultural philosophic assets, and of the major countries of the world in general, with a potentiality to develop its scientific and technological assets and knowledge, as to confront the changing economic and political visions of the world in march.

The Indian approach of this special group is truly non-traditional from any cultural exclusive perspective, yet it gives a shape to a new orientation, wherein all can participate inspite of their cultural differences.

India provides a common platform for philosophic deliberation, befitting Indian cultural process. This will be a speciality of India's contribution to the world.

This book will have also some references to other relevant writings on Indian thought both by the Indian and non-Indian thinkers only as progressive supplementary processes, to determine how far others have understood India, and to give them a right perspective of what India really is or going to be.

India, in its advanced aspect, wants that the other parts of the world must not persist with some biased traditional notions about India.

To create this ground for the change of attitude India's intelligentsia should be differently operative. They should indicate the possibility of transcending the spheres of restricted thinking in terms of one's concepts. They are processes of learning from philosophy to the sciences, with a mastery in politics, as one of the dominant trends of the contemporary world-process.

It is a scheme of freeing oneself from all ideologies, and then stepping into the spheres of transcendent insightful processes.

In philosophy the exercise is for studying the major Indian cultural processes as provided by the intellectual forms from the Vedas through Buddhism to the Vedanta and contemporary movements, in contrast to Jainism, which also pursues another alternative dominant transcendental modes.

All these transcendentalists are in contrast to the materialist trends in change and development as pursued by the Indian *Materialists* or Carvakas from century to century.

India, after the Western impact, remained aware of the historical development of European thought from the Greeks to the contemporary developments, as stretched to the Americas.

Indian thought throughout the centuries, was in relation and in contrast to the major Asian modes of reflections through Buddhism and Hinduism.

Equipped with this kind of novel non-traditional philosophic consciousness and comprehensive liberal outlook towards philosophic deliberations, as a means and not remaining confined to any restricted thinking, the new Indian mind will probe into the vast domains of advanced scientific and political ideas by transcending mere thinking and passing into transcendent insightful modes as leaders in bringing about some changes in the world's cultural processes as a whole, as far as possible, fully understanding its limitations, in a world of mutual dependence.

With an advanced philosophic disposition, and with a reforming cultural perspective the Indian mind can shape and fashion the interrelated prospects of science and politics, in international contexts, eliminating from the human mind, the suffering for dogmatic faith in some ideas or in theories as ultimates.

From the advanced Indian perspectives, all objectives are tentative positions, operative as emergent contrasts to be understood, and then can be transcendend, by a posterior efficient possibility.

They are pointers to further possibilities and further improvements suiting a country and its people, without trying to compete with other countries or cultures. Indian mind is not for creating ill-feeling or conflict as happened to the whole world in its historical developments to this century.

This capacity to be detached from the 'immediate conceptual or empirical outlook' is possible for those who are conversant with the path of discipline as developed in Indian thought in contrast to early Indian Vedic and Upanisadic thought – with a profound silence, where thoughts come back frustrated. It is an ascent in progressive insightful ways – in symbol.

India will certainly have its own choices and preferences, giving always precedence to scientific achievements with a broad humanistic attitude, it can remove of the defects of politics of the present-day countries of the world, where the objective suggestion for them, will be a general sharing of the rich scientific and technological knowledge of mankind for benefitting humanity as a whole without coming to some self-created clashes for exploiting one's rich resources or possessions.[20] The ascent in every case is a departure from positivist dispositions – as mathematics – rise to symbolical Zero by a departure from all positive numbers from 1 to 9.

In the present international situation, it has become very clear that the display of the military powers of the superpowers, as of the superior kings of the world, is a futile game – a sink into the glooms of economic bankruptcy.

Mankind should not any more waste its time and energy in vain displays but rather should come together to share the prospects of nuclear sciences in all their aspects, for the benefit of mankind in general, removing poverty and disease prevalent in different parts of the world.

There should be no desire to equip oneself with the military equipments from foreign powers, only, in place of the scientific and technological skills for the economic prosperity of the country.

Both affluent and non-affluent countries, if they have any possibility must now soar with more and more freedom, with a detachment from mere intellectualism in the galactic universe, to discover the mysteries of the outer world, or of the other types, of this multi-dimensional universe gross or subtle, for obtaining progressive visions of the experience, to give a new direction to mankind in general. The vast outer sphere or empty universe, is left for mankind as a common objective to share and exploit.

This present book, will make necessary use of this writer's earlier works to develop a prospective idea in international and intercultural contexts, with some of India's suggestions in the shaping of the future of the world in the three constructive aspects of advanced philosophic, scientific and political thinking of mankind in general, considering the precious human life in this present century to be followed soon by a new century.

In India there are two sets of mind – one refers to the really awakended seniors who pursue Indian thought in its developments from the period before the First World War to Second World War and to India's independence in 1947.

The other belongs to the period after Indian independence to the present times. The Indian seniors have experiences of both these periods within their life-history – with progressive ascents to symbols or emptiness.

Through my books (i) *Changing Phases of Buddhist Thought*, (ii) *Dynamic Facets of Indian Thought* in four volumes, (iii) *Experience in Change and Prospect Pathways from War to Peace*, (iv) *Sri Aurobindo's Vision of the Supermind*, and (v) *Buddhism and Whitehead's Process Philosophy* – I have shown, from a comprehensive study of philosophy, religion, and psychology of the major cultural processes of the world that Buddhism, is an advanced Indian philosophical and psychological attitude, in contrast to its earlier Indian traditions, and similarly, Whitehead's philosophical and psychological attitude is an advanced perspective in contrast to the earlier processes of Euro-American thought. In all there is a departure from positivist tendency.

I have then developed a study of the advanced currents of both Buddhism and Whitehead for over forty years, and formed a purely research perspective, and with a detachment found, their common points of agreements, that adhering to any ultimate or final in any stage of experiential process, from the myths to the sciences, is a wrong intellectualist attitude, fostering some form of argumentative habit, and not rising to a transcendent insightful wisdom order of freedom from intellectualism or positivism. The ascent is to a symbolic process.

These works are further supplemented by my two other works: (i) *Zero: Its Role and Prospects in Indian Thought*, suggesting also an impact on post-Einsteinian astrophysics and (ii) *The Mysteries of Vajrayana Buddhism from Atisha to Dalai Lama* – all displaying an ascent to Nothingness by progressive contrasts.

They are courageous attempts to develop my scientific-mathematical-philosophic symbolic ventures. They are also my developments of Buddhist trans-empirical psychology to suggest that mankind has finally to address itself to a transcendent psychology in contrast to the empirical, if mankind decides to terminate its man-made sufferings by too much of thinking or arguing for the sake of mere arguments – an addiction to positivist disposition.

My present venture with the above title, as triadically oriented, is a direct descent into a practical world, with a kind of survey from a concrete day to day perspective with a progressive scheme of 'insightful projects' leading finally to a non-positivist detached attitude.

A study of India's past history will make any inquisitive scholar understand, that India, throughout the centuries, in confronting the changing situations, was always operative with a reshuffling of its philosophic positions progressively improving upon its scientific knowledge and political policy, with an ingenuity.

But due to changing situations and emergent problems, India was failing or missing to realize all its objective.

This failure was not India's exclusive failures, but it was circumstantial as well, for, many problems, not thought of before, emerged after the British rule of India, and also due to the intervention of other countries, as parts of the world as a whole, created problems.

It is interesting to observe how India will be operative in this century, or in the future to come, with an increasing consciousness of the other parts of the world, in this century, in international milieu, and will make a step forward in possible progressive aspects with mankind as a whole, where the place of insightful thinking of a modernized Indian mind, is a further 'felt need.'

This is a common understanding and sharing of the enlightened knowledge and energy continuum. This new advanced Indian mind has to be progressively created by creative writings, stretching to different areas of conceivable processes, and by a revolutionary scheme of educational programme from the schools to the universities, extended to the special post-doctoral institutions. But the goal in all should be for a point with a detachment for generating a process for a common interest.

Some of the institutions of advanced studies in different places of India are addressed to this process for a common humanist outlook. It is more necessary than giving any chance to rash and unthinking activities. This statement is not an emotional expression, but a progressively constructive plan of action to change the young minds rapidly, or with a progressive speed.

By an over-all understanding of the philosophical developments and interpretations of ideas in general, this writer contends, that some kind of progress is always possible. This kind of possibility has to be effected with a new spirit, in this psychic-cosmic universe, in a progressive symbolical way.

One has to reinterpret the general consciousness of mankind as a contrasting process, entertaining some tentative standpoints or views, with no ultimate existential value, for, neither the human minds, nor the confronted objects as they appear are 'absolutes' of any sort. Hence, it is also another folly to prepare oneself, with a revolutionary attitude, declaring that nothing can be known.

Avoiding the extremes, a reasonable middle ground can always be improvised for creating a tentative solution for all conflicting attitudes, as to prepare for a solution of all problems, progressively, in terms of alternative suggestions. What is needed is a liberal reasonable attitude.

Man has always to seek some alternative ways with patience, by peaceful negotiations, between the rivals, in any critical situation, for there is no 'absolute or final idea', in any sphere of experiential process from the myths to the sciences, so there is no 'absolute God,' no 'absolute principle', in the spheres of religion and philosophy, and even so, no 'ultimate theory' in the spheres of sciences and politics.

If in physics, there is no 'absolute permanent particles' of the universe, nor even 'absolute *waves*', but all flowing to possible *fields*, or appearances then some relative tentative positions with values, or possibilities can be entertained as appearing or disappearing. We human beings, as a whole, of all ages, healthy or diseased, are helplessly placed, as intermediary phenomena, in a vast unfathomable cosmos, or in a galactic spot to perish into nothingness, with no remnant identity.

If mankind proceeds with its unattached disposition, a new consciousness, will emerge, which is fully scientific with no categorical postulation of anything as ultimate, but in a series of mathematical or algebraical symbols with stretches to *infinity* (zero).

Man has the option to pursue it, till something new can be brought out with new visions, for, mankind has to share all emergences or appearances in togetherness. This book will prepare itself progressively in its triadic attitudes, an over-all notion of the ideas as emergent in the human mind, as focussed through its general sub-title. The entire attitude of this work will be based on my earlier works, not as dogmatic assumptions, but as an useful ballast for a *forward march* to no-*thing*.

The work is projected to a selective study of books or publications, examining also some valuable newspaper materials for updating one's consciousness in process.

Both the studies of Western and Indian and other Asian thought are always reoriented by me in multifarious ways and approaches. Even Indian thought, wherein I specialize, is repeatedly recast by me symbolically, taking into consideration my progressive experiences, as luckily placed for long in different parts of the world, as a teacher and researcher.

If one attends to such article, as 'Spontaneous Dialectical Aspects in Ancient Indian Philosophy' by K.D. Kanev, in *Darshana International*, Vol. XV, July 1975, as mentioned earlier, one has to change one's vision to a new perspective, for, the Indian thought is dialectic, from its early stages of the Vedas and the Upanisads, and it becomes progressively more intensive as to reach the final intercultural stage of the present age of complexity.

Hence the dialectic, since the days of Zeno in Parmenidian Greek thought to the stage of Hegel and Marx in Germany is stretched to Russia, in Western thought and is nothing very strange, they are natural expressions of the human mind, opening up further liberal progressive views. These views are perishing processes to no views.

So India will finally stand for its "dialectic of 'neither-nor' process" not for upholding 'a specific intellectualist or empirical position' or even a 'dialectic' in an intellectualist formalist sense, but for abandoning all forms for pursuing a prospective insightful alternative in a transcendent order of *freedom*, in a post-intellectualist sense.

Some of the great Indian personalities among the philosophers of the earlier periods in contemporary Indian philosophy indicate this trend in varied ways, which will be expressed in the next and other chapters, adding to the novel shaping of the philosophic thinking of this present century, in many dimensional innovative directions, beyond any restricted West-East thought.

In the final stages of this work, some journalists' vision will be given prominance as often some inner ideas, feelings and policies are deliberately suppressed by a person or persons of a group holding power as it can be brought out like an efficient function of a midwife. A journalist's innovative scheme of questions and answers is, as we shall see later, deals with India's politics as a transition from the British parliamentary form of Government to India's complex and advanced form of democracy – a progressive transition to no 'form.'

India's prospect lies in its evolutionary Constitution in progressive change and development with *transcending* insights, abandoning mere argumentative proneness.

If the Indian intelligentsia have not reached that level, this book is an attempt in this direction, with some statements, and with some repetitions. Critics or some readers will feel some monotony, but an insightful thinker will understand the reason behind that monotonous disposition.

REFERENCES

1. History of India was written from the British and Indian perspectives since the beginning of this century. But after Indian independence, many different types of history were written. I have chosen to mention only three books on the history of India which will serve best for this text. They were written from triadic perspectives of philosophy, physics and politics. I have chosen to refer to (i) Michael Edwardes, *A History of India – From the Earliest Times to the End of Colonialism* (Grosset & Dunlap, New York, 1961); (ii) K.M. Panikkar, *A Survey of Indian History* (Asian Publishing House, Bombay, 1947); and (iii) V.D. Mahajan, *Modern Indian History (from 1707 to the Present Day) British Rule in India and After* (S. Chand & Company, New Delhi, 15th ed., 1985). The first book gives a detailed history of India through its different periods to the end of the British rule. The second is good for a general history of India with graphic details leading to the end of the British Rule, and India after independence. India now is a developing country, with growing strengths as an independent nation, with its secular attitude at home, and nonaligned attitude in international affairs. He develops also India's progress in technological and scientific aspects, and India's competence in protecting itself from the Chinese and Pakistan's invasions or intrusions. The third book deals specially with the British rule of India and also after Indian independence.

   During the British rule, the British had to face a lot of problems for consolidating its power and place, where there is a reference to the roles of various Governor Generals 'from Lord Ripon, Lord Curzon and Lord Minto to Lord Irwin and to Lord Wavell', and thereafter – after Indian independence – from Lord Mountbatten to C. Rajagopalachari to Rajendra Prasad, S. Radhakrishnan, Zakir Hussain and to V.V. Giri, India from 1950 to 1972.

2. For understanding the transition from the British to the Indian Parliamentary process through Constitutional processes – one has to minutely study the processes developed in this text from chapter to chapter of its three parts.
3. The speciality of India lies in its development of the philosophic ideas, to match with the progressive attitudes towards science and technology, proceeding to a new politics freed from any ideology. India proceeds to an advanced process of progressive democratic constitution, developing non-alignment as a foreign policy, and for self-defence, with a progress in nuclear directions, to stand on par with the so-called superpowers, if they want to retain that prestigious emblem, which is no longer there. India tries to be progressively independent of the foreign aids, but wants to step into the order of 'independence' through 'business', with scientific and technological efficiency.
4. As India's future is dependent on an advanced philosophic, scientific and political ideas, India develops philosophically, indicating its prospects through pro-Buddhist and pro-Whiteheadian attitudes – pointing out that there are no 'ultimates' in any field of experiential process where all run into an ocean of mutual self-development through continuous cooperations, opening up new possibilities. This philosophic excellence of a cancellation of categorical positivist mode has also been experienced by some eminent American and Indian scientists that there is a total zero-mass, in neutrino particles beyond the relative orders of protons and electrons. It is nothing but a discovery of a process where mathematical zero is in process towards a philosophic negation of anything positive or ultimate. This again amounts to saying that the different experiential processes, inspite of their apparent distinctions, have lost their boundaries (vide *Science*, 8 May 1992, 'Are Nuetrino Mass Hunters Pursuing a Chimera?)

5-9. General ideas of India's independence can be progressively understood with Gandhian movements and other socio-political movements, by a reference to my *Dynamic Facets of Indian Thought*, Vol. 4.

10. My book on *Zero* can be reviewed from an advanced philosophic and scientific perspectives. The search for the zero-mass of the neutrino particles has revolutionized the scientific world. If science indicates that the confronted universe in its core is mass-less then does not physics come close to philosophy, with its progressive prospects towards a complete nothingness detaching oneself from any formalist aspect of deliberations from philosophy to the sciences? The dual impacts of philosophy and science will have an effect on politics annihilating its postulation of any ultimate ideology. The present world-process is progressively moving in this direction. This process is worked out in this text.
11. The suggestion is for a new insightful scientific and technological attitude, in conformity with country's economic capability and in discerning the

priorities.

12. The suggestion is for a new insightful scientific and technological attitude, in conformity with country's economic capability and in discerning the priorities.

13 Here the reference is to some important publications on India's contemporary philosophy, and volumes on *Philosophy – Eastern and Western* – developing the processes of philosophy, religion and sciences indicating India's speciality with an attitude. India is optimistic as the world will come to that stage which India has realized by pursuing its insightful experiential trends – freeing the intellect from its progressive formalist modes, not by a dialectic in a positivist direction as in Hegel, but in the direction of Nagarjuna and his followers to this century – in the spheres of philosophy, science and politics. This text makes a reference to all of these in its three parts, as all specilists in these different areas indicate a choice for a process-philosophy, doubting the categorical character of the presented universe. By abandoning the attitude towards restricted conceptual and empirical modes of deliberations; each cultural process, in its advanced aspect, is trying to open some open spaces for deliberations for mutual advantages. This sort of advanced mode of reflection has been realized and developed by A.N. Whitehead through his philosophic-scientific works from *Concept of Nature* to *Adventure of Ideas*. The drive of the world, is beyond the intellectual forms. In this context all these trends have been demonstrated from its first to its final part.

14-16. B. Mollenhauer, 'Whitehead and the Journey of the Flame', *Darshan Iternational*, Vol. XV, July 1975 (Moradabad, India). Here Whitehead's career has been demonstrated in three stages and periods: mathematical and naturalist processes, in England, and then with a change of location in America, in the direction of a process-philosophy, discarding all ultimates from all confronted experiential processes. This prospect is non-different from the Buddhist interpretation of the confronted experiential processes. This text has carried forward the Buddhist-Whiteheadian process in its triadic pathways or avenues, as indicated in its title.

17-19. Vide my *Dynamic Facets of Indian Thought*, Vol. 4.

20. References are to my different works as mentioned in the text, continuing the choices in the direction of 'process-philosophy' – in post-intellectual direction – with progressive insightful suggestions.

CHAPTER II

About the Book through Triadic Avenues

Triadic Avenues of India's Cultural Prospects – Philosophy, Physics and Politics – critically pursues the advanced trends of these three avenues of the human pursuits of mankind in general. It insightfully shapes them to suit India's progessively developing consciousness and need, with an openness.

When India shapes its advanced philosophic attitude it is not obediently pursuing its past, but it pursues the best of its pursuits, observing the possible advanced modes of similar or not similar pursuits of the other countries of the world and delivers its preferable choices or norms, within that milieu.

Here it indicates its broad philosophic attitude, examining them with a critical eye and with insight through its representative thinkers. India's objective is to level up the processes of all major countries and cultures in a spirit of togetherness or insightfulness to share the achievements of mankind without any partiality.

In pursuing this common objective, India's claim is for giving an opportunity for an improving upon each one's existing traditions, by a detachment, with a liberal attitude.

India, in this connection, opens itself to the suggestions of the experts in these areas of the non-Indian countries, for cultivating a progressive self-improving process.

In the same spirit, India's advanced insightful thinkers from philosophy to a refined scientific attitude in development do not proceed by stressing the importance of a particular philosophic or scientific attitude, or theory, but by cultivating the different philosophical and scientific processes in relations and interdependence.

Here these exponents foster an advanced togetherness, perfecting one by the other, for developing both in a cooperative way.

Both philosophical and scientific experiential processes are deeply interconnected: often they may not be operative in specific or abstract conceptual or empirical ways, but they may be operative in a transcendent insightful direction, quitting the apparent confronted categorical aspects.

India gives a balanced stress on the importance of the objective universe as confronted, in terms of one's insightful and symbolic understanding of a vastly changing universe, where the scientific understanding of an over-all cosmology has to be determined through an advanced knowledge of astronomy, with progressive observational technological efficiencies.

It is a development of the range of human vision both macroscopically and microscopically, through the medias of X'rays or computer medias with speed and accuracy.

Through telescopes, and through satellites and voyagers, stationed in the outer space – a man extends the scope of his vision in the interstellar spheres.

Hence the earlier modes of scientific thinking were radically changed through enhanced visions leading to the prospects of an advanced physics, and other associated sciences.

The new possible conceivable ways, are no longer abstractly conceptual, but they are alternative possibilities, not in a categorical sense.

This is only a preliminary scientific attitude of entertaining the 'possible' requisites from the advanced Indian perspective.

In this area the Indian scientists have chosen the path of following the post-Einsteinian processes, beyond Einstein's conflicts between the Special and General theories of Relativity.

Here they tend to resolve the conflicts by suggesting that science as a pursuit, is not restricted to a static vision or theory. It is a continuous process and change.

They now understand a place of a specific vision or theory, for the time-being, awaiting a possible change, hence insightfully entertained with a readiness for a prospective process.

This new prospective attitude will resolve the controversies between the particles and the waves as energetic continuities, running into fields beyond both these restrictive point of views. The galactic vision in the interstellar regions makes this process a normal phenomenon, which outmodes all theorist modes of vision or examination of the confronted experience.

Thus understanding the prospects of astronomy and physics beyond all theories – the present day scientists and astrophysicists proceed towards the entertainment of possible visions, or to a transcendent basic feeling-process as a condition operative behind all experiential processes, from India's advanced philosophic-scientific attitudes insightfully.[1]

It is utilizing India's past symbolical mode through 'Zero' that governs India's deliberative ways from the myths to the sciences.

The Indian savants treat the theories as operative principles, in a broad pragmatic sense, in historical sequence, as all positive numbers from 1 to 9. But there is a vast sphere of the 'indeterminables' or 'infinities'—on both sides of the positive numbers, which can be symbolically indicated by "Zero" (0) both in the terrestrial visions or in the visions of the outer spaces.

The positive numbers, inspite of their importance, need to be transcended, when necessary.

In fruitfully realizing their prospects, beyond the restrictedness of the positive numbers or theories, Indian scientists are trying to speed up the scientific prospects of India, by going beyond all traditional scientific ways, as restrictive numbers, and quickly relate them to the symbolical aspects of technology and engineering sciences.

This is a move to the practical spheres, without continuing the obstinate process of mere discussion.

The present scientists, want to bring about a social change, as economically as possible, understanding India's limited economic assets and resources, through insightful planning and projects.

If, in an aspect, the major prospects of modern astrophysics, depend on an extended telescopic vision through the satellites and voyagers with most advanced types of X-ray equipments, then different visions of new galactic universes will be opened up, this will surely extend both philosophic and scientific visions in togetherness, nobody as yet knows in what directions.

Hence, a country like India, with its restricted economic capability has to move in the path of scientific discoveries, in these areas, with a caution and ingenuity. But India has to maintain its scientific capabilities somehow. India has successfully passed through these different stages through its progressive space programmes.

These different sciences and technologies will outmode the past insightful disciplined visions of many cultural centres of the world. It is a challenging outlook of sciences, reorienting cultural processes from a new perspective in progressive aspects.

In the political sphere too, the Indian mind is operative, in a special democratic spirit, detaching itself from other democratic countries which pursue some form of conceptually oriented democracy.

India has a democratic constitution inherited from the British, but in its transition, it formulates an over-all general attitudes of observing the differences among all types of governments.

The international policies of the superpowers or developing powers of the world have to be examined as alternative processes, as they are, but India has to be operative differently, beyond all conceptual and empirical restrictedness of the laws prevailing in a country or culture, which mankind should cultivate in a right spirit. Hence India has to be operative with the help of a possible 'internationl law.'

Any law or principle, from the advanced Indian perspective is not an 'ultimate', and hence India presents its own advanced political attitude, from that prospective insightful attitude, formulating a 'constitution' of its executive powers with a constitutional head of a prime minister, but in emergency, India will be operative with or through a president with a superior power to manage the state of affairs till another prime minister from the majority party emerges with some new policies.

Here the function of the Indian president will be like 'wait and see' given a time and in a new changed situation. He then is placed in a situation with 'alternative possibilities', and he becomes operative with a new mandate, after a fresh election, suiting a new emergent situation within India, from a secular democratic perspective.

India frankly acknowledges its failure or relative success by way of implementation.

In international contexts, too, India admits of a pervasive feeling for the governing principles of the other countries as operative, in different parts, with a detachment. It determines its policy of nonalignment functioning as an observer with an impartial attitude, it remains a pointer to the best of the insightful suggestions for solving international problems.

Indian mind is broadly aware of the frailties of the human mind, and so its suggestions should not be one of a pressure but as presenting some reasonable alternative choices, for consideration, with options to choose or not to choose, (vide Part III).

Inspite of its insightful objectives, India often proceeds with failures, knowing fully that there is a gap between the wishful idealist programmes and actual projects – and the possibility for their successful execution. Hence India's ideas and ideals are cast into the future possible actual situations, before the uncertain human frailties of mankind as a whole, creating an optimist outlook, in an intercultural milieu, knowing fully the ominous possibilities that mankind often faces.

Man has to confront two important spheres – (i) Nature and (ii) the social and international environment.

In this review of India's triadic pursuits the readers will confront the swirls of many Indian minds operative before and after India's independence, from the triadic perspectives of Philosophy, Sciences in general, and from Politics (Part III, for details).

In this internationally oriented process India tries to change the mistaken notions of non-Indian countries that Indians are all dreamers, and not critical observers, from an advanced scientific and technological perspective.

Behind this representation of Indian thought—this writer indirectly suggests that, these are advanced trends of Indian thought, but India's masses or even some of the intellectuals often miss or forget to pursue the advanced trends. This is also true of the non-Indian thinkers, both in the areas of scientific and non-scientific aspects of deliberations. Often they might deliberate with fixed attitudes, refusing to admit India's possible prospects, of a vast number of intelligentsia.

India's drawback lies in its increasing population figures, but it has to its credit a large number of general intelligentsia. Admittedly, India is an economically poor country, but no one can ignore its maximum number of intelligentsia, which outnumbers the entire number of population of many countries of Europe.

India has therefore many alternative ways of tackling the crisis-situation that might emerge, as may be devised by the teeming intelligentsia.

This book will serve as a reminder of India's special objective in international and intercultural situations. Hence, this book should be taken as having that kind of potentiality or mission, for those who are idealistically oriented, and for the scientists, who should play the role in a subtle manner, to raise India in the actual sphere of intercultural milieu, later, primarily in a scientific direction, viewing its political attitude as a secondary process, creating a complex human situation (Part III, at relevant places).

Hence India's political attitude is superbly cultural, as yet not fully associated with the magnum processes of its advancing scientific and technological achievements.

Politics is a behavioural policy in the milieu of other nations, to safeguard India's interests only by reshuffling its political policy deftly as a defence against the dominating tendencies of the superpowers, and other developing powers.

Its own inner political health has always to be maintained by progressive researches in exercising its policy of nonalignment as a forceful cultural initiative having a distant international value for the good of mankind as a whole, where India has a place to coexist, as an ally for all (vide India's scheme of using nuclear energy purely for peaceful use, for the development of country's agriculture, medical and researches in oceanography, geology and space-sciences in progressive aspects) (Part III).

Bearing in mind all these triadic processes, where India's chief force has to be understood both philosophically and scientifically, as harbingers of an advanced philosophic-scientific spirit and attitude in togetherness,—the writer of this new volume, will pick up some advanced ideas as realized through his understanding of the philosophic-scientific deliberations as developed in the fourth volume of his *Dynamic Facets of Indian Thought*, in the sphere of philosophy, as also in sciences based on mathematical symbols, as expressed in his book on *Zero*.

Hence the apparent repetition is not really a repetition – but an attempt to present the thoughts of some persons who came as reformers – from the new perspectives that will be clarified in this volume.

This writer is also for developing the advanced movements as indicated by some eminent thinkers of this century from an intercultural perspective, philosophically and sociologically, and later scientifically also.

So India takes note of its different types of people, having philosophic, scientific and sociological interests, as they are interconnected processes to prepare the basis for 'thinking' in the areas of 'politics.'

India's advanced philosophical-sociological process, can be indicated in one way by Raja Rammohun Roy, in the nineteenth century, in his insightful following of the integral reality of the Upanisads in Brahmo-Samaj.

This might be indicated also in the self-transforming processes of Arya-Samaj, as by Dayananda; he, transcended the abstract attitudes of any theistic religions, from a Vedic perspective.

Still another alternative for understanding the higher processes of religious experiences, can be determined by pursuing the best of all religions by one's own inner experience, not alone by an ecstatic experience of a Ramakrishna, but also with an inspiring spirit, meditatively acquired by Swami Vivekananda. Swami Vivekananda's method of expressing the core-spirit of religion in international religious sphere was an 'international attitude' as distinguished from the somewhat personalist attitude realized with a considerable liberal humanistic perspective by Ramakrishna.

Vivekananda prepared the grounds for participating in an international conference in Chicago, by spreading the core-spirit of religious experience of his Master, Ramakrishna, not by an ecstatic way of 'theistically oriented religion,' but it was a 'common experience' in all theistically oriented religions.

His religious experience was not excluding any theistic religion, but experimentally realizing the highest experience of all religions, Christianity or Islam et al., in eloquent communicative way addressing the intellectual masses – as 'brothers and sisters' – representing different religions at the above sessions of the Conference, and transcending them all, from that ground with a togetherness.

His religious experience was to consider each individual religion as equally 'divine', with a right to coexist in the same spirit as suggested through Hinduism. Hinduism does not want to stand as a separate religion, it ventilates a common religious spirit that is found in all religions of the world.

Indirectly, this socio-cultural approach to philosophic-religious process was shaped in an alternative garb by Tagore, from his poetic vision. This common spirit could give it another shape through his poetic work – *Gitanjali*, in addressing God as his ever-novel transcendent interpreter and a guide of life, – day and night – as a pathway. He himself was a voyager in this astrophysical universe.

In contrast to spontaneous expression of Tagore's poetic visions, Sri Aurobindo suggests a transcendence of one's 'over-mental processes'., stepping towards a 'Supramental order', which gradually comes down to the mental or other lower levels of experiential processes, only when in them there is an effort to transcend their respective restricted levels. He places before mankind a 'reciprocity' between the 'over-mental' and the 'mental' processes, in the form of 'togetherness stepping down to the final point of 'materiality', without separating it from other end-points, with a return to the highest reachable point of the 'Supramental Experiential' possibility, with pervasively present currents.

After all the transcendent Supermind is a continuously descending and ascending principle, not merely the transcendent fourth of the Upanisadic thinkers.

This transcendent Supermind of Aurobindo, is the placid order of a 'continuous freedom' a movement of the material to the spiritual, and making the spiritual come down to the material, which is experienced when an individual completely frees himself from all intellectual forms.

This reciprocity between the two orders – material and the spiritual – reminds one of Nagarjuna's insightful understanding of the Buddha's suggestion of totally abandoning the 'intellectual ways', for a 'lift' towards the 'spiritual.'

Nagarjuna's was concrete way how to be completely freed from the trammels of the intellectual ways. This is an analysis of the different forms of intellectual ways—(i) assertion (ii) negation (iii) their combination, and finally their (iv) total abandonment—by an ascent to a dialectic negation of 'neither nor' when nothing remains to choose, from the presented transcendent situation as 'there nothing disappears, nor anything appears' et al.

So what is left behind is a vacuity with a transcendent self-sufficiency the continuous transcendent basic feeling of freedom from all intellectual forms, or restlessness.

I am repeatedly mentioning about this discipline, for, man is generally fond of staying with intellectual forms – and fails to rise to the transcendent post-intellectual process with nothing to see or choose, but a transcendent sphere of continuous self-sufficient unbroken placidity or feeling.

Hence by a post-intellectual discipline, a man rises to the transcendent condition of the 'Placidity' or 'Freedom' of insightful experiential process—an unfathomable 'open space.'

In Gandhi's new political vision, a new political movement began in the garb of a non-violent form, which if felt from within in the depth of one's silence, is a 'no-choice' attitude or a sacrifice of one's egoistically oriented ideas—when applied to 'politics' one is led to a constitutional reform of the British rule of India.

His was a special type of 'nonviolent protest,' revealing, not just civil disobedience of a Thoreau of America, but it was an inner psychological change of attitude – a giving up of everything even one's dear self – a surrender – giving the ruler to do things best as he judged as right. This was a call for a supreme placidity in opponent's mind to transform himself in a new direction.

This is nothing but appealing one holding some superior authority lose the basis that compels one to exercise one's power.

Apparently it is a suggestion of upholding one's process of perfecting one's inner self, but actually carrying out a protest against the unjust British law as expressed through its 'salt act.' So Gandhi rightfully led a mass of people behind him, walking miles to the seashore, as demonstrating a civil nonviolent protest.

It was creating a 'mass awareness' by a 'mass movement'—an actualized idealist behaviour, with a potential for invisible force, touching the minds of the poorest of poors and also the men of wealth and power of the British rulers or administrators.

The reference is here to 'salt-movement', a 'symbolical gesture'—which touched the minds of the lowest levels of the Indian people, as dependent on using a pinch of salt in eating one's daily food. What a pity that also was taxed!

It was a non-acceptance of something not good for mankind as a whole. It is a superb psychological attitude, that Gandhi wanted to utilize for transforming human nature and human psyche, in international contexts, with a kind of awakening in a new direction.

The British ruling people were terribly disturbed. They were compelled to give up their policy of solving problems by taking recourse to violence.

Without going into any restrictive conceptual mode, Radhakrishnan, who began his career as a philosopher but later through a demonstration of his efficient executive power could attract the attention of the-then political thinkers of the time from Gandhi to Nehru. He was also able to touch the heart of some foreign countries as the United States, so that he was called for delivering special lectures to the American philosophically and internationally oriented persons, specially organized by James, Dewey and others.

So from the philosophical sphere Radhakrishnan's need was felt in the political sphere. India rightly responded to this elevating process by snatching him out from his academic sphere of a Professor of Philosophy and Vice-Chancellor – by appointing him to the role of an Ambassador to Russia.

When it was found that Radhakrishnan had a capacity to change the mind of an iron man like Stalin, to talk with him in human ways – with a sort of inner confession that Stalin made himself that he 'was to die soon', and 'Radhakrishnan to live longer' – was almost prophetic.

That psychic disposition in Stalin came out by a simple touch of Radhakrishnan. The Indian government was quick enough to make him vice-president and later graced him with the position of president. It was a kind of responding to Plato's rationalist attitude of reciprocity between a 'rationalist as a King,' and 'a King as a rationalist' of the highest order. Radhakrishnan paved the path of a switch from philosophy to politics, where the role of a superb personality was displayed with a novel capacity, which Radhakrishnan could acquire.

Hence naturally Radhakrishnan progressively rose to power from the academic to the political sphere, and this unique phenomenon made his place very important among the active politicians of his time.

Radhakrishnan's liberal philosophic attitude, expressed through his academic and public lectures, was not a postulation of an abstract or categorically present Absolute, but his was an 'Absolute with infinite possibilities' or 'potentialities.'

It had a wider sphere of application touching entire humanity, giving the human beings a chance to choose out of the infinite alternatives – a relativist universe – where a 'God', as a symbolical expression of a pervasive religious experience – could be assumed as a Planner, confronting a world to be formed, with alternative philosophical attitudes.

This transcendent God – as a Planner of the confronted Universe – has an option or freedom of abandoning its function after the completion of its functions. Such a God could be set aside as if as a lapsed experiential process, as not needed. Hence it lapses into the habitat of an ever-present transcendent reality or the Absolute in an insightful advanced way – as an Abode of infinite possibility.

This process of lifting oneself to a transcendent meditative experience – with an offer of a role to play – and then on completion – to lapse into 'nothingness' – was well indicated throughout his life. His role in contemporary Indian thought was of another Nagarjuna of Buddhist thought revived again. His 'power' was demonstrated from the 'wider range of philosophical thinking' to the heights of 'political thinking,' when he demonstrated that 'democracy' in the real sense 'emerges' from one's 'solitude', the same sphere which gives birth to religious and philosophic deliberation.

To Radhakrishnan, religious experience is a dynamic confronted phenomenon – to be dealt with from a broad possible perspective, without denying its presence and importance as every other confronted experiential processes. It is 'symbolical.'

But still God is not an Absolute Reality, to Radhakrishnan. The religious experience, like all confronted experiential processes – can be found as an experiential process from the myths to the sciences – but it is sheltered in another superb possibility, which is the 'transcendent habitat' of all confronted possibilities. It lapses as all other confronted experiential processes, giving place to other 'experiential possibilities' by opening up other possibilities, or pathways.

In contrast to that vast unfathomable sphere of many-dimensional possibilities man is always operative in a dependent sphere in entertaining its religious experience.

Who knows that this superb experience is one of these Supreme Orders of Infinite Possibilities of that the 'God' of Radhakrishnan is not the only possible human choice – but it is one of the possibilities, amidst many other possibilities, operative in the religious sphere.

But it is not unlike any other confronted experiential process and possibility, and that it is distinguishable from the other confronted experiential possibilities and processes, from philosophy to the sciences.

Quite naturally Radhakrishnân was considered as an attractive charismatic personality of his time, even to be chosen in the 'political sphere.' (Part II & III, for understanding Radhakrishnan's role from philosophy to sciences and politics). He was a pervasive element present from philosophy to science and politics.

Philosophic attitude was then not some cluster of ideas absorbing one's thought, but it acted as a dynamic force inspiring one's activity-process controlling the confronted situation. This quality of a personality governed the silent activities of K.C. Bhattacharyya and D.M. Datta.

In Part III of this book, I have referred to Radhakrishnan again, where he transports India's democracy' as a process that emerges in one's 'solitary' experience. Hence to him, the experience of politics, is non-different from the experience of 'religious experience' (God) as one of the confronted experiences with a further possibility.

K.C. Bhattacharyya and D.M. Datta, were operative in a similar or yet not similar possible aspects, they were just upholders of some alternative powers to represent one's proneness for thinking in terms of the final and the confronted reality, suggesting that beyond all conceptual and empirical modes, i.e. some thoughts in some sense – are not 'ultimates', but relatively potential forces – permeating all experiential processes.

Their attitudes were not radical as of Krishnamurti, who stood for detaching himself completely from any adherence to a culture – or a confronted experience by a process of a total de-conceptualizing or de-culturalizing attitude. By this non-adherence to anything confronted he claimed to acquire an unrestricted potentiality, which he thought he will miss, by holding on to any culture. This was however a very harsh attitude to hold – an unreachable precipice of elevating oneself from a ground-level where the masses travel – so that he could remain independent of any confronted culture, and shaped his personality accordingly.

But I will prefer to choose Radhakrishnan and others on his line – to give the ordinary individual to reflect a while with the confronted experiential process – to provide a temporary shelter before lifting him later to a further sphere of possibility. Often extreme detachment creates an aloofness, separating such an exalted personality from the teeming masses.

For rendering possible some form of normal educational processes, with a progressive transformation of the human nature from the depths of one's psychic nature was effectively realized by the Buddha from the ancient times, six hundred years before the Christian era. The Buddha's was a similar and yet not similar detached process, it was realized when he suggested a scheme of detachment from all forms of discursive or argumentative propensity.

The Buddha's force was still operative in the formulation of all reflective activities from the early to present century among the great scientists of India, in a gradual form, with some preparatory grounds, before abandoning all 'confronted forms.'

They were all internationally recognized, with due honors for their dual interest in the apparently confronted phenomena with a lift towards a further transcendent possibility.

Besides the philosophical pursuits, one must as well take note of reforming the general scientific attitude by following the advanced ideas of J.C. Bose, P.C. Ray, C.V. Raman, Homi Bhabha, S. Chandrasekhar, George Sudarshan, and others who gave wings to scientific thinking as an exercise with an admission of a confronted sphere – the concrete vast energy-sphere that are pervasively manifest.

Scientific experience, like all experience, is confronted in many-dimensional alternative aspects, so that a scientist could be operative with all aspects of experiential processes, from his particular scientific perspective. Like other experiences, its cultivation can change the prospects of the Universe, specially in all directions from the cosmic to the socio-cultural sphere. But this aspect of experience has to be insightfully cultivated for human welfare and not for destroying humanity by its nuclear force.

For this dual attitude for human good and prospect and choosing science and technology for human good the above mentioned Indian scientists, like the above philosophers, drew the attention of both the intelligentsia and the masses. They gave some preliminary steps to traverse, before rising to the 'transcendent fields' crossing the boundaries of all specific theories, or separating one science from another.

This new scientific gesture transforms a scientific potentiality to an appealing aesthetic experience, resolving the apparent gulf between science and aesthetics. This possibility was demonstrated by the Western scientists from Heisenberg to Schrödinger, and in India from Raman to Chandrasekhar.

These scientists, with liberal and comprehensive attitudes were able to move from one sphere to another, becoming conscious of the subtle scientific currents of the macroscopic universe of physics and astrophysics to subtle musical and aesthetic spheres of experiential process, to refine the inner nature of man, and direct mankind to real pathways of civilizing the human mind, not merely by high-soaring ideas (only for the intelligentsia) – but by getting rid of its varied inhuman egoistic propensities. They made India's science, humanly oriented, and made it technologically suitable, for the masses.

Indian scientists tapped the sphere of science with insight and in its advanced currents.

India's scientific vision is reinforcing of ideas with the necessary terrestrial energy-continua that are pervasively present and operative with relative forces, in many-dimensional ways. An access into all these spheres through scientific ideas and technologies is a need for progressive human efficiency, and raising mankind from the instinctive to intellectual and post-intellectual non-restrictive levels. It is a continuous social process. One can compare this process with some of the ideas of the book entitled, *The Impact of Science on Technology*, Edited for the Columbia University Seminar on Technology and Social Change by Aaron W. Warner, Dean Morse and Alfred S. Eichner, Columbia University Press, New York and London, 1965. The participants of this volume were some distinguished scientists as Rabi, winner of a Nobel Prize in physics, and others.

The present century should avoid playing the past of the human civilizations, subjecting oneself to some egoistic ideologies, declaring the impossibility of conquering the basic animal instincts and egoistic motives.

India's is a challenging stand, making a realistic move – an understanding of human nature in all its dimensions and aspects, from the physical to the chemical and biological spheres. Man's Nature is inextricably linked up with the progressive insightful orders, transcending the egoisic instinctive spheres. India intends to direct the entire international bodies to subtle philosophic, psychological and scientific spheres of non-violent aspects of experiential processes.

Man's nature, from a superior platform prescribes necessary food to feed the basic and higher aspirations, with a togetherness, when Man really ascends to the exalted human level in a full sense of the term, through the cultivation of a science, in the ladder of true civilization – man does not separate natural and the animal world, but treats them both with a tolerable compassionate attitude.

In this hard-earned platform of the world-process by one's strenuous intellectual effort, a refined and insightful man recommends for the entire human beings to share without partiality, all its earned capacities through knowledge and appreciation of all aspects of experience as connected.

This is not a Utopian ideal, but a concrete philosophic-scientific adventure and achievement, in togetherness. It is a suggestion for a tolerant attitude, and a choice for a life of coexistence, avoiding violence by alternative possible suggestions. It is an ascent to the real exalted order of the Brahman, as a progressive step, from 'Matter' to the 'Spirit', without neglecting any.

This scheme might sound very idealistic, and as an impossible project. But otherwise what is the purpose of philosophical or scientific deliberations, discovering the depth-processes of the human psychology, or sciences in general, or even the constitutional peculiarities of the different forms of government of mankind in general, from its primitive to the civilized forms?

Hence the new project of this volume will be first to examine the present trends of the Indian thinkers in general along these triadic lines of advanced philosophical, scientific and political developments of major areas of the world, to bring out some advanced clues for the development of the human nature further, in progressive civilized ways, in a spirit of 'togetherness,' by introducing some simple methods of education.

It is not trying to understand everything, that one has to begin with clean habits, and simplicity of choice of food and dress, and concern for others, for development of proper social behaviours, eliminating as far as possible the conflicting attitudes, spreading hatred; the choice of all will be for fostering a love for the fellow human beings.

In its three parts – philosophy, physics (including astrophysics) and politics – only advanced ideas in intercultural contexts will be taken note of, with a special reference to India's contributions to the world in general, from one's personality-process, where through each of these intellectual insightful avenues, mankind will manifest a human nature in progressive compassionate direction.

Each Part will have separate methods with its different chapters, giving an over-all perspective of the Part, in relation to the other parts, followed by the details of the subjects relevant to the chapter, with a final conclusion providing some possible acceptable ideas. The universe as presented is a possible universe, where categorical assertions have no place.

The book naturally will have some 'repetitions' about the subjects, and of the personalities for their important roles in the shaping of India's three interrelated avenues in different occasions.

The drive of this entire work in triadic avenues is entirely a secular process with a stress on a historical perspective where science and technological movement has been given a prominence.

There is also an admission of a principle of transition and advance in all spheres of experience – from philosophy and religion to science and technology from many-dimensional perspective – where there is no place for an ultimate or Absolute.

The main process that has been developed is a continuous transcendence of the conceptual, empirical, and Absolute processes to the stage of Hegel and also beyond transcendent phenomenology of Husserl, or beyond the transcendental existentiality, of Kierkegaard. There has been also a transcendence beyond realist and evolutionist trends of Bergson and Santayana, and also beyond the Pragmatist-Instrumentalist thought from Peirce to Dewey—to a socio-linguist process from the animal to the human orders.

Beyond all these confronted process there is a rise to the 'Process-Philosophy' of Whitehead, in contrast to Western thought and beyond ancient Indian thought to Buddhist negativist mode, with a further consequent processes of India's 'Movements.'

In Whitehead there is a continuous transcendence beyond the confronted spatio-temporal processes, to some basic non-sensuous cosmic-psychic process and possibility, from the 'eternal objects' and 'propositional processes' – to emergent conscious or intellectual processes – and beyond intellectual process to a transcendent cosmic-psychic order as 'God'—as a possibility in an Epocal cosmic-psychic order beyond the eternal objects and propositional processes.

All these processes from the spatio-temporal processes to God were transcended finally by *Creativity – Creativity* is a process and possibility from all *forms* to the *formless.*

But one should know that from Whiteheadian perspective – none of the above cosmic-psychic processes and possibilities are categorically oriented. They, as confronted occasions, are 'occasions' of many-dimensional possibilities as alternative processes—'may be.'

Whitehead's thoughts as concrete emergent cosmic-psychic orders are akin to transcendent insightful processes, not of the intellectual order, and even when intellectual, from the beginning, does not belong to the categorical order. Hence his thoughts are not in any sense Westernly, it is a departure from 'Western tradition'.

It is an alternative insightful order if contrasted with the Buddhist mode of deliberation. The Buddhist mode from the beginning is based on a refusal to think – hence it is 'insightfully oriented' in three triadic lines of a 'disciplined order' – cut off from the propensity to discuss. Hence it is in the new dimension of continuous concentration (*Samadhi*)—and direction of the concentrated energetic continuum to insightful experience (*Prajna*) – a continuous *transcendence.*

Here the Ultimate if any, is a continuous process and possibility in many dimensional order – flouting Aristotle's notion of the Ultimate or God as of an 'Unmoved Mover or Motion.'

If Whitehead's insightful experience is a many-dimensional creative continuity (Creativity) and the Buddhist process is a continuous cancellation of all possible 'confronted intellectual forms', by a dialectic negation of 'neither nor,' then these dual processes – the advanced Western and the advanced Indian insightful processes can be compared and contrasted from the precious move towards varied transcendence – as expressed in the book, *Transcendence*, edited by Herbert W. Richardson and Donald H. Cutler (Beacon Press: Boston, 1969).

While appraising the different writers for their recognition of the need of an experience of 'transcendence' in all orders of experience—by these participants from Huston Smith to Charles Hartshorne – of their interpretation of 'transcendence' from Whitehead's and in some sense Buddhist perspectives – I have sufficient reasons to differ from all.

To me they are neither fully Whiteheadian nor fully Buddhistic – for they are not so clear about *thought*, in the Whiteheadian sense as a pure possibility – interpreting thought from many dimensional bases, as indicated above, nor are they Buddhistic, in the sense of fully *transcending* the *intellectual forms*, in the Buddhist sense of a total separation from intellectual propensity.

They have not exhibited fully the application of the 'neither nor' dialectic of Nagarjuna.

In each of their deliberations, I find a persistent *choice* of either a tendency to analyse some of their own thoughts, or of traditional philosophic, religious or logical thoughts of the past Western thinkers. The real prospect lies in separating oneself from any form of thought, in any specific aspect, as all forms are passing phases, ending up with any final form. To claim a final form – is a similar fallacy.

One will find in all the Parts of this book a continuous pursuit towards progressive insightful experiential processes – as alternative possibilities, without making any positivist categorical attitude. If there is a pressure involved in this book, I assume that this is Natural as nothing good is obtained without Hardship. Life is hard and is a continuous suffering. But there is a 'Transcendent Possibility' in all spheres of experiential processes.

REFERENCE

1. To the Buddha and to Whitehead also, if the presented universe is not 'categorically confronted', but is 'confronted' as a possibility, then one has to detach oneself from it, by totally freeing oneself, from any argumentative propensity.

PART I

PHILOSOPHY AS CREATIVE MANIFESTATION

CHAPTER III

India's Advanced Philosophy from West-East Perspectives

As the different volumes of my *Dynamic Facets of Indian Thought* are progressive continuations and developments of my thought, this new volume, with a focal perspective title, is just a further continuing process. All these works are insightfully linked up with the developments of my philosophical thinking mostly concerned with the Buddhist thought, as a means to develop early Upanisadic thought to a further dynamic possibility beyond, to float along the Buddhist currents in intercultural contexts.

I have also built that pathway through several editions of my *Changing Phases of Buddhist Thought*. It is just not merely meditative excellence in purely philosophic direction, where the potentiality of disciplined personality was confined to the expert or his close disciple or disciples, but now with the extension of the powers in scientific and technological directions, it has become widely stretched and shared by mankind in general. Areas of disciples have multiplied infinitely.

My treatment of the Buddhist thought as developed in India is closely linked up with the cultural processes of the lengths and breadths of the entire Asian countries, and it is reaching progressively the different parts of the New Hemispheres, touching the depth-processes of their religious and philosophic ideas, through varied types of transcendental psychologies.

This extension to the new Hemisphere has opened up some avenues with some potentialities for scientific and technological study.

These books connect Indian thought from the past to the present. They will carry forward the major currents of Old and New Hemispheres in togetherness, in their advanced philosophic-scientific directions. The pathway is towards a progressive post-intellectual aspect, improving upon the socio-political prospects, if man's concern remains constant in purifying his inner character or behavioural-processes.

I have tried to indicate these processes in some details through my recent work, entitled, *Buddhism and Whitehead's Process Philosophy*.

But as my work on Indian thought is from the post-Buddhist and post-Vedantic processes to various alternative movements, and as stretched beyond philosophic speculations to concrete complexities touching human social life in the developed aspects of science, technology, politics and business, I think this entire pursuit is interculturally oriented to a continuous contrast. It is non-restrictive, but in no way it fosters any loose permissiveness.

I have always taken note of 'Whitehead's philosophic-scientific ideas, in this connection, as Whitehead gives a rightful place to philosophical deliberations, in intercultural contexts from a Naturalist basis with a prospect beyond the Naturalist spheres of mere abstract intellectuality.

Whitehead's endeavour is continuously, in progressive contrasting emergent aspects, developing the entire Western Thought from Plato to Hegel, and to post-Hegelian many-dimensional possibilities, transcending restricted conceptual modes, and going beyond the confronted restricted empirical modes to post-intellectual meditative or insightful orders. Whitehead thinks these possibilities are also Naturalistically based, for Nature in its process and possibility is a many-dimensional evolutionary process. Nature has a pre-intellectual non-sensuous process, emerging into conscious intellectual level, with its restlessness and abstractions, but beyond this intellectual level – there is a possibility towards an integral insightful order – Nature's 'wisdom-order' freed from intellectual restlessness and abstractions.

I did not blindly accept the philosophic prospects as prepared through the post-Hegelian transcendent phenomenological and existentialist processes. Nor did I move to restricted contemporary Euro-American evolutionary, realist-pragmatist process, without carrying forward a prospective outlook, when I pursued Whitehead.

But I remained occupied with developing the further evolutionary-realist philosophical and scientific processes. I followed Whitehead in his inner cores and improved upon the advanced processes of the Buddhist followers. In this continuous pursuit, I discerned a closeness between Whiteheadian and Buddhist attitudes. Both ascend towards a post-intellectual psychic condition of insightful order of 'Stability or Peace' – freeing oneself from the tautologous discursive attitude.

They realized the advantages of transcending one's restricted thought, as a freedom from 'Sufferings' from vain argumentative tendencies. This position of Whitehead, is non-different from the Buddhist experience of a complete freedom from intellectual proneness or habit. Both develop transcendence – as a necessary preliminary for indicating that the reality prepares the ground for a continuous experiential possibility, without entertaining any ultimacy to reality as an experiential possibility. Its suggesion is —'no experience is no reality.'

For this broad continuity in philosophical deliberations of Western and

Indian thought in togetherness I feel like relating these processes to lively philosophic prospects of Hiralal Haldar of the Calcutta University.

Haldar as a participant of the First Series of *Contemporary Indian Philosophy*,[1] formulated his advanced comprehensive thinking, with a title, Realistic Idealism. He went beyond Berkeley's idealism, and also beyond the idealism of Hegel and neo-Hegelians. He, by a reference to India's Upanisadic thinkers, suggested an advanced notion of astrophysics as a possible range of the sense-organs of a human organism to the infinite regions of the outer space, with a prospect of a possible visions later. His suggestion was that extension of vision means a continuous experience of experiential possibility.

Here there is not only a reference to the 'transcendent ever-presence' of the 'Reality realized within,' by every insightful individual but also the 'confronted things,' by a contrast, lying outside the individual.

Between the 'inner transcendent self' and 'the outer Existential World in transcendence', in thought, there are 'non-dual expressions, as co-existent processes.' Any attempt to separate them is a wild goose chase.

Needless to mention, I discovered the same transcendent subjective and objective remnant of a disciplined experiential continuity, through my researches on Buddhism and Whiteheadian thinking.

This is nothing but a resolution of all gulfs between the transcendent Self or Subject, and the confronted individual subjects and objects that stretch beyond a human organism.

With some insightful concern of the contemporary Indian movements of the pre-independent India, and also post-independent Planning Programmes,' of the present day Indians in general, I want to be operative, rightly understanding the needs of India from the social and political perspectives.

In this context one must not foster some mere vacuous speculative thinking. What is needed for man is an admission of transcendent feeling that goes alongside of the fundamental requisites of the human organismic life from a truly cultural perspective. Hence in the Indian philosophical deliberation, there is a 'felt need' of a transcendental psychology, which is operative in every developed personality, that all types of activities, should be disciplined and addressed to this common interest. There is a non-difference between the inner Self, and the External Confronted Universe when in transcendence.

Humayun Kabir's insightful paper 'Freedom, Authority and Imagination' in the First Series of *Contemporary Indian Philosophy*, throws a flood of warning signals. I am taking special notes of his reflective activities, when my subject of this new book will cross the bounderies of Indian Thought, and will mould or shape philosophy to touch from a comprehensive perspective, world's problems of social, economic and political processes by the scientifically and technologically oriented universe of today. I was impressed by his quick understanding of the philosophic principles, at the different sessions of the

Indian Philosophical Congress.

When Kabir went to Sri Lanka as a General President to attend the International Conference, sponsored by the University of Ceylon, he at the conference domonstrated a wonderful control of the minds of the participants, by his dominant will.

He told me that more than attending the discourses of the conference, what was needed of the delegates was a visit of the important sites and places, and mark the cultural specialities. So he used to terminate the discussion of each session by intelligently controlling the minds of the participants, and helped to show the delegates the different cultural centres, and the local folk dance performances, and the elephant procession from the Buddhist Tooth Temple (Daladamaligowa at Kandy) specially arranged for the Congress.

In tackling this complex socio-political-scientific-technological world of today, Kabir felt that one has to understand very insightfully such concepts as 'Freedom and Authority' in human-social process. Any misuse or misunderstanding of these two closely interrelated processes brings in economic and political crises, leading the world-process to the thresholds of conflicts. Finally the conflicts prepare a situation for inevitable 'war-situations,' as evidenced by two World Wars within a short interval of a quarter of a century. So there is a threat of another war of the disastrous nuclear type.

Kabir rightly thinks that the advance of civilization can only extend the scope and scale of its operations. But this operational process has to be directed towards man's change of his inner or 'innate character'.

Hence Kabir is against any instinctual explanation of human action, which is always in a state of 'unstable equilibrium'. This instability, again, is the result of the conflict between 'Freedom and Authority.'

So his suggestion was for education in a new style. It was not alone the education of the 'intellect or thought,' but also of the other elements of the human nature – Instincts, Emotions and Will, as simultaneous processes.

The human psychology is driven away from an integral whole. This is a non-conceptual project by the side of a conceptual project.

His final suggestion, in brief, was therefore, a criticism of the prevalent forms of society on the basis of the concept of a perfect and elastic social equilibrium. The suggestion along this direction, again calls for man's 'imaginative constructions.' This is a need for imagination in the fields of beauty and morality—a search for a harmonious human Nature manifest in harmonious social form.

I feel here also an inner current towards peace or freedom from mere intellectually-oriented discursive propensity—as suggested in Buddhist-Whiteheadian process.

My development of the Buddhist thought, in progressive advanced context, is nothing but carrying forward the social and cultural processes, from an earlier order of transcendent integral consciousness, to a fresh broad transcendent aspect, to which the human mentality, can be elevated, to make it gradually adept in developing progressive pragmatic schemes in the spheres of philosophic and social processes in togetherness both conceptually and non-conceptually.

Developing the advanced Buddhist thought further beyond the early two stages – (i) Hinayana-Mahayana processes in various contexts and the (ii) emergent Tantrayana processes, stepping into its Vajrayana phase, as the next higher order of transcendent-integral process of Buddhism—I remained operative beyond this relatively final process of Buddhism.

It is realizing reality as a transcendent possibility – without doubting any more; hence stopping all ratiocinative habits.

This possibility with a before and after suggests an integral process in operation, where nothing of the past is lost in the emergent ever-fresh present. The details in this direction will be evident in the Parts II and III of this work, as they together form, continuity of transcendent possibilities after developing and becoming aware of all aspects of experiential processes, and their prospective currents in interrelations.

This is nothing but a Buddhist-Whiteheadian confluence – emphasizing a need for a 'transcendent-confronted process and possibility' with a recess in togetherness. This recess is enjoyment of peace in continuity and freshness.

It is a provision for thinking with an openness, without accepting an absolute static consciousness, or an emphasis on abstract ideas – whether in philosophy, sciences or in economics and in politics, in general.

My emphasis all-through will remain for 'No Ultimate' – not as a 'theory' but as a confronted experiential continuity.

Keeping this continuity in mind this new Volume on 'Triadic Avenues of India's Cultural Prospects' will take note of the progressive ideas of some select volumes on contemporary Indian thought and thinkers.

To reflect this character, I decide to choose two volumes of *Contemporary Indian Philosophy* for deliberation – (i) First Series, edited by S. Radhakrishnan and J.H. Muirhead, and (ii) Second Series, edited by Margaret Chatterjee.

My study of these series will be by insightful selections, and not by merely pursuing the thoughts of each of the thinkers. If Chatterjee, the editor of the second series, of *Contemporary Indian Philosophy* has felt with some regrets that her second series has not given due attention to the 'philosophy of science and politics,' but still adhering to some of India's traditional modes or their exegesis – she is right. Perhaps I have responded to her wish by attempting to extend some more visions through my present work.

My present volume, therefore, makes an attempt to meet this progressive requirement, in its post-independence period – with some cautions – as will be evident from my above contentions with a clear indication of both conceptual and non-conceptual, or post-conceptual insightful or depth-processes of human development – taking advantage of the changing symbolical suggestions of 'Zero,' as influencing all subjects and experiential processes. My project will be towards the 'transcendent prospects' beyond all experiential processes in interrelation.

This book is no longer eulogistic of Indian tradition, in enforcing some specific ideas, but it utilizes the essential traditional ideas with some modernized touches, as some currents of Indian thought as matching with American Pragmatist thinkers, or with the British revival of 'empiricist traditions,' in a new linguistic garb, or even the dialectic modes of a Kant, Hegel or Marx, only from a judicious aspect. But India does not remain confined to any experience in particular.

My suggestion will be for a period of breaking of barriers among the philosophic, scientific, political, sociological and other types of thinkers, and my intention is to flow on with a common shareable international current in varied 'transcendent contexts.'

It goes beyond the breaking of walls between East and West Germany, it is breaking up other walls – territorial and cultural – of spaces all over the world. But in carrying out these iconoclastic operations, one has to formulate some grounds for 'transcendent modes of deliberations,' as merely facing the confronted empirical or conceptual elements. The confronted processes, by themselves, – are not sufficient for bringing about real change in the world, or in world-civilization. Their 'change' is possible because of a 'transcendent background' sustaining them. I am never doubting the continuous presence of a 'transcendent possibility.'

Regarding the first series of this work, I may be pardoned if I say, that I considered the philosophic attitudes of Gandhi and Tagore, in the reference-portions of this work. I took note of India's outstanding philosophers – Das Gupta, Datta, Hiriyanna, Murti, Radhakrishnan and Raju – on several occasions in my earlier works and I tried to build my edifice with a strong philosophic foundation. I would now like to add further some more ideas for tackling the progressive ideas developed in India after Indian independence to the nineties.

I would like to recast Indian thought in further innovative ways – beyond its meditatively charged continuum of experiential processes. Now the continuum is stretched to the subtlties of scientific thinking, which is also changed by meditative possibilities. Here the transition will be from the theories to no-theories – with a rise to ever-novel transcendent possibilities.

Here my reference will be only to three other contributors of this first edition of *Contemporary Indian Philosophy,* mentioned towards the end of this volume, viz., (i) 'The Evolution of My Own Thought', by R.D. Ranade, (ii) 'Dialectic Monadism' by M.M. Sharif, and (iii) 'Pragmatic Idealism' by A.R. Wadia.

They are all in the aspects of *non-sensuous* process and possibilities – directed towards *outer spaces*, with ever-fresh emergent possibilities.

They not only indicate their masterly knowledge of Western thought, but also lay stress on to India's general transcendental modes of deliberations, without any direct mention, but implicitly they are operative as something more than the confronted conceptual and empirical modes and materials of experiential processes.

They demonstrate further that the advanced Western thought has to be merely accepted as it is, but it has to be insightfully moulded with the 'transcendental elements of Indian thought,' which is no longer dominant expression of Hindu Vedic Upanisadic thought, but is a medley or mixtures of intercultural perspectives stemming out of India's other religious currents, India preciously retained.

They represent a series of cultural backgrounds of Hindu, Islamic and Zoroastrian family-environment, each one confronting the Western thought in progressive development to its modern advancements. India is a vast sphere of religious currents with a prospective social process – where each of these religious bodies has a recognition with appreciative prospects.

This will help the further development of West-East thought, in higher orders of integral psychological spheres. This aspect of the human mind, will not typically be either Western or Eastern thought, but will be a progressive human philosophical enterprise, in a novel unthoughtful direction, inviting the attention of all advanced types of people, without indicating a bias for any particular culture.

Hence, all of these papers in the above-mentioned volume are their impressive autobiographical expressions and experiences – as backgrounds for clarifying their philosophic positions. My present work is a pathway in these directions. My realizations reflect only my persistence with a 'continuous possibility' and not as an 'absolute reality.' It is an adherence to a changing and transforming reality, with a variety, not just verbiages of reflective proneness.

Their versions are not just some reflective discharges or mere mental-processes, but actual concretizations of their reflective assertions – their sufferings due to mishaps in the family either or both of the parents. Mother Nature rescuses such individuals, she has some roles in sustaining or inspiring the minds of such hapless persons.

In my text I have suggested that though there is no 'final personality,' behind a confronted individual, if one pursues the advanced processes of Buddhist and Whiteheadian thinking, still something of it remains in its disciplined aspect, as a vanishing ghost. I also behave in a similar spirit, remaining, yet not remaining soaring with 'Nothingness.'

These personal processes with commitments in ideal-real ways are real expressions of each of these 'personalities' chosen here beginning with Hiralal Haldar and his paper, previously mentioned.

Like their predecessors they also did not restrict their thoughts by restrictive confronted conceptual and empirical processes but they were all actualized or enlivened by their creative or original ideas. They were also important for me for some further reasons, as I associated with them, in many of our meetings during different sessions of the Indian Philosophical congress, and at other occasions.

I was interested in Ranade, not only because he was one of my examiners for my Ph.D. thesis, but also as he was known among the philosophers of my time, as a great Mystic. His paper, 'The Evolution of My Own Thought,' as I read it, tallied well with my first meeting with him at his residence at Allahabad, when I came with Dr. D.M. Datta, on a cold winter morning, 1946. He was found thinly attired, while we were heavily dressed with winter suits.

Looking at our face he brought for us milk. What a compassion! Then followed other conversations, with hospitable treatment of a delicious lunch, and progressive preparations for holding the examination, in a formal way.

This evolutionary history of my examination-process – confronting expressions of a compassionate attitude, with some final formal touches – is well illustrated in Ranade's observation of Western thought in progressive development of Western thought, preceding and following Plato. It was a look backwards to early Greek thinkers before Plato, and also forwards to Hegel and to Neo-Hegelians, Bradley and others, and American contemporaries finally. Ranade's was not a 'fixed thinking,' but a 'fluent process' with transient 'transcendence' towards 'all' and 'none.'

Ranade was also influenced by the great Vedantic thinkers of India, as he was aware at the early period of his life within Hindu family milieu, which was some form of spiritual mysticism, and that spirit was supplemented by one of his teachers. He was not only influenced by these double mystic trends, but also he chose to follow those trends. He became operative in his study of Western philosophy, throughout his life.

Though, unconsciously he had a strong impulse for the study of philosophy, he preferred to take his graduate degree in mathematics, and this study gave him a scientific outlook and dynamism. Later on, his philosophical studies were not systematic, but a continuous emergence with a novelty – a *transcendence.*

His early choice was for a 'Pluralistic Conception of Spiritual Reality,' suiting the inquisitive mind of his early stages, leading to his consequent process of an integral Mysticism. His pursuit was in the direction of a 'monadistic-spiritualistic vein,' as influenced by Carlyle and Teufelsdroeck, and of his own Professor who was an 'eccentric Professor.'

Hence my development of philosophic deliberations – being so influenced – is one of a 'creative manifestation,' with no attachment for an obstinate 'ultimate.'

This 'Professor' of Ranade was also known for his humorous title of his self – as the 'Apostle of Lunacy.' He was known for his irregularity, – 'regularity was not a word to be found in his dictionary.' His lectures were never given regularly. What I knew by regarding Ranade's paper, I heard the same sort of rumour about Ranade also, who was always absent-minded and never on time for teaching.

Ranade created his unique reputation by his occasional lectures. His Professor was always found in a meditative mood, in the class, as if in search of finding out 'the centre of the Universe,' in his studies of the different aspects of experiential processes from philosophy to the sciences.

I also experienced somewhat similar sort of 'method,' or 'attitude'—in one of my teachers of philosophy, Prof. Khrg Singh Ghosh, of Hazaribagh St. Columba's College. He used to teach us Psychology. Prof. Ghosh would tell each of his students to read some portion from a psychology book, that was the recommended book for the class. He will pensively pursue the thoughts of the book, when read, – but after some interval, he will, with an impulse, stop the student from reading, and remain silent for quite a while, and then would start speaking of his reaction with a serious mood, or often he would end with a summarized statement, revealing the depth of his mind.

This is a kind of mystic expression, something 'Creative', undoubtedly. Ranade's was of another kind, in the similar direction. Prof. Ghosh had also another programme of addressing the community every quarter, at the College, when the lecture hall used to be full with the people of the town, some miles off from the College campus.

At one time, Ranade's Professor came out with an expression, indicating man's interest in the centre of reality. In this search, he tried to indicate a common pursuit, in all experiential spheres. His interest was to find a unity in diversity, an order in chaos, with a further question, viz., why again, in geometry, there is a search for a 'centre in a circle', in geography, too, why there is a search for a 'centre of the Earth', in astronomy, undoubtedly, there is a pursuit for the 'Sun as the centre of the planetory universe.' Is it not a mystery? But at the same time, when a teacher starts his teaching, it is a creative manifestation, and not just a conceptual construction or reproduction of something read or understood from a past.

In all these many-dimensional aspects of reflective activity, or continuous transcendence – from geometry to astronomy – there is a search for the 'centre', or a definite 'position in space'.

But this restriction to a space is a mistake, to the meditative Professor.

His answer was that the 'Centre of the Universe', is either 'Nowhere' or 'Everywhere.' But this answer, was not an intellectual assertion, it was a form of mystic perspective – a transcendence. In the consideration of each of these philosophers I observed their ascents to their respective transcendent experiential process and a descent to the reflective level for communication to the confronted person or persons. Real philosophical deliberation is not a dwelling in an abstract sphere. It is a concrete dynamic psychic expression in a social milieu.

I remember when I talked with Ranade, for the first time about my thesis with the title, 'The Place of God in the Changing Universe' – his first question was formulated as: Have you found 'God' anywhere in this Universe? My prompt answer, without any knowledge of his philosophic mysticism – was 'No-where', Sir, but 'everywhere.'

This was an appropriate answer or a vibration, an accidental coincidence with his mystic position, dynamically entertained, and snugly placed into the ripples of the ocean of his construction – as also fitting somehow the position of his teacher. It is simply a kind of dance between two spheres of reflective activity by a dual mode of ascents and descents, as mentioned above.

Ranade's later questions and answers, in the context of my thesis fell appropriately in place, and then followed his favourable reaction—and his favourable report. But, all of a sudden, he returned to me, with a request to address his thirty or forty post-graduate students.

When I was taken to his class in philosophy – all these processes, were not just formal, nor very much connected. Hence the process was not premeditated, but it came out with the spirit of a thrill or vibration, of some mystic showers, which I till now feel, while writing.

I am now returning again to Ranade's 'Professor', for a further development of Ranade's own thought, and observation of his teacher who seems to me akin to the Japanese Zen Masters through the centuries to the present.

Then the 'Professor', after his first mode of inquiry will wait for an answer from the students, which was also in agreement with a silence followed by a consequent meaningful 'reaction' in a Zen way, however odd – a departure from any usual social usage.

This dialectic mode of deliberation – will suffice for characterizing Ranade as a mystical personality, and his natural disposition, to find out why he will not be inclined to get a positive philosophic answer to a question, his liking will be for an unusual answer which will fit in with his mystic mind, the mysteries of his spiritual inquiry and sudden creative eruption.

The effect of this pursuit led him to the depths of progressive mysticism. But my method of teaching all-though my career – by a contrast – was always extempore and continuous, with no gap in between contemplation – as characteristic of my teacher Prof. Ghosh – or in accordance with my encounter with Prof. Ranade. I had to act judiciously in appropriate ways to conform to the requirements of my authorities of a new country, and also watching the character of the confronted students, of a new Hemisphere.

I now return to Ranade's understanding of Sankara's Vedanta from a new perspective at Banaras Mutt of Sankaracharya.

Ranade could not make out this 'mystic mode' from his first study of non-dual Vedanta of Sankara, as a student, why Sankara's devoted disciples at Benares Mutt of Sankaracharya, were only for hearing 'devotional songs.' Only, after his direct experience he emerged with a 'mystic experience.'

His direct experience, later in life, made him deeply study Indian philosophy, and with this later spirit of experience, he studied, the European philosophy by a contrast. But, in both, he found out some pursuit in both Indian and Western thought, a common Mystic trend, in two different philosophic traditions.

Ranade developed an unique mode of treating Western and Indian thought in togetherness, with a progressive dynamic and dialectic spirit, but all in the direction of some mystical processes.

His was a comparative interest to discover the all-pervasive mystic element in human deliberations. His search was for some similarity, if found, in European philosophy, and not the differences. Though he found the differences, that pursuit was just a secondary process, and admission. His was a development of continuity of philosophic deliberation, crossing conceptual and empirical boundaries, and a lifting of oneself to a superior order of mysticism – an outburst, or may I say, a transcendence.

He found in Bradley's 'ultimate reality' or 'the Absolute', some closeness with Sankara's ultimate, but really Sankara had no fixed 'Absolute.' His was an experience of an ever-shining transcendent continuity. For to both – Bradley and Sankara – though 'God' was 'unreal,' they were reflecting from two different attitudes or visions.

Both found out that our souls, bodies, or confronted world of objects are but 'appearances.' Confronted 'space and time' are only phenomenal, transcended by the Absolute. This holds good of the moral distinction between the good and bad, for the 'Absolute' *transcends* 'the moral relations.' But this distinction between 'Appearance and Reality', to Bradley, was an attachment to a fixed Concept. The 'Absolute' of Bradley was the final concept.

But Sankara differs from the British Bradley and American Royce – as his was a passage into a different sort of 'transcendent mystic experience,' or rather pathways, even in his intellectualist process or proneness, which Bradley and Royce could not realize, for their attachment to some form of intellectualism – a fixed abstract 'Absolute.'

My suggestion is that Bradley or Royce had not direct experience as provided by the Buddha or by his followers like Nagarjuna and others – preparing the ground for a complete freedom from all *forms* of intellectual propensity to argue by the continuous pursuit of a dialectic, by some kind of assertion as final. The attitude of the Buddha and his followers was a total abandonment of intellectual propensity by making no *final choice* but discarding all choices by a scheme of 'neither-nor' process, eliminating all attachment for intellectual proneness by beginning with such statements as 'that there nothing disappears, nor anything appears,' or by another statement as 'Nothing has an end,' 'Nor is there anything eternal' – this is another contrasting statement with no assertion, or a claim for an intellectual form. These are statements by a contrast, making no final choice, it is a continuous process, with no end by a fixed number. Nagarjuna's usually eightfold negation, is not restricted to the number eight, it is more or continuous – symbolically speaking – an abandonment of the propensity to argue, and an arrival at a condition of placidity, by transcending the intellectual propensity, through the above symbolical scheme of 'neither-nor', where there is no place for any choice – where one automatically plunges oneself to a silence, finding no choice to make.

The speciality of Sankara was for a disciplined life, with a direct experience of the Buddha, Nagarjuna and their followers, in historical continuity, with progressive insights. His was an adherence to a 'self-shining transcendent spirit' which continues without any confronted 'restricted intellectual propensity' when confronting one's inner or outer objects – for both are replaced by a prospective experiential process, till the final transcendent experience is realized with a continuous possibility. But this continuous adventure is a pursuit always – an attitude, with a specific choice.

This transcendent sphere of discipline with a 'freedom' from 'intellectuality', totally was also the pursuit of Ranade.

Ranade, in his life and thought, wanted to manifest, this transcendent continuity too. His was a discipline, not just a philosophic activity. It was cultivated in the Indian tradition, by a way of life from the days of the Vedas and the Upanisads, with a radical change through the Buddha and his followers.

It was directed to mystical modes both at the beginning and at the end of his discourse. Ranade too following Sankara, was choosing for his reflection a boundless horizon of a disciplined mystic life. I was told that Ranade, did not take any solid food but only liquid food, juice or milk – for a period of fifteen years.

He found this flare of mystic process, also, in his study of early Greek philosophy, and from Heraclitus (Herakleitos) to Aristotle, from his Indian attitude. In Aristotle, he could see, that Aristotle found no truth, but, in his pursuit, he discerned only degrees of error. He glimpsed 'Truth' as 'one', absolute and immutable, and that was in his 'God'. It was a 'mistake'.

Criticizing, a relativist sceptic, Aristotle asks why does he not walk early some morning, into a well, or over a precipice. This argument from differences of values—was a remarkable statement of Aristotle. It follows, while there is an Absolute Truth, there are various degrees of error. But the real transcendent insight would have been, from two distinct orders of experiential processes: transcendent and the confronted, but none of these aspects are static or categorical. There is a process and a possibility in both these spheres.

In contrast, to, 'one truth', there are infinite 'errors.' So, according to

Aristotle, 'Motion' of Heraclitus, was an error, but Aristotle's 'God' by a contrast was the 'First Mover' and so was Himself 'Unmoved' as another error. But, to my mind, this was something very special for Ranade, I could not find this is my understanding of Aristotle. But my interpretation of Sankara, Aristotle or Ranade is differently 'oriented.'

This discipline may not be the same as that of Sankara, to my mind, as Sankara raised the question before all investigation, by an insightful sense of distinction between that which is 'eternal' and that which is 'non-eternal' (Nityanitya Vastuviveka). So, Sankara had a 'freedom' from any restricted 'intellectualism' – he soared in his self-shining, everfresh transcendent pathways and possibilities.

Ranade, in turn, had a special way of argument, implicitly resting on this kind of distinction between the real and the unreal.

He finds in both Parmenides and Sankara, same sort of argument denouncing the 'Idea as distinct from Being.' The 'Idea' is non-different from the 'Reality' itself, it is not a 'particular' related to a 'universal', nor is it 'corporeal,' but 'incorporeal.' Hence, Burnet's materialist suggestion is wrong, to Ranade, as he does not want to stick to a 'fixed intellectualism.'

Parmenidean 'One' or 'Being' is 'indestructible,' without beginning and end, and immovable, it is lacking in nothing. It is mysticism. It is a philosophic originality.

But other types of non-materialist thought is developed among the thinkers of the modern period, and in contemporary Western thought.

Though I would choose to develop a different type of deliberation for my special type of disciplined pursuit, I am just pursuing Ranade, in deference to his 'time' – as an alternative, but not stepping with it.

Bertrand Russell, of the present century, to Ranade, from mathematical perspective, interprets Zeno's argument against motion, as not that the arrow does not fly, but that there is 'motion' when different 'times' are correlated with different 'places', – there is 'rest,' when different 'times' are correlated with the same 'place.' Ranade takes another case: in contemporary Western thought – in Bergson – here, we have a rehabilitation of Heraclitus, and in Russell, we have a rehabilitation of Zeno.

In the case of Russell, further, it is associated with the mathematical notion of 'static change,' declaring that there is no such thing as 'infinitesimal differences,' at all – the flying arrow is always at rest, as being an illustration that every possible value of a variable, is 'constant.'

But in mathematics still, there is a process beyond Russell, which re-establishes the importance of 'Infinitesimal Calculus.' It alone can outmode Zeno's Sophism.

These varied pathways as suggested by Ranade, indicate his 'freedom' from any kind of restricted 'abstract intellectual mode' – he enjoys in himself a continuous 'freedom' from 'intellectualism.' His suggestions are, therefore, 'one in the many.' Here also not to establish a theory, but to open up a new 'Pathway' – is a further suggestion. His post-Buddhist and post-Vedantic suggestions provide another dynamic alternative beyond any intellectual form.

But as already stated, my pursuit will be different, according to my understanding and in accordance with my different interpretation of the Buddhist and Whiteheadian thinking. I am just indicating some advanced thinking in Ranade, in accordance with his time and prevalent philosophic trends. He did not want to adhere to any specific theory, and that to me has some importance, but I will not choose to follow his method, in following the present order of philosophic thinking that I am trying to develop.

From this stage, Ranade passes into the depths of Indian thought, and in his 'Constructive Survey of Upanisadic philosophy' he discusses the epistemological significance of Self-consciousness, of many of the Upanisadic deliberations, *Taittiriya*, *Kena* and *Brihadarnyaka* – that the 'transcendent Self,' with its supreme dichotomizing power of the 'Knower-Known' process – are all apparent 'dualist processes.' This is a rise to a 'transcendent mystic intuitive process,' which is the transcendent Reality, operative behind Intelligence, Feeling and Will – a creative manifestation of a contemporary Indian philosopher.

Ranade's reference to European and Indian thought, in correlation—is necessary in the interest of development of both West-East philosophy, and in their continuities, so far as their respective cultural processes, can match. His was also a reinterpretation of 'intuition' as a transcendent double-consciousness, or a mystical consciousness. Finally it was the core of his 'spiritual life' from the beginning of his career as a teacher of philosophy at the Allahabad University.

This type of philosophical survey from some autobiographical references, is also characteristic of 'Dialectical Monadism' of M.M. Sharif, and 'Pragmatic Idealism' of A.R. Wadia, and in both there is indirect or direct reference to European and Indian thought and life, as lived, in togetherness, with an understanding of the depth processes, of these cultural deliberations with elements of concreteness, of life-process, transcending mere 'intellectual abstractions' or 'reflections', and preparing one for many-dimensional scientific thinking in novel creative directions.

All these Indian thinkers were for a discovery of a superior order 'transcending' the mere advanced thought of West-East philosophical deliberations, as 'confronted then.' They display their originality in different ways. They also develop some kind of 'transcendentally based mode of deliberation,' combined with a 'life-process as lived.'

This writer's pursuit in following these great thinkers, is a reinterpretation of philosophy in further aspects, as to fit itself snugly with the demands and requirements of scientific and political movements of the world at large. Their's was a different gesture, not a mere academic form of philosophic ratiocination, as expressed among reputed Western thinkers, whether in England, America, Australia or Canada. (Cf. Volumes on Contemporary British, American Philosophy et al., with different editors.)

The indication of the depth-processes of these and other Indian thinkers in this volume is showing that these senior Indian philosophers – both before and after Indian independence – were shaping both Indian and Western philosophy in their depths and relations, never forgetting their distinct specialities, as lying beyond them, but not neglecting either.

But in contrast my interpretation and understanding of Western and Indian thought will be different, as my drive will be a historical survey of both, and then in an advanced aspect of both, I would choose to comment both on philosophy and sciences, in interrelation, and prospect, but here my interpretation of philosophy, is of some select Indian minds whose expressions were all creative manifestations of persons soaring in a transcendental region, but not cut off from the confronted universe of reflective activity.

In this context of historical survey I am aware that history is a complex phenomenon, as stretching to all aspects of conceptual-empirical processes, and it also touches the transcendental attitudes of the Asian and Western countries in general. It also examines history from one's life-history, from within and without. In my survey I have chosen the Indian 'transcendental modes' of insightful deliberations – in contrast to the similar or not similar deliberations of other cultural processes, in some ways, in the sense, I intended to develop.

M.M Sharif starts with an autographical reference to his early life in a village near Lahore. He was born in 1893. His father was not ambitious but of a mystic type, his mother was religious, her father was a Kazi. But Sharif was unfortunate, when still a student of the secondary school, he lost his mother, a sister and a brother within two years through the illness of plague. This made his life extremely pessimistic. He got over this terrible psychological shock through his own self-oriented love of 'Nature' by spending most of his time in the beautiful garden close to the school. It was the magnificient garden, known as Shalimar Garden built by Emperor Shah Jahan, the great Moghul Emperor known for his marble mosque, Taj Mahal, at Agra. His cure was a mystic association with 'Nature.' This indicates how 'Nature' could be a cure for mental sickness, and prepares one to continue one's life-process.

Sharif took the Bachelor of Arts degree from the universities of Allahabad (in 1914) and Cambridge (in 1916). His education in two different countries—helped him to combine in his thought some elements of Islamic mysticism, with post-Hegelian advanced Western thought of evolutionism and realism.

In India he studied philosophy at the University of Allahabad, and was aware of Islamic and Indian thought, and while in England, he became inclined to the Western thought of an 'empirical idealist.' In England he was directly influenced by G.E. Moore and Bertrand Russell, who studied Western thought, but retained some form of their originality.

This double impact of Russell and Moore made him a realist with a deep common sense. Inspite of his attachment to a historical past of his motherland – his cultural heritage – he was able to understand and accept with appreciation the developmental aspects of Western thought from Locke and Berkeley to Leibnitz and Kant, and also the later developments of Western philosophy beyond Hegel to the Phenomenological and Existentialist trends, and also the European realist-evolutionist thought with a terminating process of American Pragmatist modes. This progressive evolution and development of philosophic thought, depended largely from his understanding of the changing confronted universe, with progressive 'common sense,' without developing a specific philosophic concept or restricted empiricism.

To him, philosophical reflection is a continuous development and knowledge of the many-dimensional experience in process and possibilities, neglecting neither the Indian nor the Western thought – as associated with his life-experience.

Philosophical reflection, to him, is undoubtedly a 'comprehensive process,' not confined to its narrow speculative area. I had an expansive span in a complex mixed sphere. This is the comparative vision of the Indian mind which is often neglected by Western thinkers in general, only discerned in some very open-minded thinkers, both in the past and the present, but unfortunately, it is not a general or universal attitude.

In his continuous knowing process, Sharif follows Russell and his analytic scheme. To him, experience is always of some objects. It consists of two elements, (i) our apprehension of an object, and (ii) an object apprehended. Though the object is independent of the subject, it is wrong to assume their total independence, behind them there is a 'transcendent relational process'—which connects them both, and makes knowledge possible in several West-East orders, in the objective sphere, from vague sensation to well-defined perception of objects and their relations. Herein lies the 'transcendent insight' of a Sankara – beyond the exclusive subjective and objective processes – if one is acquainted with Indian thought.

Here one has to look at both the subjective and objective aspects of experiential processes in relation and transcendence. Through his 'felt experience of life'—he understood, that life and experience are nothing fixed, rigid and static. Life as lived is dynamic, fluent and mystic. It grows in all dimensions, its growth is from the concrete undifferentiated experience to more differentiated experiences. It is integrally and insightfully analytic.

It is a forward-looking, ever-growing process. Both our knowing and our experience are reflected partly in our being and partly in symbols, from individuals to individuals, and from generations to generations.

The confronted reality is known to be in time and space, when it is existential, or it may be imaginatively entertained, as subsistent or transcendental – through the passage of one's age and experience. Hence experience is a continuous change and transformation of a personality-process.

The reality as confronted or presented is known or knowable. In the perceptual situation it starts with a subject's apprehension of a sensation-complex. If one attends to all these complex modes in philosophical thinking of day to day life, one should not ever say that philosophic knowing in West-East way is not a training of the mind, in comprehensive ways of philosophy, psychology, and sciences in general.

It is a concrete expression of analytic, synthetic and insightful process of the human mind in mastering the different traditions of the world in togetherness, following the scientific process of the present atomic age.

In this advanced cultural milieu where distance has been closed up with no more creating a gulf between particles and waves, microcosm and macrocosms, earth and terrestrial processes in spheres of galactic orders of many-dimensional spaces, all confronted spheres of experiential processes are located in many-dimensional different fields. Man's many-sided visions change externally and internally. From the interests in galactic order, man becomes also interested in the geologic formation of the universe, how the universe evolves progressively through infinite orders of time – from the basic techtonic to later processes of the plates and to the order of shifts, and volcanic eruptions, with separations of continents, and becoming panoramic with the eruptions of the highest mountains in India or Africa. The varied experiential scenarios are not just presented but are also variously non-presented – but is a continuous transformation.

Here we are reminded once more of Barbara Ward's world of today with progressive ways of swallowing up of distances for a traveller in a plane or in satellites et al., by closing off gulfs between various orders of experiential processes, if one peeps through the windows, making us sing the song of Tagore's *Gitanjali*, with the sonorous sounds of his eye-opening verse: That the distance was made near, making a stranger, a close friend.

More of Barbara Ward's book will be referred to in the next chapter, relating her works with Panikkar's *Survey of Indian History*, with his concrete activities, in changing the history of India and of the world in other countries, wherever he travelled with diplomatic assignments. Here the changing suggesions are many-sided—switch from physics to history and politics.

Both sensation and perceptions are emergent processes, their basis lies in another sphere which is transcendent, it is responsive or non-responsive, when one is conscious in contrast to other conditions of the mind, as sleepy or absent-mindedness, or experiences in weightless spheres of the daring astronauts, marching towards the moon or preparing some bases in outer spaces. These expressions are from one's lived experiences and not just abstract reflections. The shifts are from ordinary psychology to the psychology of an astronaut.

These are my changing experiential processes, as I shift my attention from my studies of philosophy from one country to another, or from one science to another, or when I choose to travel beyond these areas to politics or with an abrupt shift to geology or geography.

So for rendering the possibility of emergence of experiential processes, in distant outer spheres – one must be conscious. In this, one can be variously operatve, and not restrictively operative as advocating a particular philosophic attitude or subscribing to a scientific theory, as in the earlier stages of philosophic-scientific processes of the old and new hemispheres. One's change of personal psychology is greatly transformed by the advanced scientific and technological developments of mankind as a whole.

It is a concrete dual adventure of both Philosophy and Science, during one's life-process. One's reflective activity, is not merely an acquisition of knowledge of a particular speculative or empirical aspect, with some specialization – but it is an end-process, of one's basic life-process, within a family, social and natural environment, and also an environment created by modern science and technology, and by travel from one place to another.

All these forces were operative in the life-process of Sharif, in the milieu of which he was helplessly placed.

In the life-process of Ranade, too, one can mark the influence of his teacher in his reflective activity, that made him mystic. Whether mystic or scientific, one is deeply human, and widely Naturalistic.

If one pursues Western thought from the ancient to the modern, one falls into the stage beyond Descartes' mere rationalist position of innate ideas to empirical attitude of a Locke, a sensationalist and a perceptualist, with other consequent processes, or overcoming difficultis in process of reflective activity, – ranging from one to the other fields of their interests – Descartes in mathematics, and Locke in law and politics. Now the world-process is becoming progressively internationally and inter-terrestrially oriented from the galactic soaring to oceanic dives, or even scientifically probing deep into the bases of the earth to understand the emergence of the hills, mountains, valleys, gulfs or continents and so on, from geologic and geographic perspectives. The world as a whole has become a many-sided scenarios of processes and possibilities, with progressive indeterminateness.

The empirical attitude further can be developed along the subjectivist attitude of Berkeley, or extreme empirical attitude of Hume reducing all sensible ideas, as only 'abstract impressions' with no inner or innate connections. Both these trends reflect one-sideness. But, if one studies their life-history, their special interests – Berkeley moves to Theology, of a mystic kind and Hume to legal and political sciences, from some empirical perspectives. By a contrast, the visions of the modern man, in this progressive scientific and technological universe are stretched in multifarious dimensions and ways as indicated.

Avoiding all extremes, Sharif chooses to develop the basic position of Leibnitz's Monadology, with a novel dialectic, by utilizing Kant, and then critically improving upon his own position, by accounting for the basis of the transcendent order as immediately knowable, as a possible experience, operative continuously with one's day to day life. Here he goes beyond both Leibnitz and Kant.

In this connection, Sharif, very insightfully introduces his transcendent notion of a basic entity as behind all confronted phenomena, as the ground of the individual and of the confronted universe as a whole.

Here the confronted sensation or phenomenal complexities are grounded upon several dimensional orders. These entities of an organismic sort are directly known.

Here he implicitly taps some aspects of Indian thought of a basic operation of a transcendent process – beyond confronted sensation, perception and thought, of the Upanisads, Buddhism, Jainism, and the Sankarite processes (though not mentioned) – to the present scientific universe.

He was veering towards varied aesthetic orders, as if of Dirac's beauty of the mathematical equations, or of Raman's many-dimensional scatterring laser flashes in aesthetic expressions of various musical instruments, with world-wide cosmological vibrations, with no direct mention. It is my additional suggestions beyond his philosophical deliberations.

The 'basic entity' – the 'knowing subject' of Kant – is the transcendent Self, but Kant makes the mistake of saying that it is not known in itself, but only its knowing is in the apparent phenomenal aspect.

Here he separates one's existentiality and reflective activity, and here also he restricts himself to the empirical self and its states as knowable, as a unity of apprehension or even apperception, as 'I think' – with a judgment and with appreciation, as having a 'free will'.

But, this reference to a transcendent self is still empirical only, he does not know that there is an ever-present transcendent Self operative behind one's empirical self, which always crosses boundaries, and closes up the gaps between all specific orders of experiential processes. But now with the advancement of different sciences in interrelation the human organism or the confronted universe cannot be separated from the inner vibrations of the whole universe.

Sharif was certainly aware of all these in a general way with his mystic Indian – Islamic-Vedantic background.

Perhaps Kant resists this thinking as a continuous process, and deliberatedly stays back to an abstract empirical sphere. It may be the general intellectual horizon of his time, he was not ready for any commitments.

To Sharif, knowledge soars continuously onwards from one level of basic entity to another. There is a continuous ascent to progressive 'transcendence.' This is another example of India's development of a process of creative manifestation, not found in usual European and American thought-process.

Sharif suggests also that this total separation or rather bifurcation between transcendental and empirical self has to be rejected. This will resolve many of the problems of abstract philosophy, abstract psychology and abstract logic and abstract sciences – as often indulged in by some younger thinkers of the present century, both in India, and in non-Indian countries, being specifically driven to a particular line of intellectual process, concept or an empirical situation – phenomenological or existential modes of deliberations – and arguing from that exclusive perspective. But to the really experienced thinker, with the knowledge of interrelated philosophic and scientific developed knowledge-processes – the traditional sceptical outlook progressed towards a comprehensive vision, doubting less than before.

Here Sharif chooses Leibnitz with a progressive insight and with a modification of his 'Monadology.' According to him, reality primarily consists of active and responsive monads, and their action-response qualities, the physical phenomena are first as 'subsistents,' and then as 'existents.' Then, to Sharif, this process is a transcendent continuity. The present world-process has gone beyond his time.

There are layers and layers of *phenomena* arising from the interaction of Monads – *sensations, perceptions, objects*, and *systems* of objects. Phenomena apprehended by lowest to highest orders of Monads, whose direct experience is not only on objective phenomena but also of themselves as the subjects experiencing these phenomena. It is not wrong to specialize in Western thought with a thoroughness and depth, and then come out of its horizons with a new vision from an advanced comparative perspectives. This is the new kind of philosophic manifestation, that this writer wants to express among his chosen philosophers of India, or even when pursuing some Western thinkers from a histotical sequence. But this writer will prefer always to link up advanced philosophical deliberations with advanced scientific thinking and visions in varied senses as mentioned.

To Sharif, these many-dimensional 'monads' are so-many 'energy-centres,' in transcendence, they can be controlled by one's varied intellectual and insightful operations.

Herein lies the difference between Kant and Leibnitz, as Sharif reshapes Leibnitz. Kant thought that the things-in-themselves are the grounds of phenomena, but they are unknowable, but Leibnitz did not admit such unknowable grounds. Sharif thinks following Leibnitz's suggestions, that the transcendent principles are knowable processes, but this knowability, is a continuous process. The *phenomena* seem to influence one another. These *phenomena* are the 'monads' or the 'energy-centres' of the Universe. The apparent phenomenal orders are expressions of Monadic processes. The Universe displays the dual processes of 'relativity' and 'self-identity' in togetherness. They are not merely apprehensive but also progressively *evaluative* processes, to Sharif. This scheme of philosophical deliberations, can be applied to the present-day understanding of the advancing scientific processes to advanced currents of astrophysics, if we so intend.

The monads being energy-centres, are active, moving and dynamic, they are 'dialectical,' not in a restricted sense of a Hegel, or of Marx, but it is a movement from and to all conceivable perspectives, touching all centres of experiential processes, as if teleologically oriented. These rhythmic flows of energies, are manifest in radiowaves, water or other waves, physical or chemical, psychical et al.; they are strong or weak, and operative as hypotheses for all spheres of human investigations. They are processes for closing up geographical and ideological gulfs in process, after the Russian Gorbachevian operations through his Perestroika. If science has crossed all boundaries of specific theories of each one's restricted spheres – by running into transcendent fields – politics and other social sciences must as well learn to cross beyond all restricted ideologies governing their fields.

The monads are not merely operative in the sphere of thought, feeling and will but also they are transcending or crossing all confronted boundaries to Being, with a faith in an Absolute.

In Sharif, one can find a reconstruction of philosophy and science, from a *transcendent* feeling process, beyond *apparent confronted empirical situations*, hence his is a remoulding of the whole of Western philosophy, from a touch of his growing progressive depths of feeling, the different aspects of Hindu-Muslim thought – religion, philosophy and mysticism, without mentioning anything of these traditions in particular. His was indeed a treatment of philosophy, in intercultural context, from a creative perspective. His advanced thinkings are special to him, and in accordance with his time. But I differ for a similar or not similar reason, and in conformity with my time, and the way I have been developing, reinterpreting Western, Indian and Asian thought in interrelation and prospect.

A new type of reconstruction of philosophical thinking will be found in Wadia through his scheme of 'Pragmatic Idealism,' with doubt, scepticism and criticism, facing the concrete phenomenon of 'evil' expressed in different traditions of the world, with no satisfactory answer so far. His views also are special to him and in accordance with his time. For a similar reason, I agree or differ from him, and in process – I shall be clarifying my developments of philosophical thinking, in triadic ways according to the requirements of this present work.

Wadia frankly admits that his movement of thought is on the primary basis of birth in a Parsee family and his education in India and England. In these bases, his philosophical deliberation is of a 'constructive Pragmatist thinker,' seeking for an idealist process, if possible.

In him also, like Ranade and Sharif, there is a progressive *feeling* that leads him to mould his confronted experiential process, in varied many-dimensional ways of cultural, social and political aspects.

A.R. Wadia, was born in 1888, in a Parsee family of Bombay. This family played an important social role through centuries since 1735. His life, within that social milieu, was replete with the stories of an affluent family, but his own family was not wealthy, but he thinks that was good for his life which was a mixture of shyness and pride. That psychic trend remained all-through his life, an aristocrat by temperament but grown up to be a democrat by conviction.

As a regular member of the Indian Philosophical Congress, I could experience this element, in him, in the form of a special kind of dignity in his attitude. I appreciate very much his frank admission in his paper. Later in life, I found in him a great change, I could talk with him freely, at first he kept a distance, as a senior to me, but later my association with Prof. Nikam, and participation at the Indian Philosophical Congress for years, brought me closer.

Though in his boyhood, Wadia received his training mostly in English literature, history and religion, with not much of study of philosophy, till 1908, yet in his later choice for a course in philosophy, he indicates his inner resolve to master philosophy as a subject, when he took it as an optional subject, for his B.A class. Here also, like Ranade, Wadia had an inner wish for the study of Philosophy, hence was his final choice for philosophy, even when he had chosen to study other subjects.

Ranade, graduated with mathematics as the main subject for specialization, but later became philosophically inclined, and that again turned him towards mysticism, due to his direct influence from his professor. Common factor among these chosen philosophers was their interests in varied subjects, so that they understood the interrelatedness of various experiential processes. This complex interrelational mode of viewing the confronted experiential processes from philosophy to the sciences and politics or any other experiential processes is also my continuous growing interest.

Wadia's turn to philosophy brought in him a change in his attitude towards life. This is true of me as well, as I am also changing my grounds for the study of comparative philosophy to comparative sciences, and to humanistic psychologies. I am choosing to focuss my attention to politics, as it is a subject of interest of the Indian masses and of the intellectuals. But my attitude towards politics, is not depending on heresay, but making an effort to understand its core and exhibit its emergence in the ever-fresh form of a phoenix, with a dynamic, comparative outlook. I preferred a kind of living within some political atmosphere both in Sri Lanka and at the United States.

In 1910, Wadia left for England, and stayed there for three years. In Oxford

University, he studied for diploma in economics and political science, and secured a distinction, and continued to study philosophy without any reference to examination. This gave him an opportunity to attend the lectures of J.A. Smith, C.C.J. Webb, L.D. Lindsay, A.G. Adams. C.C.J. Webb, as First Oriel Professor of Philosophy, Webb happened to be Wadia's tutor. His weekly meetings with Webb influenced him very much throughout his life, and he maintained a contact with Webb after his return to India.

His early interest was for an education in England with a spirit of an ultra-loyalist attitude for Queen Victoria and Gladstone rather than for Dadabhoy Naoroji or Tilak, known for their nationalist ideas. But this feeling changed towards the later part of his stay in Oxford, where he associated himself with a feeling for several choicest personalities of India – Hindu and Muslim. He considered his participation as an undergraduate in their associations – and this formed a legacy of his patriotic feeling.

As Wadia stayed longer in England, he migrated to Cambridge, for a graduate's degree, and he graduated in the Moral Science Tripos in 1913. There he came in contact with some Cambridge idealists like Ward, McTaggart, and Sorley, but was not satisfied with their idealist philosophy. In Cambridge he was aware of its British empirical philosophy, that did not also impress him in its proneness for reducing philosophy to logic or to science. In simple, he was opposed to all forms of reduction, which foster dogmatic one-sided thinking.

He was for developing a sense of both separateness and connectedness of different experiential processes. This led him to a socialistic differential aspect, which conforms to the world-situation of today, as an arena of continuous emergence of multiple cultures.

What attracted Wadia, the *connectedness* of experiential processes. This also attracted me, I also perceive the *dynamic interconnectedness* of experiential process, from abstract philosophical and scientific deliberations to their *inter-related concrete vibrations*, and man's increasing powers in technological efficiencies. But these changes and modifications of one's character and ideas generally do not happen, if one does not travel abroad, or does not change one's inner mind, with a progressive openness, remaining always alert to the advanced currents of scientific and political ideas.

Wadia joined the Bar in 1913, but after getting his degree, he had no taste for practising as a lawyer, but he had developed a taste for jurisprudence, this influenced his philosophical thinking in general. Watching Wadia and his changing ideas or interests – I feel that human beings as a whole are psychologically oriented, and each one's psychic frame is *indeterminable* in any categorical sense, it is just a process and a possibility, only it can be developed in different directions.

His own reading of Western philosophy during his period of stay in England, had given him a knowledge of a varied trends in philosophical thinking, and study of some traditional philosophers – Plato, Kant, Hegel, or Fichte – and in the contemporary period. He became conversant with the Absolutism of Bradley and others, but their static Absolute left him cold.

His approach to philosophy, was in accordance with his early education trom the side of literature and religion, and with a stress on ethics. But he changed in progressive ways in accordance with the spirit of his time, with a full display of his original thinking.

His was a life of moral purity, in conformity with his mother's wishes. Science he thought was not a substitute for philosophy, though he felt its need for developing advanced thinking. His study of politics and social science had added to his studies of Indian *Dharma Sastras.*

His was a separation of interests in varied fields but also in their *interrelatedness.* He confessed that during his service as a teacher of philosophy, in different universities, he did not get much opportunity or time to write a constructive philosophic work, but he always occupied himself in writing papers for philosophical and other journals. His paper, 'Pragmatic Idealism,' reflects his past, but his was a programme for a new structure, with no dogmatic acceptance of his own philosophy of Zoroastrianism, which brought in the notion and fact of *evil.*

He was critical of Zoroastrianism. He was not convinced by its solution by transferring it to another powerful principle, other than God (*Ahura Mazda*), but to an opposite principle of evil (*Angra Mainyush*), in Persian-Zoroastrian thought and culture.

To him, that was not a solution, but an invitation to a dualism, which infects the supposition of an "Absolute Reality" from his philosophic perspective

His was a search for an answer for *evil* as an operative principle during one's life within and without. The search for an answer to a concrete reality of 'evil' in human situation, led him to study different cultural processes within Indian thought.

He was therefore, for a *reconstruction* of a 'new Philosophy' by critically over-seeing the attempts of the different religions of the world. He thought some partial solution of it through the Hindu-Buddhist law of *Karma* leading to final liberation or *Moksha.*

In the law of *Karma*, he found that there was some kind of answer to human differences and human sufferings due to one's injudicious action.

Karma determines the birth of an individual with a difference. It accounts for one's some kind of unjudicious action, for which there is a punishment, within a social milieu, or with the presence of an innate intelligence, insight, defect or deficiency.

His attitude, therefore, was not *abstract* philosophy, but his was a stretch to *social* philosophy and improvement of human relations, through proper education of people in general in *insightful ways* of understanding and appreciation.

Therefore, he thought that there was a chance for every individual for a progressive *self-improvement* and *self-correction*, by a belief in a series of *rebirths*, as then prevalent upon people in general, to give them a continuous opportunity for *self-improvement* from life to life.

It was perhaps a better suggestion than letting the evil person suffer, through an eternal damnation in hell, as prescribed in Persian or in other Prophetic religions – Judaism and Christianity, in emergence, towards Islamic expressions in India's Sufism.

In all Prophetic religions – in their abstract developments—a punishment is prescribed for every individual and for whole humanity, by an absolute doctrine, from a mythical reflection of a primeval family – Adam and Eve. But this view is no longer held by the progressive developments of religions, in interrelated aspects and in the study of cultural anthropology. The different Indian religions – Vedic, Upanisadic, Buddhist, Vedantist, Jainist – in their developmental processes admit the possibility of a continuous development, and ultimately a possibility of liberation or 'freedom' from all 'sufferings,' with or without any restricted 'theistic' notion. All religious processes are ultimately directed to a 'transcendent possibility.'

All orthodox Christians grant that 'disobedience to God' is a basic heritage, and so an individual is a born sinner, and someway destined to suffer for an original sin for generations, without questioning this presupposition. Obedience and disobedience are natural processes in human social situation, and in every case there is some reason behind such responses. It is deeply psychic.

This process of disobedience is followed by the doctrine of repentence and punishment. The acceptance of this makes humanity as a sick person from the beginning, which would dissolve all spirit or tendency for a self-improvement. Life becomes a closed process, with no ventilation, or an avenue to be out of this process. So religion and morality ceases to be a cultivable process. They are not treated as processes with a speciality of their own, outside the theological milieu.

All these reflections, to Wadia, suggest, that in Prophetic religions, there is a proneness for mythical trends, without pursuing a *realist* attitude for developing philosophic, religious or moral principles in real insightful aspects or as a reflective attitude, with some progressive self-improvement, through an independent reflective activity, as in the sphere of scientific thinking, or in the development of different religions in advanced socio-cultural aspects.

Wadia's argument is convincing, and it becomes worse still if one assumes, that in the Prophetic religions, an unchangeable order would prevail from life to life – where a person has no power to get beyond the clutches of one's inheritance. He will continue this *inheritance*, from life to life, and he will fail to find out whether he has improved upon his past, or will choose a life of indifference or helplessness.

So a religious consciousness needs a *challenge* for a possible improvement, as found in the evolution of the different Indian religions.

Moreover such *religious attitude* will always go contrary to *scientific process*, and this type of diseased attitude will affect the growth of civilized processes, as could as be achieved through scientific training of the human mind, as commonly practised from the past to the present all over the world.

My experience after a depth-study of Buddhism, leads me to think differently even with regard to the common Buddhist belief in Buddhism, when there is no admission of a *stable entity* in Buddhism. Buddhism, if *insightfully* developed, denies the Buddha as an ultimate personality. There is 'no Buddha', and there is 'no Buddhism.' If so, how can there be a place for rebirth in Buddhism? So, even not accepting the 'rebirth' of the same identical person – I feel the responsibility of my activities, understanding an *insightful* change of my personality-process, not in the sense of any 'ultimateness.' It is a change of 'personality' at least in this living order.

In my pursuit of Buddhist thought – I have found that in Buddhism – two modes of thinking are simultaneously developed. There is a *conventional* attitude and a *transcendent* attitude. From the former which is not a disciplined attitude, but an attitude which is restricted by intellectual forms, i.e., 'pro-intellectual.' But there is another order also – the 'insightful disciplined order' – which is also comprehensible as a 'freedom' from any 'intellectual perspective.'

So from that *transcendent* perspective there is 'no personality' and from that perspective, there is 'no Buddha.' But there is certainly a *disciplined personality* who is freed from intellectual proness, and from the disciplined perspective – there is 'the Buddha' – 'the Meditative Buddha.'

Again, if Buddhism is followed as an advanced process in all philosophic and religious thinking – then there can be no place for 'rebirth,' for the 'disciplined personality,' as already freed from the 'intellectual order.' Such a liberated personality is no more 'subject' to that previous 'intellectually oriented' process.

But if one continues to foster the transcendent process, from the lower order of a possible intellectual demand, then one has to devise another order by way of a concession, or compassion.

There is a possibility of a rebirth only in a transcendent possible sense, where there is no more the past personality – there is a process of total *transformation* within the transcendent order, as I am insightfully feeling in this life-process. I do not have any pre-meditated thinking for the life after the present process on death.

Scientifically, also, the persons are born, from an empirical perspective, but there is no same person, to be born or not born, it is just the consideration of a possible 'birth-process,' with no presupposition of any former personality, from any scientific perspective – biological, or psychological, unless there is a switch on to some general scientific concepts, which is a material for scientists through ages, non-dependent on any one's life-process. The notion of 'cell' or 'psyche' is a general or possible idea.

During one's own life, though there is a continuity of one's birth-process, it is very difficult for one to insist, that he, as a person, will remain absolutely the same throughout his life. He changes, in body and mind, continuously, and becomes *transformed* through varied experiences if he proceeds towards progressive *self-improvement*. So one has to develop a sense of *identity*, without claiming any apparent organismic or psychic basis.

Wadia was merely following the general view as prevailing then. My changes in insightful directions, will naturally differ from that of Wadia, and in a different direction, in accordance with my speciality, and the kind of experience I am acquiring through my researches in progress, remaining detached from my past experience or an inherited experience of my traditional past, or of the traditions of other people in general.

If a Buddhist insightful thinker is thinking of 'rebirth,' he is always speaking of 'rebirth' of two types of people – (i) the first type or order, is a disciplined order, and (ii) the second type is of the ordinary, relatively disciplined or not disciplined at all. It is a continuity of a deficient dispositional process.

But in neither case, if there is no stable or fixed person, how can there be a process of 'rebirth' only there is a present birth, with no continuity, of a fixed entity, even in this life, unless one posits a 'meditative continuity.' This is a 'transcendent psychic demand for 'self-improvement.'

On the basis of these progressive experiences of mine I will continuously have a change and transformation of my experiential process. I will pursue 'transcendence' without sticking to any 'ultimacy' in any order of 'experiential process.' In this suggestion, I do not cherish any contrary view, which would contradict the scientific attitude, or any conceivable experiential process. I consider all experiential processes are interconnected in all stages and orders so that perfection of all experiential orders is a possibility. It is a progress in the knowledge of interrelated subjects.

In Buddhism its three-fold meditative disciplines – in three dimensional orders of a disciplined conduct (*Sila*), progressive concentration (*Samadhi*), and the resultant process of insightful experience (*Prajna*) – can be understood. As a voyager I shall continuously travel and progress, and cherish new ideas with a progressive detachment.

Hence, the notion of rebirth as an empirical principle, can be held by common people, but it will remain a confused statement, without any depth-understanding of Buddhism. But I will entertain the ideas of 'rebirth' – with an open mind, with a further insightful attitude, and with a detachment – of a continuity of a disciplined live in the sense I suggested.

If there is no stable or continuous personality, in Buddhism, the assumption of the ascent to 'Arhathood,' in Hinayana Buddhism, is a disciplined 'process,' but not of a fixed personality as such. For the same reason, the ascents to 'Bodhisattvahood' in Mahayana, and to'Tathagata Buddha' or 'Adi Buddha' in Tantra Buddhism – are progressive transcendent insightful stages in meditative direction – and not an assumption of an entitative notion. All these possible personality-processes simply signify some possible orders, in the disciplined sphere. Those are not restricted intellectual concepts, but continuous meditative continuities.

So, the notion of rebirth in Buddhism, as prevalent among people in general – has a place only in a progressive transcendent sphere, but still, there is a kind of persistent acceptance of it in a general sense among the Buddhist masses. But this present vision is a tentative process – in my process towards an onward march.

If the notion of a 'personality' is not real, from the highest reach of the meditative discipline, from all orders of Buddhist meditative practices – then obviously there can be 'no God,' as an 'ultimate,' in Buddhism; if 'the Buddha' is conceived as a person, that supposition is also a wrong notion, and for that matter 'rebirth,' in the context of Buddhism—is a misconception or a jumble mode of deliberation, unless properly qualified.

Really, there can be no 'rebirth' in Buddhism, nor in any advanced religion. It is just 'conventional' – a sphere for a practice, drenched with 'ignorance.' Needless to mention, advanced religious process does not depend on a belief in an ultimate reality of God, a divine personality, for, as there is no final personality, in individual's life-experience, so there is no personality in the context of God. There is no ultimate as such, all are progressive possibilities, if taken as confronted. The notion of possibility – is distinguishable as *intellectual* or *meditative*.

In contrast to the above notion of rebirth as a reasonable experiential possibility, with doubts the notion of an 'original sin,' as cherished in Christianity, is a solitary myth, known to a localized class of people. This is not known to other parts of the world, but the notion of 'Karma' not in any specific entitative sense, has a place. It has a practical or a conventional value for retaining one's sense of responsibility for any action, during the duration of one's present life, but not in the context of so-called rebirth. Here a broad scientific attitude can be entertained, for a future inquiry, as a possible experiential process. But advanced notion of possibility, I am upholding as post-intellectual.

'Karma' might be considered as a general principle of activity, with a resultant process with effects – merit or demerit – good for life as it exists, or good even for posterity as historical records for emulation, whenever good for humanity as a whole is contemplated. There is a place for 'unknowability' in every sphere of experience from the myths to the sciences – if construed as abstractly intellectual.

If finally there is a transcendent continuity of a feeling process, without any entitative implication, there is a continuity as an energetic possibility, even when the process is not directly perceived, for the confronted universe is broadly a 'seen-unseen' possibility. This is my sense of post-intellectual meditative possibility.

There is such a thing as a universal process of Man's activity in different pursuits of life, but that activity-process is nothing ultimate, as a thing, but as a perennial possibility, in an 'if-then' sense. The net-result is a dynamic assest for humanity if insightfully entertained and maintained – in a post-intellectual transcendent sense.

One can just admit the notion of process and possibility in a non-entitative sense, in many-dimensional indefinite senses.

If viewed scientifically, in its advanced developments, it is nothing but a notion without any fixed scientific theoretical sense, where the role of a 'transcendent Nothingness' (*Sunya*) in a 'Symbolical-Philosophical sense' – has a meaning, or in a symbolical mathematical sense of Zero, is a reasonable scientific expression.

Wadia was in search for reality of various orders of experiential processes, with no admission of reduction of one into the other, they are all distinct orders of experiential processes. So 'evil' is a distinct type of experiential process and character, to be understood from various psychical, moral and social perspectives. It is a *demon* in its core, with protean expressions, tending to be *magnum opus*, if indulged in. It has a place in social, cultural and international situation, as a concrete experiential situation, not in a final abstract or ultimate sense. It is nature-centric or Man-centric in a variety of senses or in experiential contexts – in possible cosmological processes, and possibilities – so long man is locked in abstract intellectuality.

The individuals, for their separate 'inner evil disposition', prepare the grounds for conflicts and war, but it is not a fault of whole of humanity, as thought in Christianity.

These events of conflicts and wars are due only to group actions, mischievously directed, affecting the destiny of mankind in general, there is no dispassionate search or pursuit for an answer, in the prophetic religions, in general, unless they are deeply meditative in the advanced order of the Buddha or Whitehead, as I have developed so far through my publications.

Every activity-process is psychically oriented, so progressive insightful activity – in all spheres of experiential processes – has a progressive value, and is worthy of cultivation, deviation from this path brings in evils.

I am now reviewing my search for philosophic pursuit, as below with my understanding of the different religions of India and of the world in general, with some references to the philosophic position of Wadia, wherever relevant. But here my ideal is with a possible brevity. If in this process, the readers get into a monotonous feeling, they are at liberty to abandon this portion.

In Indian religions, since the Vedic Hinduism, the evil in a broad sense, arises from man's inner egoistic disposition to think that one's powers rest on the individual alone.

But really any power has a *transcendental* basis, for, the confronted sense-organs are just instruments, without the *transcendent* continuity or element of power, or self-manifesting character, which equips a living organism with varied capacities behind one's intelligence, or behind the operations of the different sense-organs.

To assume that a capacity, or power belongs to the organism or sense-organs, is indulging in one's own pride of power, an evil of a sort, which, being a confronted aspect of experiential process, really is a borrowed power, but lacking a sense of indebtedness from a basic transcendent power, or continuity of experiential possibility. One's so-called power is a confronted or emergent borrowed power. The real power is post-intellectual or transcendental.

Without recognizing the transcendental basis of all powers, if a man brags of his borrowed power as, self-originating or one's own power, is a misplaced concreteness.

Such a dogmatic assumption amounts to egoism, and therefore it is entertaining an evil. It is vicious, and an expression of ignorance.

The Upanisadic thinkers carry forward the Vedic tradition differently, suggesting that the powers behind the different sense-organs are not *in* the different sense-organs, but they are *transcendentally* based in one's 'Self.' The power is not placed in one's *confronted* sense organs, but in the *transcendent* process, which supervises the confronted operative powers, without being affected by them.

From this transcendental perspective, man's organismic powers, of any cosmic emergent levels – from the physico-chemical levels through life and mind – are borrowed operative powers or emergences. The lack of this knowledge by the confronted powers is ignorance.

The Upanisads, therefore, declare that the confronted sense-organs lack the insightful experience of the double-consciousness (or consciousness of consciousness), by mistakenly thinking that the sense-organs are operative by themselves. The sense-organs are projected processes, from a transcendent self-shining background. Here also a frank admission of a transcendent power from within, every moment is necessary.

Not to admit the place of operation of this transcendent power fosters one's egoism, producing all sorts of evils in psychic consciousness, or disposition, and these evils are transmitted to social processes, making the world-situation a vast sphere of evil processes. From social process the evils travel to the sphere of politics, and in international situations, infecting the cultural processes of mankind.

This transmission, is like a virus behind everything, flaring up unseeable mental, social and other types of diseases. The evil is associated with one's inner dispositions, but, from the naturalist perspective, it is associated with Nature also.

In naturalist sphere, the referrence is to the unnatural occurrences in the forms of volcanic eruptions, earthquakes, floods, storms et al., or any calamities or destructions, epidemics et al. But these natural processes may not be directly good for human beings, but Nature has its own processes, which can be understood scientifically. Scientific perspective is differently oriented. But it has also a post-intellectual insightful transcendent possibility, as becoming manifest through modern researches.

But besides these, following from human nature, or from Nature, there are criminal activities, or terrorist activities, in human situation, followed by wars et al. These are human creation of *evils* of diabolical sort. An individual, in a cosmic-social situation, is placed in the milieu of these situations, also, which manifest such characters.

These disciplined life and thoughts are continued to Buddhism – which denounced all argumentative tendencies, for, generally, the argumentative tendencies are the roots of all possible *evils*, if not controlled insightfully, because the insightful *transcendent* process prevails, or predominates.

The transcendent process is in a different dimensional order, dependent on the process, of conduct, and continuous meditation, waiting for the emergent order, in insightful experiential processes.

The mere argumentative tendency has to be distinguished from real and pragmatic investigative ways – as fostered in post-logical periods of Indian thought with a comprehensive philosophic attitude, during the centuries of the varied, Buddhist, Vedantic and post-Buddhist Vedantic period of India, leading to complex socio-cultural movements in contemporary Indian thought.

I have indicated this process, of the two-ordered universe, in some details in the third Volume of my 'Dynamic Facets of Indian Thought,' with an insightful understanding of man's continuous possibility for one's development, in progressive 'insightful orders,' through progressive elimination of *ignorance* of some form whether in thinking of one's capacity, or in the thinking of one's activity-process.

This kind of myth-free detached thinking is not found in Western thought and religion, as an explicit disciplined way, in its earlier stages, it was involved only in some abstracted conceptual way, in philosophic and scientific thinking in Whitehead, Einstein, and post-Einsteinian scientific thinkers like Heisenberg. Schrodinger, Dirac of the Euro-American milieu, and Indian contemporary scientists – C.V. Raman, S. Chandrasekhar and others.

When the Western thought steps into the many-dimensional perception in the Naturalist sphere, and with this development Whitehead discerned that in the orders of Nature there are initially two orders already present – (i) one order which is basic and widely *relational* and non-sensuous, and the (ii) other order is the transcendent order – a post-conscious insightful or wisdom order, within Nature in transcendent 'may-be' processes and possibilities, with no claims for a static existentiality.

Then from this understanding of the three confronted Naturalist order of process and possibility – one is able to discern, by an inter-cultural perspective – that this 'post-intellectual insightful order '– is non-different from a similar and not similar order as cultivated in Buddhism.

Not to be aware of the Buddhist process, which historically preceded Christianity is an unpardonable 'historical ignorance.'

This kind of thinking is developed by Whitehead in his *Process and Reality* and *Adventures of Ideas*, and in *Religion in the Making*. Whitehead courageously refers to these, unlike other Western thinkers.

This avenue of deliberation, as set up by Whitehead, runs alongside of Buddhism as an alternative insightful current. Indian philosophy,-religion, psychology and science – as preceding, and following the Buddhist process – by a contrast as transcendentally oriented – with disciplined practices – leads one from a lower to a higher order of concentrated experiential levels.

The disciplined processes are exercises, projecting intellectual process to a transcendent insightful order, which is a continuum of consciousness.

This book develops an independent thinking-process in all aspects of experiential perspectives, without the assumption of anything as ultimate. The transcendental process and possibility is a concrete insightful experience, not as an ultimate finalist sense, but as a process and possibility, experienced before and after all confronted subjective and objective experiential processes as all possible continuities or possibilities.

There is no such thing as 'there,' in a categorical sense, but as an 'if-then' process of 'may-be' in a symbolic sense.

Hence, from this perspective, philosophy is an expression of creative possibility – from a transcendent – to a confronted empirical world of interrelated processes of knowledge in possibility. Here the supposition of a transcendental element, before and after, is symbolically entertained, without any claim of existentiality.

Without the triadic disciplines as mentioned, in the context of Buddhist and the discriminative attitude of the transcendent (*real*) and apparent orders – the elevated Vedantic forms of life as a post-Upanisadic, post-Buddhist life – are not possible.

In both Buddhism and the Vedanta, in Indian thought, there is a deliberate detachment from discursive disposition, in one's insightful thought and activity, otherwise, no real solution can be found in philosophy, religion, scientific or political experiences of mankind.

Too much abstract thinking, without the development of insightful process amounts to a continuous suffering, from vain discursive habits, with no final solution or freedom.

The exclusive empirically oriented discursive philosophy or science, ultimately falls into the traps of empirical modes of continuous discussion, with no solution.

The effects of indulging in too much discussion disturb the entire judicial system of the United States. The judicial system, indulges in intellectual discussions – it tries to find out different types of murders, committed by a culprit. He is vainly placed under trials in several courts on murder charges, though he has committed several murders, and also in some cases such criminals exhibit even cannibalistic propensities. Some psychologists and lawyers indulge in studies on such criminals. Such criminals even are allowed to preserve the heads of murdered human beings in a refriegerator, and they seem to eat or taste the mutilated body.

There are defence lawyers and even professors appointed at some universities, to study such cases, to prove that the culprit, in question, is not a culprit, but a sick person deserving compassion and pardon.

Such expert lawyers and professors must understand that the culprits also are adepts in exhibiting their capacities in varied deceptive ways to outmode the intelligence of the learned lawyers and professors.

There is also a psychological specialist in employment, recording such cases for research, trying to find out psychological reasons to prove the culprit's innocence.

But in no sense, it is a worthy psychic phenomenon, where a person, in question is genuinely conscious, he should be disciplined in the highest reaches of meditative processes, as cultivated in the different orders of Buddhism, to the Zen Buddhism of Japan, in contrast to the early and present stages of Indian thought, or alternatively in the case of Whitehead, in contrast to the early and present state of Western thought in development.

The culprits of the United States, are conscious of general psychic frailties of the people of the United States, which is evidenced by their frank admissions.

Such self-conscious or super-conscious criminals are generally pardoned as suffering from a mental disease, as if mental disease is an illness as any bodily disease, whereas an ordinary criminal, suffers the duration of his entire life in jail. What a pity!

Here what I think, is that the public in general, needs to be educated to cultivate a disciplined life, which is an ascent to progressive insightful orders with an acquired capacity of progressive discriminative capabilities, or learning – otherwise, any statement is a simple excuse, or a discovery of a mental confusion between different types of behaviour, and not a discovery whether a man/woman, in a particular case is culpable or not.

An individual is simply normal or abnormal, without any transcendental insight or knowledge, so one needs to be trained in the sense of real 'Wisdom.'

It is not an abstract scientific discernment, but also a determination of a different behavioural process, with insightful discriminative attitude: whether a particular behavioural process is social or anti-social, or simply an expression of wickedness, obstinately nourished and developed, not only by the society at large but also by any individual, with a plea for specialization on a subject.

If such wrong behaviours are not corrected from the beginning, or from early childhood, then no civilized training is possible. The correction of one's behavioural ways, is only possible when a person separates or detaches his/her basic personality from one's confronted activity-process, where a person can feel that he/she has an option to do or not to do an offensive action.

It is a discretion or an option, as a Master of himself, one has sufficient power to control one's activity-process. Without the cultivation of controlling one's activity-process – a man/woman falls into the clutches of his/her misdirected activity-process, when he/she earns the disposition to go by his/her disposition or impulse.

World's moral code, which is broadly a symbolical law becomes more powerful than the transcendentally oriented, natural or psychic laws – as expressed in the Vedic *Rta*, which is at once, psychic, natural and moral.

This is India's transcendent philosophy, with a confronted creative possibility, not restricted to any exclusive sphere of experiential process, with an element of a fixed existential character.

I do not find any justice in such confusion between one's decision and the impulse, compelling one to act on the spur of the moment. In such impulsive action a man loses any sense of real value.

But Hinduism and Buddhism, too, inspite of their merits, are not always studied with an insightful scrutiny by ordinary undisciplined or semi-disciplined ordinary person.

Instead of a strenuous spirit of real research, human beings entertains some wrong reflective insightful processes, and face some social problems, which have emerged in process, due to no fault of disciplined thinking of their philosophic positions.

Hinduism is confronted with some form of static caste, initiated by disciplined social structure. But here proper disciplines are not always observed by men or women in general, so there are some deviations, and the resultant processes of non-observance of the caste rules come into the social structure.

Hinduism takes into consideration those natural processes, and makes provision for accommodating such other social structures. But there are lots of other types of persons living in society who do not come under the category of the disciplined order.

They form their own community, such orders of people in India, are generally taken as out-caste people, in the sense they do not belong to the Hindu four-fold social structure, but are born of a different social structure, and hence they are disciplined in a different sense, and this does not amount to saying that they are undisciplined people.

Besides such people, in India there are lots of tribal peoples in various locations of India. They are outside of the Brahmanic, or Buddhist structure. But according to my judgment, given these communities as different structures, they are treated with due respect, but within their social milieu, they are all subjects to their social rules, and there are deviations from their respective groups, as in the groups belonging to Hindu, Buddhist or Jaina structure, or generally among other religions of India.

Hence the social structures in Indian soil, have to be interpreted, not alone from the perspectives of 'Hindu caste system,' but they are recognized groups of different social structures, as indicated above.

They are not social *evils*, but emerged in social process, not with *evil* motives.

The discriminative attitude is not determinable alone by the caste-system of Hinduism, or those who are the deviants from the Hindu mode of life. To tackle India's problems, from the perspectives of Hinduism alone is not only narrow, but wrong, or only a foreign intervention, as this outlook ignores different multifarious other social structures prevalent in India, from the ancient times.

A comprehensive sociological study, is not just an exclusive discourse. This is becoming evident from various kinds of publications from the Readers Digest or from National Geographic Society et al.

If this discourse leads one to unnecessary digressions – then one might cut this discourse with complications in different ways, by polluting human motivations, so that the past basic principles would be completely lost, and would become a jargon of labyrinthine opinions, as the human mind maintains some notion of distinction, a distinction of people according to one's distinct professions.

There is no natural process, in any cosmopolitan social structure of a country or culture. The Hindu caste-system is not merely a general social structure, but it is psychologically oriented structure as a depth-process, in terms of one's inborn dispositions aiming at progressive self-perfection. But it had a historical degradation and dissolution for its progressive courses towards impurities, as common to all human processes.

The solution of the problems of the different emergent societies lies in a different perspective or approach. The answer, if any, can be provided from insightful perspective, and not by irrelevant arguments.

A thoughtful sociologist has to accept those social orders as they are – from very early times, even before the historical period. Besides the followers of different religions, there is a large number of people who do not follow any specific religion. But they may be *disciplined* in their own ways. Hindu caste-system was originally based on one's disposition and a natural choice of profession according to his/her capacity.

Indian societies, as emergent processes, from various religions or even followers of no religion, as a whole need the impartial facilities for education in the line of the disciplined aspects of character-formation and also learning to master various branches of learning for adding to their efficiencies as members of India's multi-cultural social structures with their varied forms of historical past and with skill, preserving all the time a sense of purity of one's inner self, to be operative with others of the same spirit. This smacks idealist way of life, but without it there is no palpable hope. It might bear a faulty notion, but not less pragmatic.

First by an admission of the so-called 'caste-ordered' society, if a Hindu, on par with the different other caste, communities, religions or no religions—is evaluated, then his position is non-different from any other cultural societies of any country.

The so-called 'depressed class' of people, is an offshoot not alone from Hinduism, but they emerge in all possible religious or non-religious groups of people, from all social or political structures, with democratic structures of any variety, throughout the centuries to the twentieth century – among the Hindus, Muslims or Christians, or of any other religions.

Too many books have been written so far against Hindu caste system, now my suggestion will be for distracting one's attention from this long-drawn repetitious mis-directed cultural criticism. One must keep oneself separate from this hackneyed choice, or monotony.

All these varied orders of people—are equally important or present among all orders of people in the over-all frame of India's community structure. All these emergent processes, are subjects of a sociological situation, in human society in general.

The so-called suppressed or depressed classes of people of any social structure, are to be lifted up to the more or less same status giving facilities for training and impartial job opportunities by progressive education in human and specialized sciences.

Admitting that the study of Indian society is a complicated phenomenon, any arbitrary solution often amounts to a sort of coercion, or unpsychological project, just as the study of the criminals of the United States is a complicated subject-matter of the judiciary of the place, dependent on Freudian psychology, or false notion of a compassionate attitude.

In that country also, though there is no apparent class difference, there are professional differences and gulfs between the rich and the poor, people coming from the different parts of the world, Europeans, Asians, Africans and South-Americans et al. They may be employed or unemployed, or may be totally homeless. With no education, they are all empirically placed to their impulsive natures, if there is no suggestion for some disciplined ways of life, through proper education, according to their religions or understandings or religion in their essential cores – then there is no provision for the emergence of a higher cultural process to follow.

There is also a gulf between the rich and poor classes, in the world as a whole, business community and the government officials and the hapless unemployed numbers. There might be a conflict always between extremely rich grouped together and the Government.

It may be that discourse, in the context of Wadia's problem of *evil* in society has been unnecessarily carried long, but it was somewhat deliberate to indicate that 'evil' is an emergent process, arising from all perspectives. It is a century-old problem, and is found everywhere in the world – past and present.

But Wadia does not go very much into all these details mentioned above, he gives only a general reference to this, with a sympathetic understanding and awareness of the so-called out-castes in Indian Society.

He was only worried by the *evil* as a phenomenon of life, in all its phases. But he was courageous enough to indicate the fallacy of the postulation of a dualistic religious basis as ultimate in Zoroastrianism, his own religion. His examination of religious process was from an insightful perspective.

An advanced religion and philosophy, of Buddhism, too, suffers for remaining static in exclusive forms of Theravada, Mahayana or Tantric disciplines, without trying to continuously develop the deliberations, in each stage, as evolving towards further advanced stages.

This writer's study of Buddhism, is by a distinction and also from an integral insightful aspect. It is a development of a dynamic process through Buddhist disciplines, throughout his life-long research, to a *transcendent* – 'No Buddhist' process – which to his mind, is the goal of all religious bodies of the world, as indicated in 'No-Buddhist process.'

Similarly, one can cultivate a 'no-Judaic' process in Judaism, or similar such transcendent processes in many other religions of the world as to flow in a common international stream of religions of the world, as shown in many of my works through Buddhism.

Everywhere the choice of study is an individual's choice, taste and capacity, just as in any scientific or technological study.

The Jainas, too, to overcome the *evils* of social discrimination, are trying to take resource to an extreme form of ascetic life in prescribing the Digambara form of life, of a total nakedness – so that from within and outside – a Jaina abandons all desires and aspirations.

Hence this form of a total discipline or asceticism is almost a life of *isolation* away from any society. Within Jainism, also, one would choose to live with a moderate form of discipline, and hence it claims to eliminate all forms of evil to generate the possibility of a 'liberate condition.'

The Jaina Svetambara class gives some concessions to normal householder's life, and also upholding the theory of possibility for the females to attain full perfection, or a place among the Jaina Tirthankaras. But there is also, an admission of a female liberate. She was Mallinatha. Here also, my prescription will be for a different type of social process, and a process of elimination of *evil*, more or less, in a Jaina society, by a scheme of moderation.

But, one must know, that the Jainist demands are too difficult to conform to the social and political processes in development of a present-day scientifically oriented world.

The present-day social structure of the traditional types are greatly affected by the political ideological schemes as fostered in the present century, which out-modes all past-processes.

The future of the world depends on the existences of the *ideologies*, or a complete evaporation of any *ideology.*

The world-process is fast proceeding in this *dissolving* processes of *ideologies*, it has now reached a stage of a complete vanishing of any *intellectualist process* as in the advanced aspects of Buddhism, and all other advanced religious processes of the world, if one follows Whitehead of the present world – so philosophically and scientifically oriented – in its advanced processes in togetherness.

This is admitting a transcendent order of experiential process as enveloping from all sides of confronted orders of experience.

Wadia does not subscribe to this kind of advanced religious or political process, he is not also sympathetic to the extreme ascetic life, of the Jainas, or even of the moderate process of Jaina life, prescribing fasts. He cannot also approve of the social processes of the Hindus, Muslims or Christians, for different reasons as might be involved in Wadia's mind.

His suggestion is for a *new* social-structure, some idea of which I have attempted to clarify, but not in any specific details.

Wadia stands for a novel principle of his, 'Pragmatic Idealism,' by eliminating the defective processes of the different social orders. He illustrates an attitude befitting his time, but I have gone beyond to the present order of socio-political evolved conditions.

For eliminating *evil,* Wadia suggests, a total reconstruction of Hindu society and other Prophetically oriented social structures. His interest, therefore, was naturally for a new social structure not by a movement for politics alone. His suggestion was for a changed moral character from within, through some educational schemes.

That was one of the reasons why in his early student days, Wadia was not so inclined for associating himself with any immediate demand for political movement, then prevalent during his time, though later he changed his mind. What was true of Wadia was also true of the many Indian intelligentsia and the masses in general.

Wadia's understanding of the religions of the world, was from a broad moral perspective of eliminating *evils* from all social processes.

Speaking of his inherited Zoroastrian ethically oriented religion, he suggests, that its teaching is anything but ascetic. Zoroastrianism does not advocate fasts or celibacy.

It does not look upon life with sickly eyes. It has faith in a righteous God and believes that in order to succeed, man has to try courageously to rise above *evil* temptations.

Like the ancient Greeks and the Persians – he believed that Man is *most natural* when he is *most developed*, which is a life of civilization, where arts and literature, science and industry flourish.

If the Absolute Spirit lives in all its parts, and directions then there is nothing which it needs to repudiate as alien to itself except the *evil* which hinders its life of harmony.

According to Wadia, however, the life of the Spirit is found in the interplay of individuals within a social life around us. This sort of life, has to be intelligently matched with a life of reaonable detachment for progressive self-perfection, but it is not a totally solitary or exclusive life. solitary or exclusive life.

Wadia thinks that mere cherishing a theistic attitude is not sufficient for a cultivation of religious life, man has to orient himself in the progressive direction of philosophic, artistic, literary mode of life, as suggested by Plato, who combined all these processes of human nature. Comte, a positivist, in post-Hegelian thought was a pathetic example of denyingGod, but was driven to raise a *new* God, which is *Humanity*, to give vent to one's emotional cravings.

Russia's introduction of Godless Communism, ends up with an adoration of a Leader, and a Gorbachev emerges with a programme through his 'Perestroika,' he wants to modernize or remodel Russia to democratize its entire constitution of communist union to an unforseen process towards a market economy to be approved of by the Russian countries as a whole, which is a *sociology* of many countries and cultures. This project is intelligently projected to the Europeans, Asian or other developing countries, to show that its constitutions also can be changed to take any complex shapes with alternative choices, suiting the needs of the different countries and their peoples – considering everywhere the balanced budget, according to the capacity of each developing country.

My view is different, for, the *social porcess* has evolved so much, that nothing really can be totally eliminated, one has to be operative neither by accepting nor by negating, but one has to proceed by a progressive scheme of *transcendence* which becomes acceptable by all without hitting any one at all.

In European Absolutism, the highest category, cannot be a person, without sharing all the limitations of a personality. Different *forms* of mysticisms, from Shammanism to highest Monotheisms, speak of the highest religion's mystic experience as having one ultimate value.

Similarly the Christian Catholicism or Hindu Vaisnavite or other types of mysticisms clearly show the purely personal origins of many of these mystic experiences.

My attitude after the depth-study of Buddhism, in all its dynamic aspects—is towards some conceivable stages of personality-process, and then by a detachment to no personality process.

It is a process beyond *Vajrayana*, still staying with some personality-process of a *Lamist* movement. That must also has to be abandoned, but retaining an attitude of reverence as a ballast.

Religion is fundamentally an attitude of reverence reacting to Reality as Truth, this may be variously symbolized as God, or a Living Presence that unites each to All.

The God is the personified aspect of the Absolute, the mind in which everything lives and moves. This *symbolical* God is within and around us.

In the words of a Medieval Muslim-Hindu Mystic – Indian Kabir – there is *Musk* within the deer yet he thinks that the fragrance comes from without, and he hunts in the forest vainly in search for it. So God is within us, but we mortals miss Him, and pass Him by.

To Wadia, religion as applied philosophy is a life in *harmony* with the purpose of the Spirit, through different lives, till the individual soul merges in the Spirit that pervades all.

His prescription to the Youths of India is to turn their attention not to an abstract Metaphysics, or to any *ism* in any sphere of experiential process, but to a New Social Philosophy, reconstructing it in a non-traditional spirit, uniting all classes of Indian people to a unified social life, as to be strongly placed to face all invading foreign powers.

But how this 'uniting' is possible. It is not indicated in a right integral direction. My suggestion by a contrast is preparing the human mind to a continuous advanced process by progressive *transcendence.* It is a constant preparation and a heartfelt willingness.

Wadia's is not a distribution of powers to men alone but to the emerging females with the enlightened processes of comprehensive learning and understanding, not with dogmas or rituals, but by a development of the advanced scientific and technological ideas.

This new orientation of the men and women of this land, should interpret philosophy as a pursuit towards no barren speculation, but a search for illuminating vision of Truth which inevitably prompts self-culture and social service. This pursuit also can be non-different from my view of *continuous* transcendence without any stop.

Wadia's study of philosophy is for a new reconstructed social life, by an elimination of *evils* from all misdirected traditional religious, philosophic or mystic pursuits, by the symbolical suggestions of his paper, 'Pragmatic Idealism.' This paper, is a departure from any traditional abstract ideas behind them, but is a progressive reconstruction of all with an insightful attitude of self-culture in the devotional spirit of a social service.

With this as ideal, one can formulate or reconstruct the prospective project of this volume in the direction of emerging scientific and technological world of today with varying political ideas circling round the world.

In this departure from the past I have also continued the subtle lessons of the past without abandoning its core.

Coming to the end of my four volumes of *Dynamic Facets of Indian Thought* several editions of *Changing Phases of Buddhist Thought* and *Four Principles of Whitehead's Philosophy from West-East Perspectives*, and other books in process – I have in my mind some ideas. These ideas, as below: That India before independence from the British with her varied political leaders – became mainly operative through the Indian Congress.

They constitutionally prepared themselves for a parliamentary democratic system of government, as inherited from the British, with a willingness, and also with a reluctance. It was an inevitable possibility – which the Indians had to accept – with a non-violent attitude. This was a unique historical achievement, as some form of friendly parting of the British out of the Indian soil.

Now it is the responsibility of the Indian people as a whole not merely to accept what it is, but with intellect and insight, try to improve upon it, not with clashes but with schemes of willing acceptance, maintaining an inner peace and order, to rise in strength of a nation within, with an international outlook, without. India welcomes all who come to India, and settle down in India, if they so decide.

India should not follow the violent ways of the West in general, but should proceed with an art of compromise and mutual settlement of problems with progressive understanding, tackle its own country's future and also the future of mankind as a whole.

With this as an objective, this writer will pass into two of its later parts to reconstruct in a dynamic and creative way, West-East philosophy, science with technology and politics – with a new and integral vision, getting rid of the details as far as possible, for, most of the things have already been mentioned.

This is not again an inconceivable structure, but replete with progressive exchange of ideas, for a genuine progress in human nature and civilization, modifying the historical currents non-violently as far as possible.

Here also as in the formulation of a new attitude towards philosophical thinking in continuity and prospect, one has to review the scientific and political ideas and objectives of today with scrutiny, criticism and appreciations of progressive achievements in each of these dynamic areas in interrelations, admitting the roles of each of the parts, as useful and necessary.

Be it noted that in this transition to science and technology, or in the stepping into the process of politics – only the core-principles of these areas will be considered from two general perspectives, from India's own perspective, and also with an eye to maintain the interests of all countries and civilizations as can be possibly known, according to this writer's limited knowledge and understanding.

My objective in its subsequent parts of this book will be a new approach to India's Planning in progressive improvement of the Indian masses through gradual education and through the best ideas of India's culture, in progressive secular directions.

My final suggestion will be: No more by ideologies, but by direct progressive technological ways, paying some heed to reshape India's political future through liberal ideas, as may be clarified with alternative suggestions.

In the next chapter a vision of India in rapid change has been presented through the writings and activities of India's great historian – K.M. Panikkar.

REFERENCES

1. The reference for this Chapter will be from the perspectives of some relevant books suiting to this chapter. Hence from the point of view of this writer – for this reference – is not an usual affair.

   Taking note of this I am referring to the chief book I have chosen to refer to in this chapter *Contemporary Indian Philosophy*, Edited by S. Radhakrishnan and J.H. Muirhead, first published 1936.

   I have chosen from this volume only five well-known thinkers – Hiralal Haldar, Humayun Kabir, R.D. Ranade, M.M. Sharif and A.R. Wadia – as some of the thinkers who did not interpret 'Philosophy' as an abstract thinking scheme but as something which one creates from one's study of philosophy from an inter-cultural perspective, not by a following, any, but by creating an avenue with insights to express one's philosophic attitude in confronting all other aspects of experiential process. It is an exercise with an insight and application of an attitude towards shaping one's way of life with a 'freedom' from all 'abstract thinking.' In this particular volume, I have not considered many other India's philosophers, only for some reasons suiting the purpose of this present book. In other works of mine I have considered most of them from varied perspectives.

   The present work is in no sense, encyclopaedic, it is a restricted work with a defined objective. When dealing with the philosophic deliberations of the above selected personalities of this book, I have made enough references from their papers concerned, in the body of the book itself.

The speciality of this volume is in requesting Mahatma Gandhi to answer the following simple questions: (i) What is your Religion? (ii) How are you led to it? (iii) What is its bearing on social life? – lest he would simply excuse himself by referring to his busy political life.

Indian philosophy from the early Vedic seers to the modern Logicians (*Naiyayikas*) is a presentation of a contrasting outlook from the Vedantic integral basis to a progressive analytic process. It is a similar contrast in the analytic direction – from the mystic Neo-Platonic and Patristic tradition – to Kant and his followers in Western philosophy.

The drive of the Indian mind, as will become evident to the readers, will be one of tolerance in every sphere of experiential process from religion to politics, and the Indian attitude will be international in spirit.

India's ideal is not pessimistic as supposed by some earlier thinkers. Indian philosophy very efficiently tackles the existence of *evil* not denying it, as not an ultimate phenomenon, it is a perishing process, as a confronted phenomenon, and so it can be transcended by freeing oneself from one's narrow egoist motivational habits in one's behavioural process, whether in arguments or in the ways of life.

How can *evils* be transcended in one's day to day life, including social and political spheres, can be found in short answers that Gandhi gave to the above questions.

I, as a writer, will try to further summarize Gandhi's answers, as below: (i) If Gandhi admits that his religion is 'Hinduism,' then, as it is, it is non-different from the 'Religion of Humanity.'

Answering the second question, which is considering 'religion as a process of life,' Gandhi gave an appropriate answer by saying that he was led to his 'religion' through Truth and Non-violence,' which is love in the broadest sense. So instead of saying God is Truth, he would choose to say, that 'Truth is God.' Elaborating his ideas further he said, 'Denial of God' we have known, but 'Denial of Truth' we have not known. Even 'Truth' is known to the most ignorant among mankind. We are all sparks of Truth. Gandhi's mystical attitude is expressed when he said that the sum-total of these sparks is indescribable, God symbolizes that Unknown Truth. He was pursuing that path by constant prayer.

Answering the third question 'the bearing of religion on social life' he said, lies in one's daily 'social contact' or 'social service,' losing oneself totally in that process. So, there is no escape from social service, and it has to be extended to all aspects of life, and to all types of persons, with no discriminative attitude. In this lies man's happiness.

In contrast to Gandhi's answers about his religious experience, and how he was led to it, or its bearing to social life – one will find a detailed exposition of religious experience of Tagore of his time, and his was a creative expression or shaping from his poetic and artistic vision.

I will not go much into details of Tagore's thinking, but will simply bring out the religious experience of the time he was born. The reference is here to his family background. The different gifted members of his family, for their revolutionary religious ideas – were ostracised but this 'freedom' provided them to develop India's inherited ideas with a novel attitude confronting the Western impact.

His family was very much influenced by Raja Rammohun Roy and his Brahma Samaj. His father Debendranath Tagore was an advocate of a slightly different religious attitude of Raja Rammohun Roy.

Besides being associated with a religious movement, his association was also towards the innovative linguistic revolution brought about by Bankim Chandra Chatterjee, Bankim departed from the 'ponderous forms of Bengali language' locked with Sanskrit heritage, he lifted its dead weight with a touch of his magic wand, and aroused in Bengali literature a new life releasing it from an agelong slumber.

In the currents of religious, social and literary movements, Rabindranath Tagore became operative, by refusing to go to school. Academic training, he thought, will not suit him, but he began his own way of becoming directly conscious of European language, specially German, through his study of Dante through the help of a Missionary German lady. He almost mastered the language, the core of which he could not know before, through English translation.

But finally he chose to pursue knowing his mother language through his studies of Vaisnava lyrics, and Sanskrit literature in depths from the Upanisads to dramas and poetries, probing into their rhythmic innovative expressions.

So inspired Tagore switched his mind off from the materialist attitude of Western philosophy and scientific ideas.

He became engrossed with the depths of Nature's beauties, from dawn to night, and through the transition to different seasons, and in different places from the high mountains to the flowing rivers, so that he could hear the music that runs everywhere making him a perennial traveller with galactic visions, leading with him from the finites to the infinity.

Tagore realized his philosophic thinking, as entering into all aspects of experiential processes, and made it a creative expression with a freedom.

This integral consciousness has been expressed by me in the fourth volume of my *Dynamic Facets of Indian Thought*, and in this book too through its new title, for, I have not kept philosophy, soaring, in some abstract aspects, but as flowing through the sciences and politics, breaking all-through its abstract boundaries.

I have not only found this creative, connected expression in Gandhi and Tagore, alone, but also in some of the chosen contemporary Indian thinkers in this chapter. Their attitudes were not strictly confined to some abstract philosophical ideas, but they had distinct approaches to philosophic studies.

They utilized their education, whether obtained in India or in some foreign countries – with a freedom – and finally they expressed their philosophic ideas relevantly in confronting the problems that emerged in the context of social and political life, by a reference to their respective times.

To me the confronted universe is so variant, that experiential aspects cannot be integrated from any fixed idea philosophic, scientific, social or political – it is dominantly 'Naturalist' governed by the laws of Nature, pertaining to its various aspects, nobody knows finally how Nature will behave. Similarly the 'psycho-social' and 'political' universe, is not determined with any final certitude. Therefore only some possible general ideas are given with no claim for a categorical assertion. This book, therefore, finds only some relative 'Truths,' in Gandhi, and also some other 'Truths', in Tagore, and some more in other, taken note of, in the book chosen for consideration.

I have not deliberately mentioned about all of these contributors, but I have developed most of the ideas of these thinkers, or of their general backgrounds, in Indian thought, through the different volumes of my *Dynamic Facets of Indian Thought*, culminating with the Fourth Volume of that work, where I have touched the movements of Indian thought in the present times, freeing philosophy from mere abstract realms.

Besides the main reference to Gandhi and Tagore of *Contemporary Indian Philosophy* there is a reference to some other select personalities as mentioned above, and their papers of this book, dealing with their philosophical ideas as also related to all other aspects of experiential processes, transcending mere abstract ideas of philosophic thinking.

Though in this chapter there is no reference to contemporary philosophers of other major countries – America, Australia or America – they have been treated in general as developing Western thought in general. From that background only Whitehead has been chosen, and from the Indian and Asian background, only Buddhism has been chosen as expressing advanced Eastern thought – for carrying forward West-East thought, but the main emphasis of this volume – is to demonstrate that philosophy is no longer an exclusive deliberative process, but is also related to other aspects of experiential processes.

PART II

PHYSICS AS BREAKING UP BOUNDARIES

CHAPTER IV

India's Independence Symbolized from Philosophy to Politics: With Prospects towards a New History

In my *Dynamic Facets of Indian Thought* in four volumes – I presented Indian thought in many-dimensional facets from the Vedas to the contemporary Movements.

The entire development of Indian thought was from the Vedas to the Upanisads, and from both – to Buddhism, and to the post-Buddhist Vedantic processes.

After the impact of Western thought in India, the advanced Buddhist and Vedantic thoughts were developed in alternative forms of West-East philosophical movements – in comparative philosophic directions.

These mixed complex processes of Indian thought, in an aspect also touched Whitehead's process-philosophy – developing all philosophic thinking as a continuous creativity, separating oneself from all restricted *forms* to the *formless*.

This progressive movement towards a many-sided alternative process touching American thinking could be discerned earlier when Swami Vivekananda went to America to attend the Parliament of Religions at the World's Columbian Exposition in Chicago, in 1893.

Vivekananda's was a carrying forward of India's Buddhist-Vedantic current in its mystic ecstatic experience of an enlightened person as Ramakrishna – who inspired Vivekananda by his religious experiences of the major religions of India as non-different in their cores with the religions of the West.

His was also a demonstration of finding the divinity in every individual – man or woman – so that there was no gulf between God and Man, and between Man and Woman.

Both Ramakrishna and Vivekananda were messengers of international spirit, and that becomes an international movement when it was carried to America.

This kind of mystic-philosophic experience could be found in the deliberations of William James, *Varieties of Religious Experience* and in the Zen Buddhist experience of 'Satori, as carried into America through Suzuki and others, from Japan.'

The book, *American Mysticism from William James to Zen* by Hal Bridges (Harper & Row, Publishers, New York and London, 1970) represents a movement in intercultural contexts.

William James carried forward the transcendent experiential process of Emerson and others who were influenced by India's Vedic-Upanisadic thinkers and the German Indologists.

James also carried forward the novel pragmatist-instrumentalist modes of Peirce, Mead and Dewey departing from the restricted conceptual and empirical trends of post-Hegelian, Neo-Hegelians, Phenomenologists, Existentialists, Evolutionists, Realists. He and others like Peirce and Dewey prepared the grounds for Whitehead's process-philosophy as a further transcendent mode – touching the advanced trends of Buddhist-Vedantist processes of India.

But this prospective outlook was exclusively emergent from various dimensional aspects of philosophic thinking as prevailed in India's cultural past of historical periods, with a general reference to other experiential processes as prevailed in India, Asia and in the processes of Europe and America.

Now these complex processes have been focussed in Indo-Whiteheadian Euro-American process from an intercultural perspective with a dominant movement in Buddhist Whiteheadian form of process-philosophy in a common intercultural direction of insightful peace.

My move will be, therefore, from a basic philosophic aspect of a traditional past to some aspects of concrete and changing situation before and after Indian independence. It will be a development in the aspects of economics and politics, with projected visions to the present day universe of science and technology. The resultant possibility will be a change of history in a different direction – not pinned down to a 'form' but released from all restricted forms. This aspect of world-process from philosophy to politics, in concrete realistic scientific-direction has been indicated in its total uncertainties, with a penetrative insight in the book, *International Politics in the Atomic Age* by John H. Herz (Columbia University Press, New York and London, 1962).

India has to be fitted into that dynamic frame of reference and process of the changing international milieu – in a non-categorical direction as a new multipower system to emerge, bringing in a new order, of philosophic thinking with a stretch to scientific and political thinking, and with a progressive separation from any restrictive forms of conceptual or empirical perspectives. It is a prospective outlook matching with the indeterminateness of scientific deliberations transcending the restrictive modes of particles and waves of physics, and moving beyond the restricted visions of the galactic universe, to a series by progressive transcendence – in 'supernovic unpredictable aspects.'

All these expressions finally point out a dawn of a non-violent age, which mankind as a whole welcomed to see with a broad realist outlook. Here mankind has to abandon its old fallacious habits holding obstinately some *ultimates* losing sight of the transcendent prospective alternative insights of infinite varieties.

In this context one may advantageously consult a cooperative work, *Peaceful Change in Modern Society*, edited with an Introduction by E. Berkeley Tompkins (The Hoover Institution Press, Stanford University, Stanford, California, 1971). The participants of this volume – from Anthony Eden to Robert A. Scalapino, Bernard de Jouvenel, F. Hilary Conroy, G. Warren Nutter, Josua Lederberg, Bernt Ingvaldsen, Sir Percy Claude Spender, Sidney Hook, Charles J. Hitch and William G. Pollard – were widely known personalities during their historical periods for their many-sided specialities and roles, they observed the resultant processes of the two World Wars, and demanded a peaceful coexistence and amicable settlement of an international organization more powerful than the existing ones.

My present work is an adventure in alternaive insightful directions, it admits that the continuous sufferings of mankind, are due to attachments to some abstract unreal historical processes.

Now India will stretch itself to other countries from the old Hemisphere to the various parts of the New Hemisphere – wherever Indian thought, including Buddhism, had gone. This process is not only philosophic but also scientific and political.

In my present volume on, *Triadic Avenues of India's Cultural Prospects – Philosophy, Physics and Politics*, I would like to utilize the general ideas of my previous volumes of *Dynamic Facets of Indian Thought* as basic, from a traditional perspective for proceeding onwards, insightfully and selectively, in relevant aspects to a new type of deliberation, where philosophical deliberations as detached from all restricted forms, have been stretched to the advanced non-formalist scientific thinking, and from both to an advanced political thinking, where all forms of restricted political thinking in terms of a fixed ideology or Constitution have been abandoned.

Just as all advanced philosophic thinking have departed from all formalist modes, even so, the advanced scientific thinking have crossed all the boundaries of theories. Similarly, also, all types of political thinking have crossed the boundaries of restricted ideologies and fixed constitutional forms of government. We have come to an advanced stage in every experiential aspect, and have developed advanced insights in contrast to the earlier historical processes.

The world by a departure, in all aspects of experiential fields will be prepared to look ahead to direct itself towards further insightful orders of experiential processes.

The stretch will be towards the avenues of new philosophy, science and technology – as grounds for placing an order of political consciousness – to build up a possibility of international peace. It will be a movement with a stretch of confronted experiential processes. None of these aspects of experiential processes are to be considered as ultimates. If perfection is a demand of all aspects of experiential processes their prospective pursuits are to be understood with an interest. None of them are more important than another. One runs into another, as interrelated processes.

The transition will also be to the new bases of this new project of the present work, in further dynamic ways.

Reflectively the present work in development – will be in the currents of progressive contrasts – treating the earlier volumes as so many centres of past views, philosophically oriented, having a retrospective value, in contrast to the ideas presented in this present volume. This is a prospective attitude in all directions of experiential possibilities.

Here the stress is on a further advance in other aspects of experience and thought, as will be clarified in process.

The pattern of treatment of the Indian thought in this present chapter will be to draw attention to some major stages of Indian thought of the different periods of before and after Indian independence. None of the early periods of Indian thought are separate from the present day socio-philosophical and political movements.

Here the thought is not exclusively philosophical, but it touches generally other aspects of experiential processes, directly, though beginning with India's political independence from the British rule from the early periods of the present century.

Why, for example, I consider the different stages before Indian independence as only possible philosophical *views*, with not much reference to other aspects of experience in a concrete sense of the term, even when India, in its different historical periods was conscious of other experiential processes, with a scientific and technological efficiency will be progressively clarified.

Such a judgment on my part was entertained because then those *views* had some values as preparatory processes, but they could not become potent enough till the Indian mind, became actively operative, with a freedom and strength to show other types of capabilities, besides its knowledge of philosophy with efficiency.

The awareness of these experiential possibilities became gradually patent when the freedom struggle began, and the British rulers felt that the progressive intensity of the Indian reactions, compelled them to quit the British grip from the Indian soil.

It was a detachment from the rule of India, from its political supremacy, after finding out India's efficiency in that political sphere, also, with a progressive consciousness, and also in other spheres, suggesting India's capacity in military aspect indirectly demonstrated by Subhas Chandra Bose's organized Hindu-Moslem-Sikh military potential in the form of INA, and aggravated by the proposed trial by the then British Commander in-chief.

India's many-sided potentiality – philosophic, political, military and also scientific – was for long forcefully covered up, or suppressed, as India was a dependent country, and the British interest was primary.

It was till then, the dominant Indian personalities, inspite of their efficiency in the understanding of different experiential processes had no opportunity to express their talents. A least expression of it will amount to terrorism, trial and punishment, amounting, to banishment and execution. India had all these experiences to pass through in process. This accounts for a novel emergent process, judiciously decided upon in favour of a non-violent attitude.

Naturally only the philosophical aspects of experiential process through its different stages could be expressed from the early Vedic period to later nineteenth and early twentieth century, in a ground-consequent way. India pursued this process in the spirit of a non-violent expression, wherein really the potentiality could be manifest, reaching the depths of human understanding. India's movement was a continuous historical process, and the operations of intelligentsia and teeming masses in togetherness.

These philosophical processes were just potentially operative processes, without actually realizing their objectives. After the British left India, with a partition of former India, into New India, and a Pakistan (East and West) – with the same sort of coexisting cultural processes in development – India was free to proceed in a new direction. It had to face its divided bleeding apparitional form, but had to emerge, with a many-sided consciousness to build itself in progressive strength, maintaining its cultural spirit, expressing its force as based on its transcendent ever-inspiring energetic continuum, which does not become restricted to any 'emergent' possibility of the confronted universe.

In spite of its truncated character, one has yet to feel that it was an achievement, for the Indians who fought to achieve this objective as 'freedom fighters,' but the result was not a complete success, it was an achievement, with a segmented part of a living organismic creature.

This segmented India – if insightfully viewed – was an inheritance with the scar of a political intervention upon its basic geographical process amounting to a geographical dissection of India, having two parts opened to each other in their respective borders—with rivers flowing from one country to another, leaving them to the mercy of its impoverished individuals, to play with. Each had to learn the ways and means to determine each one's competence to exist with the deficiencies of its respective truncated limbs.

It was a definite violation of the ecological principles, disturbing Nature. It was a kind of new birth for both countries, after serious surgical dissection through a political scissor of the cultural and geographical processes. The two vulnerable children were left in the hands of unkind step-mother.

It was a beginning with a frail political consciousness, of a democratic secular state for India, but for Pakistan (on both sides) it was a confused process, towards a possible religious state, with a linguistic and cultural difference, and messed by some undefined dogmas.

But becoming Independent – India could express its many-sided efficiency, as India from its ancient period developed with a knowledge and efficiency in all aspects of experiential processes from the Myths to the Sciences.

The history of two truncated parts – India (impoverished) and Pakistan (with two far-flung disconnected wings) – became operative with further separations as two political entities (units), with an emergence of two separate cultural and linguistic expressions. Now with partition, the two wings of Pakistan – East and West – were joined forcefully by a religious affinity. But they had to suffer from cultural and linguistic differences.

There remained many other differences, which mere religious affinity between the two extreme parts – West-East Pakistan – could not solve. Even in the religious process there was prevalent difference in custom, language, habits and place.

These two parts of Pakistan were linguistically, and also culturally wide apart. The East Pakistan was closer to the Bengali culture of West Bengal—and West Pakistan was also linguistically closer to the language prevailing in Delhi, Lahore and other adjacent places. The West Pakistan linguistically was Persian-Urdu process – with a mixture of Hindi as an effluent from Sanskrit and Pali with local variations.

For the time being, the two parts of Pakistan were excited by their religious affinity, but it was lost by a new political enforcement of West Pakistan dominating over East Pakistan.

But when India was one unit – India's religious process was of an advanced order, it was one cultural 'continuity', and was expressed through a 'unitary advanced religious process, with expressions in various aspects of commonly shared aesthetic experiences – of music, songs and dances of many varieties, throughout the lengths and breadths of Indian soil, where the Muslims did not differ from the Hindus.

This early Indian religious efflorescence was always indicated in unified processes of Vedantic Sufism, and Vaisnavite expressions of no religion, or no caste, a freedom from both the processes were exhibited in concrete possibilities of Yogic and Tantric practices in East Pakistan. These elevating processes were accepted both by the Hindus and Muslims – controlling man's mental and bodily processes.

All these complex processes were focussed in the higher transcendental goals, as manifest in Nathist and Baul movements, in the form of 'Sahajiya or naturalist cult,' and as expressed through their songs. Life was a smooth sail and expressed through in 'Vatiali' tunes of a boat-man's song in both the places of West and East Bengal, before partition. These advanced syncretic processes I have developed in the third and fourth volumes of my *Dynamic Facets of Indian Thought.*

In these advanced processes religious expressions were realistic movements, where science and technology of various subtle kinds were manifest.

Similarly, language in development, specially of East Pakistan – being Bengali – had become an exchange of common feelings, between traditional Indian and the present East Pakistan, through Tagore's and Nazrul Islam's mind-penetrating songs touching the depths of one's heart and feelings. A new type of Bengali language also emerged with the Muslim Persian touch, adding a new glamour besides the touch of Sanskrit.

Indeed the common expression of the elevated human heart was always expressed through musical instruments of many native varieties. So these realistic interrelated processes were dominantly present between India and the present emergent Bangladesh. India was on a way towards an emergent advanced religious possibility.

With an enforced political dissection, after partition – things flowed, but lasted for some years with a bitter end, resulting in the independence of Bangladesh from West Pakistan.

In West Pakistan, Sindhis and Hindus, too, expressed themselves linguistically in the efflorescences of Urdu, with some touches of dominant religious currents of Sufism in mixture with the syncretic processes of Kabir, Nanak and Dadu, exhibiting a mixture of Islamic Sufism with Hindu Vedantic syncretic processes. There was an under-current of a common linguistic and religious procèss, which was broadly a process of Indianization, of many currents that came into India from the Greeks or the Persians, Central Asians et al. India flashed with its specialist features in its dynamic developmental process.

A synthetic Islamic-Hindu religious confluence with a new linguistic flare and international outlook could be found in Muhammad Iqbal – lawyer, jurist and poet – of this twentieth century India. His cosmopolitan mind has been well expressed in the 'Foreword' by William O. Douglas, Associate Justice, Supreme Court of the United States. This 'Foreword,' is of the book entitled, *Iqbal – Poet-Philosopher of Pakistan*, edited by Hafeez Malik. (Columbia University Press, New York and London, 1971). For further details, the readers are advised to read the Preface by Hafeez Malik to know the reason why Iqbal chose a separate federation of Muslim provinces, and yet claimed for himself the ideal of a philosophic humanist.

It was however somewhat tragic, when Iqbal who totally identified with India with a song 'that India is mine' (*Hindustan Hamara*) could switch on the process of division for the Islamic minority. But really the Muslims remained in India as the second largest community in divided India. This number of Muslims in India, constitutes half the population of the United States.

India absorbed them and waited for the British mode of life and culture, to assimilate in course of its historical evolution with the introduction of British mode of education. Here also 'Indianization' of the British life and culture got mixed with the Indian elements.

After partition an enforced Islamic process was fostered, ignoring the traditional Indian influence. So the West Pakistan remained in a kind of suffering with a dual current, of a painful partition. But that past feelings were often reflected through the beautiful 'Gazal songs' by expert Muslim or Hindu singers. These songs are in the unitive aspects of religious experiences, where there is no gulf between religions.

Each suffering Kingdom – West and East Pakistan – existed with internal pains of having lost something – which each needed as transcendent requisites to compensate each other's deficiencies.

Hence it took a long time to stand or heal the pangs of the artificial wounds, by a series of conferences, controversies, leading to conflicts, resulting in the dominance of West Pakistan over East Pakistan, which was a heartless political issue.

Here because of the dominant process of politics, the basic process of life was disturbed both geographically and culturally, which expressed in its dual aspects both in the perfect expressions of a unitive religious and linguistic process, with freedom and heartfelt satisfaction.

The climax of an inhuman political movement was the invasion of West Pakistan upon East Pakistan, with an impatience and ferocity, and when mass subjugation and killing became a veritable inhuman eventuality.

This chaotic situation, created after the British departure, with a political independence – ultimately needed the intervention of the Indian Government, then under the leadership of Prime Minister of India, Indira Gandhi.

She wanted to prevent the possible attack on India by West Pakistan. When Pakistan attacked India, India's the-then Air Commander-in-chief, took the responsibility to stop the progress of the West Pakistan's military might, to cross the borders of East Pakistan to India.

The final success of India with the help of the rebellious group of East Pakistan under the direction of Mujibur Rahman, was possible when he sought India's help.

This resulted in the emergence of a new country – Bangladesh. It is a step into a novel process of an international and inter-political situation, which was not seen before by the politicians in power. West Pakistan lost East Pakistan – in its hazardous game.

But here also both these countries could not enjoy the highest forms of transcendent experience of 'No Religion' of Vedantic-Sufi and Vaisnavite process, of the earlier 'United India,' with expressions in Nath-Baul forms, or the final expression of linguistic fluency expressed through Tagore-Nazrul songs, in their transcendent feeling aspects.

Due to one's dominant political attitude, nobody pays any heed to India's mixed population – Hindus or Muslims – and the Muslims as an emergent process, with elements of a few foreigners – Turks, Central Asians and Moghuls, mixing with the Hindu women – constitute the Muslims of India. India's Muslims really are about ninety percent Hindus, with Hindu cultural processes, than Muslims of Central Asia and the west-Asian cultures. India's was a new emergence.

The emergence of a new Bangladesh was not a solution, as both the countries earlier were divided by enforced political principles. It was a process of psychological maladjustment, and not choosing the best in a progressive cultural aspect. Only by an exalted super-religious process, can outmode the political difference between 'Bangladesh' and India in that corner, I believe in that common prospect.

Through India's help, East Pakistan could eliminate the dominance of West Pakistan, and its invasion, but it could not step into an advanced order of cultural possibility. This would have been a kind of possible freedom of this new country to choose from several perspectives—political, economic and linguistic. But that further prospect is not excluded. It is still operative in depths.

It could have been a process of return to the lost self of the great Bengali community with its Bengali heritage, beauties and novelties. But the geographic dissection remained with problems, as scars, unless the two Governments become impartially interested to solve the problems as if the two countries were one, from all poirts of views, with a transcendent outlook, not lacking the highest religious and linguistic expressions through heart-reaching songs eliminating all politics.

Keeping these general ideas of lost souls in mind, I would like to suggest that the prospect of 'Independent India' or of 'Independent Bangladesh' lies in the development of the three major objectives of a new advanced philosophy, as detached from any restrictive ideology – proceeding towards scientific and technological developments – with an insight, taking care of their respective economic resources, with an attitude of a detached political consciousness – without restrictive ideology, towards a democracy with a secular attitude – and approaching to all solutions from advanced scientifically oriented geographical process and river-water controls.

What is needed is a convenient life for all types of people in both countries – without creating a political or geographic barrier, or rifts.

The present volume is not so much a pursuit, in its traditional inherited direction, as it is with a new progressive elements which are expressed through its comprehensive title – *Triadic Avenues of India's Cultural Prospects: Philo-*

sophy, Physics (Astrophysics) and Politics. They are interrelated processes, with novel prospects.

Each of these processes vibrates in togetherness. This, I will have to reshape through my ideas of philosophy, as a creative manifestation, differently, with some strings of the past and with attachments and detachments, by stretching its suggestions, realistically, to science and technology, as stretched to the advanced ideas of (i) astrophysics, with a developed role of advanced scientific process of Indo-Euro-American turns, and to (ii) politics, with some novel touches – by progressive contrasts and improvements. The move shall be a choice for a commonly shared existence, with an eye towards a dynamic prospect for both. It is nothing but welcoming a 'New History' for mankind as a whole in the direction of an evolving and developing aspects of experiential processes.

This passage will be from the ideas entertained, to a concrete actualized consequent processes, for, what is politics after all? It is the acceptance of a frame of reference, for the good of the people, removing all types of problems, sufferings and barriers.

Politics always depends on philosophic mobility to face changing new situations, by an ingenious operation with effective scientific knowledge and technological efficiency. India is progressively heading towards that efficient philosophical disposition of a secular Republic – progressive human life with progressive compassion. It is a new move towards a benign political process, progressively freeing oneself from past antiquated political diplomacies, leading to conflict, tension, cold war to real wars. Now is the time for cancelling the past to manifest one's dignified nature of progressive cultural prospects.

Hence, the entire resultant process, is another complex-process, in contrast to my previous volumes on India, developing its many-dimensional cultural processes through the centuries. India does not want to confront once more the various environments, spoilt by repeated foreign invasions and interventions. In India's historical past, India was suffering from a condition of *deficient actuality.*

In India's philosophical tradition, there was very little dogmatic touches, India was continuously trying to foster a progressive social life, only it insisted on a continuous operation through transcendentally oriented meditative or insightful processes. One has to get this 'advanced cultural spirit' through insightful processes of educational schemes of New Emergent India.

This spirit was lost through political movements, the right attitude towards a triadic confluence is a possibility through a re-education, in scientific-technological ways for a concrete materialist prospect, as will be suggested in this work. India is ready to abandon its mere idealist suggestions – and proceed by a new channel – in the realistic orders of advanced science and technology. Mere idealist suggestions have no value without scientific and technological efficiency, the advanced process should be in the aspects of self-defence of the country, with a novel exercise with political ideas – dignified with a secular policy at home and nonalignment policy in international milieu. The call is to a new emergent history in contrast to the bitter history of the past.

In the present changed frame of philosophic-scientific background I would like to take some help from the insightful expressions of Barbara Ward's well-known book, *Spaceship Earth* (Columbia Press, New York, 1966, third printing, 1968), as expressing a realistic attitude of the present world.

What Barbara Ward suggested was by a general reference to the advancements of science and technology for the benefit of man for a common good. Now that possible prospect has been blurred by contrary observations, when one examines the contents of John H. Herz's book, *International Politics in the Atomic Age*. He draws one's attention to the consequent processes of the advanced researches in the arms race of the superpowers and also the minor powers in the same directions, where both are for destroying the world.

The world is heading towards a new multipower situation, where the 'Space-Ship's life' has been jeopardized thousand-fold, by progressive lifts from the atom-bombs, as baby-bombs to giant hydrogen bombs. There is no more any place for conventional power or conventional weapons. All have become old-fashioned fictions. This accounts for the indefiniteness of the age and the uncertainties of the future.

That apprehensive situation has moves in a different direction, after the Gulf War, and the dissolution of Russian Union. The world is slowly getting rid of political ideologies, and after breaking its walls it is proceeding towards another process of a market economy where every country and its people have a common interest. After all the over-all movement is psychologically oriented. Hence a disciplined psychology – with insightful possibilities – is to be progressively devised, which alone has the path, for peace. This emergent process has to be earned with proper insightful educational process all over the world.

I will also take note of India's K.M. Panikkar's *A Survey of Indian History* (Asia Publishing House, Bombay, 1947), for what is needed, he insightfully contends is a diplomatic negotiation, by a preliminary improvement of a country's defence through advanced technology, through the development of scientific knowledge of nuclear energy, for self-improvement, discarding the motives for fight or war. Then alone negotiation with strength, will be possible with neighbouring superpowers or developing other powers. India addressed itself to all these, through scientific and technological processes, soon after India's independence. India's motive is not a soaring with mere ideas, but making oneself *nuclearly* strong and progressive, also making some gestures for a world-wide *peace*, through cultural proliferations, through films, songs, dances et al., stringently indicating the value of nonalignment as a political policy in international situations.

It is a political game with a political paradox – an effort of search for progressive power and yet not exhibiting any power – but showing with an artistic gesture towards a frantic search for shelter of a child, who has lost his mother. It is not just soaring with an idealist hope, it is facing the real with its drawbacks, but with insightful wisdom process – preparing the ground for a possible real-ideal process, with progressive strengths, with some scheme of educational process in a world-wide form.

It is a process with some hopes as expectations, but these hopes are as problematical, drenched with uncertainties. This is a tragic phase of human advancement, but mankind cannot stop this onward march towards progress.

Panikkar, through his advanced understanding of India's historical processes in evolution and developments, suggests a unique direction for India to be clarified later in this chapter. Here the role of Panikkar all-through his life was a contribution to India, from a realistic perspective, placing India directly to an international forum, with a great potentiality – through diplomatic exchange of political relationships.

Barbara Ward had a very insightful perspective of each one of the countries, with one another, and also of the old with the new hemispheres. Her's was for expediting the process of relation between countries – through improved processes of science and technology, to close up the gaps of distance of space and time.

Aided by satellites and voyagers, traversing the outer speces—the present world will speed up its prospects with progressive communication facilities, with progressive peaceful gestures.

Barbara Ward's project was for restoring a reasonable balance of power between continents, with understanding and tolerance between world's rival creeds. Her objective was to create a world community – with cooperation, closing up gaps in power, in wealth and ideology. Her hopes were somewhat idealistic, but she felt its necessity, for the crews of *Spaceship Earth* must have some confidence to go by for its survival and navigation through infinite space.

Our life is one of speed, as if dwelling in a space-ship craft, with a speed and its vulnerability. But Herz announces to nullify all these attempts, if military powers, of the world—proceed with a breakneck speed. But one has to combine hope with a realistic possibility. The present world fosters a double-outlook, finding truth in both – but it has to learn to proceed towards a prospective outlook which is a real educational process.

But the world has already moved forward, beyond the suppositions and apprehensions of the above thinkers – Ward and Herz. Panikkar's foreign policy, might be utilized as a realistic step towards an advanced political process.

The world has now come to a situation for appeal, if at all the so-called powers of the world pay a little hearing for the safety of humanity as a whole. It is a plea for a coexistence, facing the common fatal destiny for all, choosing peace as the best of choices, getting rid of the phobia of superiority of a powerful nation. It is an understanding of every one's – *efficiency* or *power* – in a relativist sense, as contributing towards a world power. Hence each one's power is relative and contributory to a *magnum opus*. It is an age of sharing and distribution of powers with progressive tolerance, and extension of faith among nations. Safety for each is a safety for all. Hence United Nations has to be based on a universalist attitude, spreading education all-through the world by schools, colleges and universities.

Panikkar had an eye directly to India's prospect with an insightful attitude, contrasting its advantageous position after independence when India could act independently and with the force of an Indian Union, even with a partition.

But India could not take full advantage of its independence for its inheritance was one of a divided India, with not powers for self-defence.

India, after independence was confronting two wings of Pakistan, and also within it, it had to confront several millions of Muslim population, second in number to the Hindu population. It had also to face four-hundred or more of Indian princes, within India, ruling over the Indian people, from the days of the British rule.

The British, when left India, declared their independence, leaving them with the option to join either of the parts of the divided India.

The problem of the princes was quickly tackled by the iron hand of Deputy Prime Minister, Sardar Patel, with a speed and insight, so that India became a Union, with only a problem for Kashmir but its Hindu ruler joined the Indian Union. Repetition of an earlier event was needed for this writer to continue the important role of Panikkar in bringing about some international change. His was a suggestion for internal coordination of power for an Indian Union.

As a historian Panikkar described the historical situation by referring to its chaotic condition that India was left with, it was open to an ideological universe surrounding it.

France realizing the changed situation negotiated a settlement with India in 1954, by which her possessions were peacefully handed over to India. The liberation of Goa, Daman and Diu from the Portugal's rule was effected later through India's military, without any loss of life in 1961, when people joined the Indian Union.

India was threatened with varied economic and political differences and problems. In this complex situation India could have been easily trapped by any type of ideological, or dogmatic frames of religious processes, stretched from Pakistan to the Middle East countries, or it could pass into some form of Communist ideas.

The ideological dominançe of communism could be found in different stages after India's independence, from the stage – as formulated by Lenin, Trotsky, Stalin, Khrushchev, and Brezhnev and others in Russia through several decades. They could have grabbed the minds of the people of India in general or of some intelligentsia.

Before India's independence, M.N. Roy was drawn to Lenin, but with a different vision.

But then India, under the leadership of Nehru and others, preferred the transmission of the Parliamentary form of Government as passed through Lord Mountbatten, the negotiator on behalf of Attlee's government. India accepted with the drawbacks of partition and, after decades of varied confrontations – has come to the present stage of historical process, with some kind of liberal visions for enjoyment of a freedom mixed with some kind of 'compassionate attitude,' totally eliminating the totalitarian spirit of Europeanized Communism.

If the earlier forms of Communism had continued, in Russia, with some tolerance and insight – it could have tolerated the free expression of a comprehensive religious experience, and it could have prevented the psychological malady of a vicious expression of death as happened with Father Alexander, in Russia, whose only fault was to declare that God is within one's self. Reference is here to an illuminating article by Lawrence Elliott, with the title: 'Murder of a Russian Priest', *Reader's Digest*, September 1991. But this did not happen, for its extreme stringency.

Human being has an inner obstinacy, and it refuses to change quickly on time, and so it exhibits cruelty and suffers for nothing.

Who knows, alternatively, India could have been captured by some fundamentalists—of Hinduism or of Mohammedanism, if India had not pursued a secular attitude in its political sphere?

But luckily India could escape from both, and pass into its hard-earned secular attitude of religious tolerance and development of religious experience through every possible religious expressions of mankind in general, of traditional India, preparing a path of an exalted transcendent political process with progressive secular ideas, crossing the difficult processes to its democratic expressions of an advanced politics.

This new current with sufficient changes of the political schemes of enactments and amendments – will be symbolically indicated in the progressive development of Indian politics, as a process to advanced peoples' Government, crossing the boundaries of ideologically oriented conceptual or merely empirical politics. India has now to be operative in this present advanced international political phase, with a possibility, for opening up the grounds for a 'new historical process'. But this process has to be learnt by progressive educational processes in India and all over the world.

India's democratic expressions, as reflected in the framing of her Constitution, in an aspect, is an advanced expression beyond some restricted forms of conceptual and empirical forms of reflections as prevalent in Great Britain, France and the United States.

But till India became independent, India was in a state of suffering and also is, in a sense, a going beyond many crisis-situations in its progress towards the making of further perfections. This is a dependence upon historical process with dynamic changes within the country of its people, and international politics with changes within and without. (Vide, Wallbank, T. Walter, *A Short History of India and Pakistan*, A Mentor Book, The New American Library, 1958).

All these processes can be left to a natural course of history, or it can be expedited by a progressive world-wide scheme of innovative educational policy. I shall always remain for the latter process.

Barbara Ward is a noted English economist, with a degree in philosophy, politics and economics, hence very appropriate for enriching my present work, with its triadic perspectives of (i) philosophy, (ii) physics (astrophysics) and (iii) politics. But now we have been placed in a further advanced historical stage, as already mentioned. I am not merely writing but I am as well writing with some suggestions for some innovative programmes for the progress of the world as whole.

I agree with her viewing present mankind placed in an unprecedented stage of historical process, where modern science and technology have created so close a network of communication, transport, economic interdependence – and potential nuclear destruction, that planet earth, on its journey through infinity, has acquired the intimacy, the fellowship and the vulnerability of a 'spaceship' (preface). Admitting her statements with some basic truths, I will have to make the necessary modified statements even by a reference to India's nuclear developments and advance towards space projects through satellites to the outer space.

I also fully agree with her contention that 'the gaps in power, the gaps in wealth, and gaps in ideology' – hold the nations apart and these make up the abyss into which mankind can fall into annihilation. It is on these disproportions that the world policy has to concentrate, so that its three interrelated processes of philosophy, economics and politics, could efficiently express its complex processes of relativity, among the nations of the different parts of both hemispheres of the world. India is proceeding fast with its growing intelligentsia and technologists. But these ideas should not just be entertained or merely be left for floating, but need to be taken up as educational items for mankind to cultivate in proper contexts.

An internationally oriented possibility lies in restoring a reasonable balance of power between continents, with a reasonable balance of wealth between the two hemispheres.

Further also there is a need for a reasonable balance of understanding and tolerance between the world's rival creeds, with ideological and economic controls.

Her conclusions are also very impressively expressed with a dynamism and balance, in all directions, when she says the following: 'Then when the gross inequalities have been remedied, there can be more hope of building the common institutions, policies and beliefs which the crew of Spaceship Earth must acquire if they are to have any sure hope of survival' (preface). These general statements have truths but some further advanced ideas have to be furnished for fitting them with the advanced situations that have emerged after some historical sequence, as mentioned. The move now is not merely through ideas but through the developments of science and technologies.

But the world has already the chance to display the strength of the super-powers (America and Russia), in terms of their nuclear power. Now we are in a new historical period beyond the cold war, and there is an abandonment of the process of any 'superior power,' or the claim for a 'superior power' has gone into the underground, ready to be expressed in different insightful ways. The world should now move forward towards a process of coordinated unified world-power with a new faith drawing attention to the prevention of contrary processes, which damages the world-process, with a threat of total annihilation. No one power is ever sufficient. UN has to be progressively strengthened.

The unity of strength between the superpowers was often disturbed by their separate interests as holders and controllers of 'oil powers' (Saudi Arabia, Iraq et al.), or powers emerging from 'dogmatic religiocity,' of Middle Eastern countries (Israel, Syria, Iraq or Iran).

The conflicts among them resulted in the consequent processes leading to the Gulf War, with no possibility of solution of world-problems in any near future. But the international situation changed to take the present shape, as it is now as already referred to.

Extending my thinking of the ultimate process of Einstein's special and general principles of 'relativity,' as running into transcendent field processes – I am disposed to think, that the different power-centres, with some ideological displays will someday pass into the oblivious process of 'no-theory' or 'nothingness.'

In that prospective world-situation – when science, will gradually lose its specific theories, with the emergence of a new type of intelligentsia, then, it automatically will close the gaps between *particles* and *waves*, and run into *scientific fields*, closing up gulfs among themselves. Politics, in turn, will lose its scientific strength and also its wrong visions of distinct ideologies. Now politics is, in the way of losing its ground and is in distress, and is ready to shake hands with various opponents. It remains for the insightful scientists and politicians to shake hands for coming together with progressive projects for mutual welfare in all fields of human life.

They should stand for a new *transcendent current* to flow into an ocean of an *internationalism*. At that time, *history* will change with no more of proneness for *abstract* ideological choices, and setting up the conditions for war-situations.

Then what will prevail – with insight, the common pursuit for a concrete demand for an economic solvency by an open market for all nations. This is a soothing advanced psychology.

In the determination of which, I have also perceived through my studies of the journal *Foreign Affairs* – America and the World 1990/91 – a new current or project for change towards a common internationally oriented market.

I found that Gorbachev preferred a process of solution, by a switching on to a change, where he found that continuing the expensive military superiority will become ultimately ruinous or a draining out of Russia's economic solvency, so he was compelled to quit that process in his political thinking, he pressed for his socio-political change through his *Perestroika.*

One good thing has emerged, viz., that now all can see, is the force of the *historical process* – that awakens a person to understand the spirit of the international situation – and Gorbachev represents such a person, he was quick to understand this *historical* process.

In the context of India, it was a kind of Britain's progressive decision for quitting its colonial powers, not only in India, but also wherever it became expensive to maintain.

The representatives of the superpowers are understanding the superiority of the *historical forces* over mere *ideological forces*. So *history is changing*. But all these complex processes are to be communicated to people at large through some such ways, as I am trying.

The demand for an economic solvency became the primary concern for a developing nation. Russia stepped down from its previous choice for sticking to 'Superpower' status.

We have come to the stage of Herz's multiplicity of 'superpowers' flowing into the processes of multiple 'minor-powers' with potentialities to destroy the world in many spheres.

Different nations of the world—willfully chose a moderate order of holding an international power, with a prospective solvency in opening up varied channels for a competitive progressive market economy. Instead of forming a party of world powers with a restricted number of international powers, it will be a wiser move to admit other powers to come with an willing attitude to foster a *consolidated* world power through UN.

Gorbachev threw the entire East European nations into this dynamic flood and turmoil, with a struggle for its own existence, flouting the possible threats of the French and the British – from an emergent Germany after the unification of the East with West Germany.

This is my way of observing the Russian preference for an alternative shaping of political frontiers of this world of various complexities. It is nothing but a realistic stepping down from a Platonic living in an *abstract* otherworldly sphere. It is a kind of judicious prevention of an early economic collapse through a craving for a militarist supremacy, which finally goes by a process of annihilation of others followed by one's own annihilation.

The process of the world today was tremendously disturbed after the Gulf War and the world has gone further in historical process, as touching a consequent period in *political history*.

Politics has now emerged, with various *psychic abnormal* expressions which so-called civilized personalities vainly try to demonstrate, but is not quick enough to accept the *meditative transcendental* path of a *psychological possibility.*

Here my reference is to progressive development of Buddhist transcendental psychic postures and *possibilities*. Only such avenues can rescue mankind from a crash.

When Barbara Ward wrote the world-situation was tolerably morally oriented, but now it has taken a new turn, it is polluted by terrorist exhibitions and, in that *phase* of *human history* – man is passing through a stage of *moral* degradation, with some concern and apprehension, everywhere. It is expediting its collapse further, losing all human values.

Now human life has lost all *values*, and sense of humanity has crossed the earlier boundaries of historical crises and calamities. The world has stepped into further complicated situation – terrorism et al. with operations towards 'human bombs'.

World has also learnt the hard days, and has become more vulnerably miserable under the sway of any individual villain, or a group of such persons, bringing in more of human miseries, due to man's growing knowledge and power, losing sight of the transcendental meditative possibilities.

It is now a plunge into a stage of the world-process, after the victory of the major European powers with the American leadership, but its glory of the Gulf war is a temporary phase of the human history.

This new advanced order has been won by a defeat of Saddam Hussain of Iraq, and these powers are again for crushing the threat of nuclear capability of the defeated Iraq, the world is threatened with another possibility of war.

Again, there is a problem to be faced with measuring the strength of Gorbachev, who is trying his best to quit the process of Communism, in favour of progress towards a new market economy. But market economy, is a complex international phenomenon, dependent on each nation's economic stability, business skills and the trade-facilities provided by major and the minor powers competing in that field.

In this milieu of complex international politics and economics one has to observe the continuously changing problems – in the Russian areas, viz. how ably Boris Yeltsin, the Russian President, can tackle its internal problems, within its shrunk 'Union.' Can he also manage to maintain Russia in tact, in relation to other emergent powers, cutting off connections with the 'Old Union'? Russia has also to retain its prestige confronting the varied nations of the world, where the superior role of the United States stays as a historical process.

Hence setting aside the emerging problems that are happening among the emerging countries which are gradually proceeding to their respective process of progressive independence, they are still someway linked up with the last snag of Communist control of Russia. In this growing complexity China is left to play its roles in different ways.

Russia is taking some form of democratic attitude, but China is still holding to some sort of communist or totalitarian power not admitting the place of democracy at all. China, is still obstinately demonstrating its underground nuclear tests, as heard from on TV on 22 May 1992, for example.

The world is patiently waiting for the collapse of all political ideologies, with some kind of triumph of economics over politics. What will be China's political future none can yet say.

Singapore, in its political evolution has given a precedence to market economics, as if suggesting the emergent Hong-Kong after 1997 to take its decision – in that uncertain future.

Russia, in facing the dissolution of 'Russian Union' – is moving towards a market economy. It is also confronting the liberated European countries the other powers as the British and the French, not excluding the controlling power of the Americas.

Who knows – the Chinese will have to face, a similar collapse of its abstract disposition in 1997, as mentioned, with the emergence of a new Hong Kong, there will be such an alternative possibility with its non-traditional emphasis on 'Economic Forces,' abandoning any specific politics? It will be stepping into another 'New History.'

The world is waiting to see the dissolution of old ideologically oriented politics. The future of Hong Kong after the British leaves its sovereignty over Hong Kong in 1997, will give birth to a new kind of politics, which I feel will be economically oriented from an advanced multifaceted perspective, but its longevity will not be very long, if it is not moderated by some broad-based democratic-humanist attitude even in the execution of its business skill. This is a new phase of world history.

The vigilance of the world powers will become more and more subtle with the possible scientific knowledge and technological success. It will be world of satellites and voyagers.

Why, a large number of disoriented individuals are coming to Hong Kong, not alone the Asian people, but also the people from Scotland or from Africa or from any other place of the world as a whole, to begin a new form of life – suffering all hardships – nobody knows. Is there any mystic principle involved in that particular place of the world?

Possibly, it may be that the immigrants have a hope to choose some possible prospects, even sleeping in expensive bunks, it is a 'possibility,' as can be found mentioned in the wonderful paper, with the title, 'Hong Kong Countdown to 1997', by Ross Terrill (*National Geographic Society*, February 1991).

There are obviously some more gaps, which Barbara Ward could not dream of for a possible solution. Several historical fronts have emerged, but the possibioity of a further historical front is not closed down.

As history of mankind is linked up with the empirical psychology, without making some attempts to accept some form of the best elements of transcendental psychology of India in development – there is again a possible continuation of the mistakes to be repeated in history.

Conceptually and empirically, oriented man, always fails to transcend its basic animalism, and wrongly chooses a psychology that indulges in demonstrating the lower orders of human Nature, and continues the human history, bragging of some kind of attaining a higher order of civilization. This is my considered opinion that haunts my mind still. So I feel history will repeat itself – in an intensive potential order.

The term *view*, I have coined or improvised, with a new meaning to suggest, and to incorporate in my mind such complexities, that it is not simple, only suiting a particular 'time', and the-then, 'socio-cultural' milieu. This new historical front is many-dimensional with many fronts.

Human psychology, left to itself in the Naturalist level – often over-reaches its segmented boundaries to multifarious flashes. It hangs without any sheltered habitation. But enjoys life, with confronted phantasmic visions, with claims of 'Civilization.'

To clarify this point of mine – with a project to create a new enriched transcendentally oriented scheme of thought-process by constant writings – I will reconstruct the main ideas of the processes of the 'Movements' of my fourth volume, of *Dynamic Facets of Indian Thought* – by some advanced general ideas, and by some concrete references to the roles of some outstanding personalities of India, who helped to change the India's *intelligentsia* and the mass-minds, to confront the varied modification of the minds of the British Rulers, controlled from the British Constitutional frames as historically developed, with operational capabilities and from further perspective outlook.

Here my main reference will be to the two volumes of *Sources of Indian Tradition* edited by Wm. Theodore de Bary (Colombia University Press, New York, 1958). It is a little deviation from the results I obtained in my analysis of the different ideas of the fourth volume of my *Dynamic Facets of Indian Thought.*

I am now proceeding to find out the actual results derived from the operational scheme adopted by different Indian personalities, sensing the general processes of historical understanding as patent to their minds. Presently my search is deeper than before.

After the continuous insightful studies of Buddhism, as a contrasting philosophical deliberation beyond some of the entitative concerns of the Vedic and the Upanisadic thinkers, that there is an ultimate Reality (*Tad Ekam* of the Vedas and *Atman-Brahman Process* of the Upanisads) – I find a dominant role of Buddhist drive towards Nothingness (*Sunya* or *Zero* when mathematically expressed). Philosophically and religiously too, it is towards no abstract philosophy, nor towards any abstract religion. That was why the Buddha was bold enough to declare there is 'no Nirvana,' only there us a pathway, with no traveller, as such, no 'Personality,' 'no Buddha' even. What a detachment? This transcendent vision, can alone open up 'perennial pathways' or 'possibilities' – beyond all specific philosophies or religions, and for that reason beyond all specific politics – pregnant with a possible new historical front.

In the process-world of the Buddha or of the process-philosophy of Whitehead there is no ultimate – but we are in continuous process of change and transformation, so that there is no ultimate personality, or there is no ultimate theory in philosophy or science. New history is confronting the same prospect. There is no ultimate reality now.

In Buddhism, there is 'no person' as a permanent entity, but only a collection symbolized by a name. In denying the existence of the so-called person, Buddhism symbolically assumes that 'Misery 'doth exist', 'none miserable'. The 'Path' towards freedom exists, but not the 'traveller' on it (*Visuddhimagga*).

This attitude of Buddhism is in conformity with the advanced scientific notion of Heisenberg, which can be symbolically expressed as really there is no ultimate *particle* or *wave* but only transcendent *process* or *field* – implying *energetic continuum.*

Philosophy and science no more can indulge in any theory, or ideology as ultimate – or as an ultimate personality. All objectives are in the order of confronted processes – they are in a 'going or proceeding order', with a possibility of perishing and emerging.

In this unrestricted process, there is a transcendent possibility, in infinite alternative ways. There is finally no Buddha, no Buddhism – no Einstein, no theory of relativity, but, it is in an 'if then situation' with a dynamism and progress – all are treated as steps towards a further prospective possibility, with infinite indefiniteness. This 'New History' is a perennial possibility. It is a possibility with infinite uncertainty, with a basic certainty that the multiplicity of human mind insightfully could determine. Categorical certainty has been cancelled by the realistic understanding of the possibilities of 'if-then' process and variety.

Gorbachev's trip to US, with his wife Raisa brings in a story of change of historical current in a new direction. Gorbachev permitted the Berlin wall to come down and the Eastern European satellites to go their anti-communist ways.

He also points out confronting a cheering crowd of 15,000 in San Francisco, that the major centres of world-politics, have today been made in favour of peace, cooperation, interaction and over-all security.

To continue this process, he feels the need of the expansion of the role of United Nations as a world peace-keeping force and suggested that Japan, Germany and India, be included in the Security Council, as permanent members with the previous groups of superpowers, for introducing some changes in world-process, in progressive realistic directions.

This advanced consciousness will become my basis for the study of the philosophic position of some Indian contemporary philosophers or of the scientists. Some advanced ideas, both in philosophy and science – I am already indicating in process, and this process will be intensified continuously in this book.

So no more I will indulge in a 'final', religion, philosophy or science – as all religious now are placed in a mixed international order of comparative religions, comparative philosophies and comparative sciences. Let UN take the lead in every affair, where let every nation should have an assured place. In any emergent situation right action is first, before all discussion.

This advanced idea or outlook, with alternative changes and transformations will be expressed in later chapters of this work from progressive meditative perspectives with a relativist outlook, but not as a halting post.

This will influence the formulation of my new politics, also, which ultimately will result in peoples' rule with progressive insightful experiences.

Here I feel the need of a different educational scheme for the different parts of the world. The world must try to abandon the wrong traditional ideas, or faiths in defunct cultures or religions or even some radical changes in the scientific or technological ideas and attitudes, always. One should wait for changing insightful experiences of the enlightened personalities, freed from the expectations of any ultimate in any field of experiential processes. The experiential processes in general are like floating *stars*, helplessly placed in the galactic universe, with possibilities of *supernovic* eventualities, over which human intelligence has no control. It is a process beyond focal fields – a new continuum.

So, we must respect the past and the past achievements, as platforms for a forward outlook by a contrast – from the Myths to the religious expressioos of the world in general, but must not follow as they are, but with insights, with symbolic suggestions, and stretches along the 'Wisdom Paths' of Buddhism or of Whitehead's process-philosophy – and with no cry of ideology or *ism. Ism or ideology* stops the continuity of infinite possibilities.

In the study of these two volumes edited by De Bary, my main interest will be with a development of the evolving ideas of the second volume, mainly, with symbolic suggestions – in the background of the efficient king's rule by the Mughal Emperor Akbar, as mentioned by Akbar's friend, companion and a great historian, a scholar and a mystic, Abu'l Faz'l, of the sixteenth century A.D. as an example.

Abu'l Faz'l joined in influencing Akbar against Sunni orthodoxy and in obtaining the assent of Muslim doctors or law to a declaration giving Akbar the deciding voice on religious questions. Akbar was also greatly influenced by his great grandson, Dara Shikoh (1615-1659), who was influenced by Hindu philosophy and lofty ideas of the Upanisads and later Indian syncretic Hindu-Muslim mystic Sufi-Vedantic confluence.

Dadu's liberal ideas were stretched further from the processes of Kabir to Nanak. Kabir was for a synthesis of Hinduism with a vision of the mystic Islamic Sufism, and Nanak's was a resolution of gulfs among Hinduism and Islam, by the emergent process of Sikhism.

Dara's position was a further possibility of religious experience in interrelation, and not of a particular sort 'Hinduism' or 'Islam,' as the continuously transcendent religious process, is an 'energetic continuum' of religious experience—in perishing and emergence.

But this advanced gesture had a sad end. Dara was condemned as a heretic and executed by Aurangzeb, the last great Emperor of the Moghul dynasty. That emperor obstinately stood for a static or ultimate religion.

This was nothing but a repetition of the Muslim reaction of the dogmatic Islamic process, heartlessly expressed.

My move is to recognize the best of the past but follow or pursue the prospective luminous insightful suggestions of philosophy, science, religion or politics, as indicated in different dimensions of energetic continua. I cannot cease to indicate the prospect of the world, in a developed religious consciousness where there is no place for an ultimate.

This disappearance of an ultimate has already been indicated in all spheres of experiential processes – philosophy, sciences and politics of different varieties. Only after this abandonment of an *abstractive absolute* from the domain of each aspect of experience, there is a place for continuous transcendent possibilities within each experiential process, with a prospect towards an interrelational possibilities in all aspects of experiences.

I choose to move forward in the line suggested above, with no fixed propensity, but with a progressive developing insight.

My main reference is for developing the evolving ideas of independence and its final realization in a form after actual independence, from the British rule.

The move here is not merely philosophic or religious, but an access into the concrete spheres of sciences in general as to reconstruct the processes of innovative politics and economics as continuously dynamic – perishing and emerging – that rule and control the world with scientific and technological prospects.

I am shaping here the processes of a dynamic market economics all over the world among the different nations of the world, operative with different sorts of ideologies, controlling wealth and military powers, by some benign ideas, changing with a transcendent insightful way.

Here alone the core of civilization lies – in the transcendent feeling aspects of human nature. This I found today in graphic expressions by Charles Carol, while reviewing Walt Whitman's *Leaves of Grass* in honour of his centenary. (19 April 1992).

The stress is not on *isms* but on transcendent meditative feelings, with a tenderness or softness, behind all orders of experience. Life is a progressive transformation from Being to non-Being.

After India's actual realization of some form of 'Nationalist status,' in a different dynamic level, India dreamed of playing its special role in international sphere, rescuing itself from the British colonial grips, but facing later China's intrusion into India, and repeated attacks of West Pakistan.

India gained relative experience and foresight for defending itself from any kind of future attacks, in a realist direction, with a strong 'defence programme,' so that the appeal to the foreign powers was not merely in words or ideas, but with a scientific and technological force, without any obvious demonstration.

It was just pursuing a political game.

When this process was actually realized, or will be set up, with full defensive strengths – India will then be fully able to show its potentiality for its operative political scheme of nonalignment, as a workable principle.

The ideas of nonalignment influenced some European countries as Yugoslavia, under Tito's time or in Egypt under Nasser – during the time of Nehru's Prime Ministry. Nehru was occupied always with ventilating the ideas of nonalignment as realistic, in idealist symbols.

But Nehru's gesture then was still idealistic, as it should be as he was fully conscious of India's vulnerability, but the later Prime Ministers, with a progressive change in the defensive directions – naturally expressed it with progressive realistic gestures – previously with some foreign helps, but later with less foreign helps.

This relativist change with progressive strengths from science and technology to politics – as could be understood in different stages – India could demonstrate its progressive strengths. In Nehru's time, India faced the Chinese attack.

India, could foil the intrusion of West Pakistan during Shastri's time, India halted Pakistan's movement with a positive strength with some Russia's friendly gesture, supporting India.

A full show of a victorious sort was demonstrated by Indira Gandhi – when she helped emergence of East Pakistan as 'Bangladesh.' She then openly repulsed the attack of West Pakistan of East Pakistan – and also of India.

India's progressive hard-based politics of the present Prime Minister Narasimha Rao – has facilitated the first visit of the Indian head, President R.

Venkataraman, to China, and to solve the complex border problem in a patient and systematic manner without resorting to violence.

China's leaders, in turn, stressed the need to improve economic ties between the two nations, with border issue, through dialogue.

China's General Secretary Jiang said both countries were neighbours, with long borders, and hence, it is judicious to agree with India's President – that both the countries will maintain tranquil and friendly relations.

Then ensued the process of post-Independence period of India, when it became actually operative going beyond just *views*, playing its important roles, as an independent country, through UN. India's role, in that platform through the years was important, and detached.

At that stage, it became incumbent on India to reconstruct India's foreign policy of nonalignment with a real potentiality, in a scientific and technological way, from all aspects of experiential processes.

India's foreign policy remained confronting the two powerful foreign countries – China and Pakistan – and the South-Asian countries from – Sri Lanka to Indonesia.

India's gesture remained with expressions of some good feeling for south, north and central Asian surroundings – to match itself appropriately with the whole world finally.

India's attitude all-through remained transcendental, with a non-violent diplomacy and democracy, coming finally to the stage of the dissolution of Russian Union.

Not alone in the sphere of philosophic and scientific pursuits – considerable patience is a requisite quality. In the sphere of politics, too, patience is needed, to let history evolve itself, letting *events* to happen taking its rough course, of repeated fallacies.

This historical event as observed confirmed the concrete reality of some form of potency of the transcendental reality, that never stays *static*.

India's goal remained for helping non-independent countries to rise to a status of transcendent detachment, from any superpower, controlling the dependent countries, mostly in Africa.

Everywhere an individual or a nation tries to cultivate a transcendent unified feeling-process beyond all abstract principles.

Now is a concrete phase of abandoning the craving for some kind of superpower dispositions as evidenced in the journals issued from the scientific papers of American Association of Science, to be mentioned in relevent occasions, when it is a fact that all nations of the world will become aware of all phases of scientific developments.

Nothing will remain secret or private. It will be a period of mutual participation and sharing of powers. The world is on a *new* move towards a common 'scientific development', and shaping politics from a new historical perspective. For a comprehensive understanding of this complex historical stage, one may conveniently use the popular work by K.M. Panikkar, *A Survey of Indian History*.

Panikkar was educated in Madras and Oxford. He was a scholar of Christ Church and later called to the bar (Middle Temple), in England. After a period of experience as a professor at Aligarh Muslim University, he became editor of *Hidustan Times*. Later he went to Patiala State as Foreign Minister, during which period, he was also a Secretary to the Chancellor or Chamber of Princes. He then went to Bikaner as Foreign Minister, later becoming Prime Minister of that State.

After Indian independence, he was called for more responsibilities and represented India at the 1947 session of the UN Assembly. In 1948, he was appointed India's Ambassador to China, and then to People's Republic, recognizing the new regime on behalf of India to desist from carrying the Korean War beyond the Yulu river. After this successful diplomatic role, he went as an Envoy to Egypt in 1952. He was appointed a member of the States Reorganization Commission set up, at the end of 1953. He was also India's Ambassador to Paris, and a member of the Rajya Sabha. He was for sometime Vice-Chancellor of the University of Jammu and Kashmir and till the time of his death he was Vice-Chancellor of the University of Mysore.

He was a personality of the superior level, having an access into so many experiential centres, displaying a versatile capacity everywhere, raising the importance of India's independence.

With such an all-sided experience both of India, and of the world, in international contexts—Panikkar's was indeed a dynamic approach to India's history through the different periods to the end of the British rule, and after India's independence.

Here one can see, that 'history' and 'politics' change their shapes by the role of some extraordinary personalities. Panikkar was such a personality. It was for, him the 'history and politics' of India and China *changed* to their favourable orders in course of time as noted below:

Any well-read scholar conversant with the depths of historical, scientific, economic and political processes in interconnectedness – can gain a clear perspective of progressive steps before India's independence, and of the processes immediately after independence, till the territorial encroachment of China in the Akai Chin area of Ladakh. Then there were further movements.

At this stage, the controversy ensued regarding the boundary between India and Tibet, as it was undefined with consequent pretensions to grab parts of territory on the Assam border (known as the MacMahan line). This led to a growth of a hostile feeling between China and India.

This and other problems of China's progressive intrusions into different Indian territories on the Himalayan borders reached a climax, when the Chinese made an attempt to penetrate the McMahon line, the frontier forces between India and China, came into open clash, bringing the two countries to the verge of an open war. The Chinese after some initial success withdrew due to international pressure.

With Panikkar's experience and his diplomatic roles accorded to him, by India's Government—he paved the path towards a solution of the present border problem with a strong neighbour as China.

This intermediate period of India's history after independence had flouted the past Buddhist affiliations of these civilized countries, they faced the poisons of modern political facets.

But Buddhism remained, in its transcendent sphere alive, as will be progressively given shape through my further researches in that sphere from an advanced comparative international perspective. But politics changed, with India's progressive defensive strengths. Here philosophic ideas needs a supplementary process of progress in science and technology.

Hence the progressive developing processes of India, after Indian independence presuppose a preliminary passage through constitutional changes during the last stages of British rule of India.

The British introduced the evil principle of communal representation, which was the first expression of two-nation theory. But due to India's progressive schemes of political evolution the members of the imperial legislators under the system were no longer Maharajas, Nawabs or other noble men but India tried to welcome political leaders, like Gokhale, Srinivasa Sastri, Madan Mohan Malaviya and Surendranath Banerjee. The progress was in the Constitutional aspects only, in the absence of any scientific and technological aspects during the period of the British rule, but this situation took a different turn with considerable 'Constitutional' changes, after the Indian independence.

The progressive recognition of gradual Indian power was through the Minto-Morley Reform, in 1906, it was an introduction of democratic government of India. India's persons of exceptional capacities as Sir Bhupendra Mitra, a member of Viceroy's executive council, and others, later were appointed to seats of effective power, all in the direction of progressive democratic process.

The further evolution towards a pro-Indian democratic process – breaking away from old traditions to a genuine democratic process – was effected through an Oxford don, Lionel Curtis, whose letters to the people of India on responsible government, facilitated the process of administration, which was placed under Indians, responsible to elected legislatures.

And after India's participation in the First World War for the British, the Parliament in August 1917, defined the goal of the British Rule in India as a responsible government.

The Montagu-Chelmsford Reform was first step in this direction, which introduced the system of diarchy in the provinces, under which subjects like 'education, local self-government, public health and industries' were placed under Indian Ministers, responsible to the legislature, while subjects like 'law and order, finance and land revenue' continued to be under members of the Civil Service, responsible to the Secretary of State.

There were further processes of gradual Indianization through this reform India rose to genuine powers, departing from any idealist touches of philosophic deliberations.

The weakness of the scheme was in the government at the centre: but even here three major achievements can be claimed of the Montagu reforms. The first and the most important was that it led to the foundations of the parliamentary system of government even though the central cabinet was not responsible to the legislature.

Men like Motilal Nehru, Mohammed Ali Jinnah, Vitalbhai Patel dominated the political scene through the mastery of the practices and procedures of parliament and this made it possible for independent India not only to establish a constitution based on parliamentary democracy, but also to work it efficiently and smoothly.

What this author tries to emphasize here is that India's move at this period was no longer a traversing in the path of pure speculative philosophy – but it was a step along the avenues of concrete political intricracies, that render a country gradually democratic, in its operational aspects with schemes. This kind of evolution and development in the political sphere was a progressive change in the attitude of the British administrators and India's efficiency as shown by some Indian personalities, operative with the British administrators. Hence the resultant political situation was an emergent process, with a post-British possibility. It was a kind of earned property.

It was no longer a soft movement, it was a realistic process and achievement – in an efficient democratic direction. The present Ministry's movement was a further consequent process of India's progressive strength in all spheres of activities and enterprises.

Secondly the central legislature was able to secure a measure of Indianization of the office cadre of the Army and also to create a separate Indian navy: thus making the transition to independence easy. In this process, one should give necessary credits to Narattam Morarjee, for his shipping agency and its maintenance even to this day through his able son and his wife.

Thirdly it established a system of discriminating protection which enabled India in the period between the two wars to create an industrial basis for her economic life, and progressive defence programmes.

Inspite of these advancements of political changes, India's far sighted nationalists became operative, though interrupted during the first great war, and through other wars, where the European powers demonstrated their varied ideological expressions with consequent processes, developing a kind of civilization with ostentation of their scientific powers with armamental efficiencies, bringing in the glories of materialist civilization.

The British Indian Government, after the first World War tried to put down by repressive measures the widespread discontent caused by the war. This led to an explosion in the Punjab. The British ruthlessness culminated in the massacre of Amritsar (April 1919). This unthinking action was the turning point in the history of Indian Nationalism.

The revived Indian National Congress, under the leadership of Mahatma Gandhi embarked on a movement of non-cooperation, the object of which was no longer mere reforms but a move for a complete independence (*Purna Swaraj*).

The effect of stringency of rule led to independence process. India, chose a path suiting its fair traditional heritage, but with a progressive realist insight and gesture.

Panikkar, in referring to the non-cooperation movement of Gandhi from 1920 to 1947, interpreted this movement as a method of revolutionary mass action, without resorting to violence. Though apparently it was simply a withdrawal of cooperation from the British government, it was a revolutionary movement as it took the claims of nationalism from the urban classes to the masses, touching the villages.

This brings about a radical *social change* by the abolition of untouchability and establishing equal rights for women. This movement became operative and developed through three stages: (i) from 1920 to 1923, it became operative with the Muslims organized under Khilafat Movement. Then Gandhi himself conducted with the masses, the protest against the salt-tax, which affected the poor man's daily food, by the *symbolic* violation of the salt act (*Salt Satyagraha*) through 'Dandi March' in 1930.

The third resulted in the Quit India Movement as a spontaneous protest of the entire Indian people, against the British Rule in 1942 during the period of Second World War – a period of crisis not only for the British, but for the entire nations of the world.

At this period, I was a professor of philosophy at Rajendra College, Chapra, when its principal announced that the college will stop its function for an indefinite period. I had to take immediate steps for earning a little money for my family, I met the need by buying some ricksaws, and renting them out.

A reflective mind can clearly see, how this progressive movement, was a period of training of the entire masses to think in alternative ways with a consciousness of reformed social, political and economic processes, and with no academic education, but with a common sense.

At this period, a scheme for a general mass education, became a programme for education launched by schools and colleges, for contacting the masses, by visiting their houses. So, this transition from the British to the Indian rule, with a mass-consciousness was a gradual evolution – a learning of a possibility towards a fresh form of Nationalist Independence, and transference of power with a more or less peaceful or friendly manner.

But this process was not easy. The changed British mind had to be operative with the Indian minds with caution, patience, foresight and yet with a civilized gesture – with progressive realist touches.

It was a meeting ground for both sides – a unique and novel achievement on both sides. The world still was morally oriented.

Panikkar was finally interested in indicating the emergence of an independent nation, with a totality of a 'mass-awareness.'

But this adventure unfortunately, brought in a process of partition of the Old Indian Empire into Pakistan and India. It became an inevitable process, when the leaders of India, differed among themselves, and had to choose from the British, with some disadvantage an alternative political solution. (For an alternative interpretation of India's transition from the British to Indian constitutional form of Government, vide, *India from Curzon to Nehru and After* by Durga Das, The John Day Company, New York, 1970).

The subsequent post-independent activities lay in the role of India's prominent nationalists and to the majority of peace-loving Maharajas, Nawabs and Noble-men's resolution to join Indian Union. They were gestures of recognition of India and its masses. It was also a reaching out to the order of the organization of the Indian village masses – to an advanced evolutionary socio-political consciousness.

Panikkar with a great political, and as well a sense of social unity of his time, presents those processes where one could see the readiness of the Maharajas and Zamindars in quitting their hereditary rights, as quickly as possible after the British left them with their independence.

Here there is also a reference to India's reorganization of the structure of Indian constitution, by a quick legislation to reshape Hindu caste system, its marriage system, and its inheritance principle, and reaching the masses by abolishing the discriminatory attitudes towards the untouchables. This process should have been stretched to all Indian religions.

India also gives special attention to the tribal people in different parts of India, as members of an integral whole. New programmes were introduced to develop fully the social life of tribal people giving them incentives to develop through education and providing job-opportunities to place them snugly within the inner core of the Indian Republic.

It was a reduction of India as a whole to one continuous social community process, with a dynamic cohesive force towards progressive possibilities of an advanced socio-political continuum – by continuous five-year plannings to expedite the steps of industrialization of the country through educating the Indian masses gradually. India has reached its eighth planning stage to begin from 1992 to 1997.

The scheme fostered was also in the direction of character formation with an emphasis on humanist curriculum, as also through science and technology, for, India has a tremendous population growth, which cannot be tapped properly without judicious steps with a breakneck speed, enforcing uniform social system and education.

Panikkar's critical insight for 'insightful integration' of Islamic people of India, not by looking at Pakistan or any other country – was for a kind of parental protection. It was a felt need for an integral growth of India, fostering the process of a total union with a secular spirit abandoning its chances for wrong step of a Hindu Raj to exhibit a counteracting force, to Pakistan, for its wrong choice for an Islamic State.

Panikkar's reference is to the development of this 'secular process and development' of India, through constitutional legislation. It was linked up with many-sided scientific and technological projects. His was a programme of reaching out to India's 80 percent village community through large-scale irrigation schemes with such projects which commenced with Nehru Ministry by such projects as of Damadar Valley, Hirakud, Chambal, Koyna, Tungabhadra and other projects, and progressive industrial development through planning and change of educational policy through University Grants Commissions.

These schemes were for progressive developments of improving upon India's economic and political processes on concrete realistic ways, maintaining in tact the fundamentals of India's transcendental attitude, and not falling into the traps of discrete visions of merely empirical outlooks towards life. In this adventure one can clearly see the evolution from abstract thought or philosophy to a concrete social-humanist political process – with a progressive scientific-technological gesture.

It is indeed a novel phenomenon, if India can successfully execute in its historical process, preventing population growth by a restricted marriage system.

Since Panikkar's time, this project has been furthered by several such projects in Central and South India. In his records, one can find a dawn of a *new history*, with a combined philosophic and scientific insight, to shape the future of India with a strong *political vision.*

The life-process has to be based on compassionate and loving attitude towards the children, women and older generation, sick, depressed and of the different cultures and linguistic and other differences of the country's population, taking care of all, as other possible types, efficient or non-efficient people at large.

The world-process is a possibility, with differences, and its possibilities – are not in the hands of an individual, but, on all minds as dependent on cooperative caretakers, with sincere responsibilities.

In this connection one may examine a failure of the restricted linguistic attitude of another country as of General Franco, when he allowed only Castillian to be spoken in public. But at home native speakers nurtured the Galician, Basque and Catalan languages, which along with various Spanish dialects, survived a ban of almost 40 years. Responding to demands for increased local autonomy, the 1978 constitution divided the country into 17 regions with unprecedented power and responsibility.

In contrast to what happened elsewhere – India from the ancient times allowed linguistic freedom to all spoken languages, both urban and also tribal. This freedom led to the emergence of many mixed languages between two or more adjacent states.

Recently India has been for the progressive developments of tribal languages, for, language in interrelation is an evolving and growing process. This evolution of interrelated languages fosters the varieties of *cultural evolution.*

India was or will remain awakened to this process, without fostering any individualistic trend which is suicidal. India passed through a very bad period, in trying to impose a nation-wide language scheme through Hindi. It was a fallacy, so it had to be corrected, but its venom persists, and has spread through South Asian countries. Language, like all other experiential processes is an intellectual process, with a growth, evolution and development. It is not akin to social process of population growth – creating problems.

It is wrong, however, to think that the infection was from India. In entire South Asia, from the ancient to the present, historically many languages from the different Asian countries have gone into these places with varied European impacts, Spanish, Portuguese, Dutch, and French, and lastly English. With all these, there were also some impacts from Middle-East Asian trading peoples. Hence in these Asian countries there are many prevalent languages. But with the end of the British rule, and also of the other foreign European rules, when the different countries became independent, each faced linguistic problems, due to wrong visions of narrow linguistic policies, in these Asian countries, leading to fatal results in Sri Lanka.

India's religious freedom is patterned with a tolerant spirit all-through the centuries, the growth of religious process allowing each religion a free access into India and tolerating and fostering their developments with a cooperative spirit ultimately enriches cultural growth, just as free access into a country of the scientific ideas – helps the development of a country's civilization, from century to the century. Religious process is associated with a social and cognitive enlightenment.

This *tolerant process* stretches to the present century. These progressive understandings of religion, and of the other experiential processes including science and technology, give mankind an opportunity to learn things from various alternative perspectives.

Difference lies in the core spirit of all religions, that sustains them with a speciality – but that should not be construed as materials for conflict. The difference casts into a religion its enlivening spirit for presenting a different outlook. This inner core in development keeps all religions active with a progressive growth.

The world as a whole is not categorically placed, it is placed problematically and as a possibility, in every spot, perishing and emerging, in alternative and many-dimensional ways.

Accepting this novel *Dynamism* I am reviewing the main philosophical ideas of my Fourth Volume of *Dynamic Facets of Indian Thought* with considerable relevant changes to pinpoint many of the ideas, not focussed before in that volume. Here, in this volume the philosophical and scientific ideas will be developed progressively, with a togetherness, in the advanced aspects of Buddhist non-entitative symbolic-insightful ways, going beyond any earlier boundaries.

In this aspect, this writer will always take the Whiteheadian spirit in development, for a parallel development of philosophic and scientific ideas, beyond Einstein.

It is a change of categorical to many-dimensional *non-categorical* possibilities. Though all philosophic and scientific pursuits are social processes in aspects, yet the pursuit of religion is often controlled by some faiths, beliefs and customs leading to dogmatic attitudes. But philosophy and science can easily be freed from such restricted processes of faiths, beliefs or dogmas.

From the days of the opening of India to the West – India became open to various foreign contacts, in different parts of India – and it passed through the evolving processes of early contact of the East India Company of the British trading companies.

India, then, was placed in different possible alternative situations, where the East India Company was gradually gaining in power and its supremacy – leading to India's accession to the direct rule of the Queen Victoria and subsequent process of British rule of India.

These led to various evolving and changing processes towards the ideas of a developing political consciousness, all in favour of building the grounds of a new empire. India passed through the pro-British rule, replacing the Moghul rule of India in political historical process.

This present historical process of complete independence of India, to the state of an Indian Republic, through a process of transference, as mentioned, was an achievement for, India, when it became operative with a novel foreign policy of nonalignment.

India made itself felt in the community of nations, its special position, by a rise to a position of a detached observer of the activities of the other nations and countries. It is an emergence of a novel socio-political-cultural continuity.

But India reserves the right of remaining separate, without involving with any. These were idealistically oriented, earlier as referred to, but now it has reached the progressive order of a *realistic order*, facing concrete changing situations. Now India is in this advanced stage of its independence-process, indicating a new emergence of a political consciousness.

This acquisition of a unique position, is in no sense, neutral, or an indifferent attitude or a gesture of a progressively powerful nation. It steps into an international process, with a progressive outlook.

India's suggestion for a detached attitude will give it a platform to be operative in an active way, and influence international situation with a heritage of a transcendent power in a novel elevating aspect.

This transcendent power, if manipulated with insight, will bring about a revolution in the political sphere, which will progressively create conditions for a balanced experience in inter-personal and international contexts, with a detachment.

This becomes a possibility for India, because that special foreign policy with a capability issues out of India's transcendental philosophy and psychology.

It is an understanding of other aspects of experiential processes, as confronted processes, from the myths to the sciences and technologies, as progressively evolving towards prospective processes, as so many evaluative contexts of confrontations, to be faced and properly dealt with.

To represent all these contrasting complex processes, with accuracy – in contrast to merely empirically oriented mode of life and thought, of the Western world, in general – India will always refer to something 'non-seen'. Therefore it has a reason not to plunge itself always by associating with other nations, when they restricted themselves to mere conceptual or empirical modes without detachments.

The other nations inspite of their varied experience cannot totally free themselves from uncontrollable proneness for conceptual-empirical modes of deliberations. As the Indian attitude, is detached, it confronts all experiential processes, as *relational*, and never as an ultimate, so it can move from one experiential process to another, without any *restrictedness*. Hence India provides a ground for a possible new emergence.

India also sometimes becomes influenced by the foreign impacts, but it maintains its *independent attitudes* and views, as will be found in the following cases mentioned below.

The first Western impact, if historically viewed – can be traced back to Ananda Ranga Pillai, a merchant of Pondicherry, when Dupleix, was the French Governor of the place. He established a friendly partnership, and identified himself with a loyalty with an empire-builder as a sponsor. But as a religious man, he kept himself separate from his business associations.

With Dupleix he enjoyed a mutually profitable connection or business relation. At the sametime Ananda Ranga remained a staunch and orthodox Hindu, never violating in the slightest, the rules of his religion.

European culture remained for him, as something 'external' in contrast to his own traditional culture and belief, which were transcendentally oriented. His was an evaluative perspective and a tolerant appreciative attitude. But now the attitude towards any experiential aspect is a cultivation of both adherence and detachment with a transcendent insight.

Abu Taleb, born in Lucknow in 1752, of Persian and Turkish descent, who served under the Governors of Bengal and Oudh, during the Muslim period – was forcibly retired by a jealous Prime Minister. He then sought employment with the British, and assisted the ruling British in putting down a rebellious Hindu Prince.

He moved his place to Calcutta, and learned English efficiently and had a scottish friend, that enabled him to travel to England and Europe for three years. He on his return to India, wrote his reactions to English life in London.

To him, his European experience was a superficial standard of life-style in contrast to Indo-Persian culture, of India, so he also indicated some dislike for the Western materialist attitude and culture.

But this mixed process was religiously oriented, with not much of advancement of transcendental process. It is just an expression of another attitude.

But this writer is waiting to develop India's cultural process, progressively and with speciality, with the *changing environment*, beyond these two cases as referred to. He switches on to another example, of the nineteenth century, when the British education had become widely prevalent.

Then there is a reference to a short-lived Anglo-Indian scholar, a Christian poet Henry Louis Vivian Derozio (1809-1831) of Calcutta. He received the best of English education available in Calcutta, and he was appointed assistant headmaster at the famous Hindu College.

Here the brightest young Bengalees were flocking to learn from him, the new knowledge from the West. In two years Derozio received the credit of appreciation from the minds and hearts of his students.

He inspired his students with patriotic ideas, but at the sametime he was against idolatry. It was for that reason he was denounced by the orthodox people as spoiling the minds of the youths. But Derozio's mode of knowing and learning was evaluative – a different order of expression with equal stress on the importance and interrelation – a balanced insight with a detachment.

The College Board demanded his resignation. Here one is reminded of Greek Socrates encircled by Sophists and some dogmatic social background. We know about his tragic history.

Derozio's advanced nationalist ideas influenced Michael Madhusudan Dutt, a pioneer of Modern Bengali Literature. Derozio's premature death through cholera, was an end of the finest flower who sowed the seeds of ideas of India's independence through independent thinking through English language – a prospective religious process, beyond all orthodoxy the young mind of India. This period of Indian history, was socially, religiously and politically important – as providing later grounds for India's shaping of the transcendental processes. He expressed an advanced attitude, and opened for India a new avenue for progressive insightful reflection towards nationalism. These various ideas as emergent possibilities keep the Indian minds always open to a new, casting off the past dogmas.

This writer is not always defensive of India's every thing, he is critical of those aspects of civilizations of India, which have to be criticized by reference to later developments of India's nationalist ideas as also internationalist ideas. The progressive path lies in examining the confronted situation, and also developing a prospective vision, as none of the presented situation is 'ultimate,' but 'a possible situation' with *alternative* possibilities.

Here an advanced philosophical thinking is a need. This becomes manifest in diverse philosophically oriented social and political processes. There is here also an elevating process beyond mere nationalist to internationalist ideas and attitudes, as found in Gandhi, Tagore, Aurobindo, Radhakrishnan and others.

They became not only interested in comparative philosophy and religion, but also comparative politics, sociology and psychology. India's scientific studies also was from an intercultural perspectives, where the Indian scientists became associated with the eminent foreign scientists of different areas, but could as well bring out something very special resulting out of India's cultural background, as could be found among a series of Indian scientists as J.C. Bose, who could even discern the life-process in plants, becoming perhaps conscious of Jaina and Upanisadic thinking. He was able to relate the confronted physical universe with the vibrating vital processes. His was an association with physical and biological sciences with philosophy.

P.C. Roy, could find through his knowledge of chemistry something good or health giving process in his chemical analysis of the elements for the extension of the longèvity process, as already realized in Atharva-veda, in the advanced aspects of medical science. Roy's was associating chemistry and biology both with philosophy.

C.V. Raman could discover in his analysis of the radiation process, a clue towards its multilinear scattering process, the vibrations of a musical instrument, paving the path for linking up the physical universe with the aesthetic processes both as *seen* or *heard*, connecting the *optical* with the *hearing* – the sounds of musical instruments. Raman connects the audio-visual processes with music and aesthetics diving deep into philosophy.

Meghnath Saha, Satyen Bose, Homi Bhabha and others discovered something special fitting with some aspects of Indian cultural process, linking up the cosmic with the mathematical. They in different ways paved the path for going beyond Einstein's physics and mathematics in other possile ways – indicating philosophical transcendence as deeply symbolical.

These are just examples to indicate some general ideas of the religious attitudes of some persons and of the Indian masses or groups in general, to philosophical, political, scientific and internationalist attitude among the intelligentsia, as all experiential processes are interlinked, so religious experience is not totally excluded from other experiential processes.

In this interlinked process, the liberal internationalist attitude in learning different foreign languages, wherever felt necessary becomes another avenue for a further movement.

India needed at this moment a personality to unify these processes to a prospective direction in all directions.

But with progressive Western impact these attitudes changed to concrete philosophical and religious movements which could be socially and politically considered as dynamic and changing processes, which could be utilized with some detached attitudes, giving vent to various dynamic attitudes, as to bring in some international change as will be indicated below, going backwards and forewards to indicate the prospective attitude from a background.

Then one may go backwards in history to the outstanding contributions of Raja Rammohun Roy, the Father of Modern India, a social and religious reformer, from an advanced transcendental perspective. But this was a movement to a really complex *intercultural* process and possibility. He wanted to be operative with the British Masters to bring about a change of India by education through English, inspite of his vast knowledge of the advanced process of India's Sanskrit, and Arabic and Persian heritage with a masterly grasp of the advanced processes of Indian religions then prevailed. His was a pointing to a *new direction* of developing one's cultural processes through mastering a new foreign language, English as associated with other foreign languages. His was not an exclusive attitude.

His was a vehement protest of the activites of the prosletyzing Christian Missionaries. But this objective he could realize by effecting a socio-philosophic recast of Indian thought. It was an emphatic interpretation of Upanisadic transcendental non-dualism, removing emphasis on idol worship, but not forgetting the transcendent ultimate reality behind such idol worship. His was a correction of India's customary religious process, by a revival of the philosophic process of the Upanisadic transcendentalism behind the religious customs. His was also a confrontation of the impact of the prosletyzing policies of the Christian missionaries, pointing out that Indian religious consciousness is neither an idol worship nor a theism. India does not subscribe to a fixed Absolute.

Roy was for reforming the wrong social practices involved in 'Sati' practice, as of compelling the helpless widow, after the death of her husband, binding her with a rope, and mercilessly throwing the body on the funeral pyre, to symbolize the love of the wife for the husband by dying with him.

What a heartless circle of people who allowed this bad practice to continue in a civilized social process? He also encouraged widow-marriage, when he observed that a respectable male could marry several times during the life-time of one's first wife, several other ladies, only to give a prestigious touch of his family as belonging to a higher order within a particular caste. Whereas an unfortunate lady has to sacrifice herself at the funeral pyre of her husband. Is it really a 'Love'? Society had to be reformed immediately. Roy stepped into that advanced process.

In his recast of religious process, from some specific religious and philosophic ideas, there was a transition to a 'movement', when it could be operative as a power to cancel the prosletyzing influence on Hinduism by other religions, taking advantage of some of its drawbacks. India could not rise to the condition of an evaluative process from an intercultural perspective before Rammohun's treatment of Hinduism as philosophically self-sufficient, and also supplementing it by removing some social evil practices within Hindu religion.

Rammohun found a gap between high soaring 'philosophic thinking' and the wrong culpable 'social practices.'

Rammohun held the view that without this philosophical and social reformation, and balanced attitude, of Indian Society – India cannot protect itself really from a superior moral, religious, social or philosophic perspective – from the operative impacts of the conversion by the Christian Missionaries, both from outside or from inside of India, as they offer many privileges in their social system.

Making the country healthy and balanced, from an advanced religious, philosophic and social-moral and legal process—Rammohun wanted to improve the condition of India, with a novel spirit, in confronting inter-cultural impacts.

He effected the abolition of the 'Sati-practices' (of mistaken notion of a faithful wife) through the help of the British legislation, even by an appeal to the British authorities of England through the-then Governor of India.

Similarly he introduced the education through English, as a need to elevate India to an internationally recognized current of the world as a whole, both for elevating India's greater future in obtaining the advanced scientific and technological knowledge and skill.

It is wrong to suppose that he was not aware of the rich Sanskrit and Persian heritage of the time, but he wanted to be in the spirit of the time, and suggested that India should take advantage of this opportunity of educating itself both linguistically, and scientifically, through English. He had the foresight of seeing the value and potency of English language as an instrument for mastering scientific and technological skills and harnessing them for political and business aspects – in man's drive towards mastering of the international relationships in various spheres of experiential processes, gaining progressive controlling powers from the ocean beds to the outer spaces.

This is not possible through a development of the knowledge through Sanskrit or Persian alone, as was in practice before. Mastery over the ideal world is one aspect of human efficiency, but there are other progressive *confronted* historical processes that claim priorities, opening up the possibilities for proceeding in varied intercontinental directions conquering the gaps between space and time, and making things accessible within moments. The present world is a world of 'speed' and 'quick prospect,' and not for lagging behind.

The world has already ascended to a status of speed, scientifically indicated through the speed of light, and now it is waiting for the world-process towards a faster process as indicated through 'Tachyon' (Greek word for speed). This process of unification of scientific attitudes were effected in this century by India's outstanding scientists – C.V. Raman, J.C. Bose, P.C. Roy, Bajendranath Seal, Meghnath Saha, Satyen Bose, Homi Bhabha, and by other recent scientists of the types of S. Chandrashekara, a Nobel laureate, and George Sudarshan, and many others.

These scientists are not exclusively Indian, they are internationally known and operative, stationing themselves in India, or in any part of the world, according to their discretion or interest.

To my mind this was a right judgment, though, in this new drive, there was no suggestion for not developing one's own native languages, or the classical languages of Sanskrit and Persian for the development of other varied aspects of improved possibilities of literature, art, architecture, music and other refined cultural processes.

Efficiency in computer science is another ascent to speeding up one's mental capacities, in all spheres from philosophy to sciences and technologies – by mere observations, and stretching of visions, with an inconceivable speed and accuracy – curtailing the limitations of thinking by aiding it through a machine, which mankind can easily utilize, as if climbing upon a new horse of infinite speed.

The role of computers, in their varied aspects, has given speed to all sciences and technologies. In an aspect, the whole world is computerized or computer oriented.

The decision for giving prominence to a particular regional language, abandoning English or giving English a secondary position, of the post-independent India's constitutional policy – was a colossal fallacy which infected the future of entire South Asian countries, with confusions, dissecting the countries to their cores, and severing their unified cultures, by linguistic fights. Introducting such retarding process into the social and political sphere is injecting the human system with a virus.

India missed for a while, the superb *transcendent element*, that connects all languages together as the different sciences perfect to a height all human adventures, with no restrictedness, by a wrong linguistic policy after Indian independence. Between two insersecting states, two languages mix and emerge in two clear cut wings becoming another, signifying that language *is* a dynamic living continuum.

The restricted understanding of a single language as an exclusive process creates a situation of seclusion of its people and culture from other adjacent linguistic family and cultural processes by creating a ground for antagonism.

This partial attitude leads to brutal killings of innocent men and women. Such a possibility will spell a losing of the evaluative aspects of human life, in the linguistic sphere, and in every other aspects of experience from the myths to the sciences.

Human mind often gets confused with the mass minds, which is often not effective, as confused by the operations of multiplicity of chaotic uncontrolled minds.

It is wrong to think that 'language' is everything. Language is an expression of man's mind in an aspect. This one-sided exclusive vision of a particular language or of a particular religion et al., loses sight of cultural unity and reciprocity, and fosters bigotry and dogmatism, creating a polluted atmosphere.

This wrong vision spoils the traditionally acquired knowledge through the refined processes of religions, philosophies, psychologies, et al., in interrelation – from a hoary past, obtained through local languages and then translated or communicated into English. This is nothing but an extension of oneself to a new power or potentiality.

This practice is now adopted belatedly by all advancing nations – Asian or European. India had a priority due to British rule. It has a mastery over the English language so much that Indians – in general in different fields – from philosophy, religion and literature to science, technology and business – have reached such an excellence that an ordinary Indian – can reach an efficiency of an ordinary British. Both have experts and they are comparable with equal efficiency.

So, in any historical period mind has to be kept detached from any confronted orders of experience – from language to sciences. Herein lies the cultivation of a detached attitude. This is transcendental insight, that has fertilized the growth of English language in India. India should not lose this status in international market of cultural exchange.

This unified nationalist process, through English – was carried on during the British period, through education, in the whole of South-Asian and South-East Asian countries.

At this stage of the cultural change with a progressive development from international context one may conveniently refer to different papers of the book, entitled, *The New Asia*, edited by Guy S. Metraux and Prancois Crouzet (A Mentor Book, published by the New American Library, New York & Toronto, 1965). For getting a general idea of an insightful change of a country from no specific perspective – geographical, religious, social, cultural, or anthropological – but from all possible perspectives – if one considers the European impact of the differently located Asian and European countries from Turkey to North Eastern Siberia – one may get some broad ideas of the change as resulting out of the European impacts from the different cultural backgrounds of the different countries.

In this connection one may select for study two papers on Turkey: 'The Impact of French Revolution on Turkey', by Bernard Lewis, and 'Turkey: From the Oriental Empire to Modern National State', by E.Z. Karal and also a paper on India 'The British Impact on India', by N.V. Sovani. All these papers present advanced outlooks due to foreign impact. It is wrong to think that to accept the best of foreign impact is not patriotic, but is a gesture of acceptance of what is best in others with a detached attitude.

The change that came to Turkey was made possible by the extraordinary insight of Mustafa Kemal (alternatively known by another chosen name, 'Ataturk' meaning 'father of the Turks). His was a culmination of a secularization of the Turkish State from the prior stage of 'Ottoman Empire,' through a constitutional process, not by just purifying of one's soul, by an adherence to an Ultimate God, but by a cumulative body of progress achieved by mankind in the technical, scientific, social and moral and linguistic spheres (replacing defunct Latin by Arabic), with no fanaticism, where freedom was both for males and females, and through nation-wide education, modeled on European lines, devoting considerable attention to Arab culture.

If one surveys through the detailed history of British impact from the period before and after the coming into India of East India Company to the shift of India to a rule by the British crown and the spread of British education of India throughout India, by pursuing Sovani's paper, one will find how the final process of a 'parliamentary form of government' emerged from the British heritage in India where several Indian minds became operative to bring about a change in a cooperative way in the social and political spheres. They are not just nationalist movements – but pierced through and through with internationalist ideals in conformity with the advanced philosophic-religious ideas of India's cultural processes. The world is heading towards international orders of personality process.

Here, also there is a display of concrete advanced ideas of Western philosophy, science and technology. These have been judiciously taken, unlike the different alternative choices made by other Asian countries. India will always stand for a 'secular process' in all aspects of experiential process, cutting off the 'ultimates' in any process of experience as they hinder the process of lift to understand transcendent experiential process that shines through the presented or given orders of confronted experiential processes. This is not a sphere of argumentation but a sphere of freedom – insightful or penetrative understanding – a new mode of advanced outlook of life to live.

The Japanese, after Western imapct, chose to perfect themselves more towards technological aspects of sciences, and the Chinese riveted their primary attention to the military perfection through nuclear sciences. Those were their chosen prospective attitudes suiting their cultural processes.

By a contrast India's choice after the British rule remained for a transcendent 'Peace,' beyond all 'argumentative propensity' – as the only objective to realize.

Now what is needed is an international attitude, through language – by its unifying transcendental possibilities, as human beings are community-units in interrelation. This efficient linguistic process, in international contexts I would suggest, is good for the whole world, which China and Japan are belatedly trying to accept, as they are also becoming more and more commercially conscious, and reviewing their frailties in communication skills.

India, should preserve its advantages obtained from its British rule and its heritage of English languages. This also helped India to obtain efficiency in science and technology, and other subjects that prepared the ground for a civilized status for a developing country. Its further progress can be obtained through its transcendental ideas and pursuits, through the centuries, to this period of history. Its inheritance of a 'parliamentary form of government' was from the British, but it has to be modified and improved upon through its insightful cultural attitudes in varied transcendental directions. This has been indicated in the third part of this volume.

The world is for advancement not alone in business, but in all orders of experiential fields reaching the technological skills for controlling also the outer spaces, and remaining alert to observe the activities of the different foreign powers to control or challenge their unlawful activities from the perspective of international laws and computerized frames of reference. India fosters a many-sided ideal with a farsight through a mastery over the outer space with technological efficiencies.

India will feel the need of the efficiency through English, and also through other foreign languages of the world, to rise high in the competitive business world of today, to extend its communication capabilities, when the world-force is directed to the efficiency in business, as a scheme for controlling the world-market efficiently both from the perspectives of politics and economics, as also from some languages, wherever they have some roles to play in any foreign country.

The present world is gradually understanding this. Both China and Japan have to depend on the American or the British linguists to render their thoughts into English or to learn English directly.

This process is also taken by Russia with a speed utilizing the computerized and fax facilities. But India had a basic advantage. Now India is thriving in the world at large with its computer and fax facilities, and communication through English and other foreign languages.

But India had been wrongly interrupted by a temporary misdirected emotionally ideologically oriented political scheme of a 'language' preference, tinged with a wrong sense of a restricted nationalism. This brought in narrowness, and truncated nationalism pierced through and through with the venoms of inhuman elements.

India is a union of its past cultural heritage of different varieties through linguistic, religious, philosophic, and scientific stretches, and now with the prospects of further discoverable experiential fields. It must be remembered that the progress of a country depends on all-sided developments. A country has to rise with a stable sense of the pursuits which bring in the insightful sense of one's 'values' in the exalted sense of the term. A personality of this type has to be entertained.

Man should not be allowed to losehisbroad-based human feelings that bring togetherhis neighbours, now stretched further with the advancing technologies, planes and telescopes, satellites et al. We live in a world of an extended community-process of improved technological efficiencies.

It is for the understanding of the interrelated possibilities and for the future prospect of India from all possible aspects that I am trying to shape India of the present and also of the future in the triadic aspects of philosophy, astrophysics (or sciences in general) and politics, by a possible perfection of all these processes, and in unision, keeping in mind its prestigious position both inside India, and outside in international contexts through an advanced outlook mastering all areas of experiential process as far as possible.

India, with its linguistic efficiency of English can with strength and ingenuity – bring about progressive changes of the Constitution, suiting the need of one's own country and culture, in relation to international situation. An ideal and yet a real personality is heading progressively towards this process and prospect.

Within the country it can be operative with its various other languages prevalent within India, and finally it can rise in progressive cultural spheres in appreciating peoples' government, in the best sense possible, by improving upon varied social relations by travel and enjoying the film-products of different provinces.

But it is possible only through a mastery of English language, with a possible skill of learning as many other foreign languages, as far as possible, and other languages within India, with full sympathy and interest.

Rammohun's suggestions, so dynamically interpreted, lead one to his immediate and distant followers, who played some important roles in facilitating the process towards Indian independence, through linguistic efficiency.

Through English, it was possible to rise to a progressive reconstruction of India's philosophy, stretching it to social reformations, and perfecting country's efficiency in scientific and technological directions.

India has to rise to political, linguistic and legal efficiency, and in other conceivable spheres, deepening the depth-processes of benign principles that sustain and foster the liberal humanist ideals, of real growth of humanity as a whole. Mass or even the general intellectually capable person, cannot have mastery over all subjects or areas of experiential process, or can have proficiency of professional capability, so it is wrong for the narrow-minded politician to make a mass-movement by some restricted ideas.

Indian masses should be trained in this interrelated aspect, in a general way so that they will learn to safeguard their interests insightfully with a many-sided attitude.

If India loses its evaluative understanding of cultural processes of its own, with a full understanding of cultural processes of the other parts of the world – India's advancement, according to this writer, will be an adequate evaluative achievement.

Here also my method will be a quick survey of the roles of many others not alone of Bengal or Bihar, but of the other parts of the sub-continent of India.

The world can make a wider stride of progress through computerized world, aiding all aspects of learning from mathematics to various linguistic spheres, with a transcendental scheme of reference, as I am trying to express myself through my books.

Referring to Rammohun's incentive to learning, through English, one finds that this suggestion had tremendous effects upon the Indian intelligentsia, who came in direct contact with the whole corpus of Western learning.

Through English he could raise voices for developing the interest for the study of the past of India and other ancient countries of the world, as a policy for self-improvement. With this attitude, he could also show his interest for the different cultures of the world.

India later could compete with the British rulers of the period through proper and efficient education, by choosing to join Indian Civil Service, through competitive examinations in Indian educational and administrative services.

Eminent Indians could also participate in law and business, where some technical English terms and expressions were needed. Indian efficiency became an efficient stubborn fact, which the British during its rule, had to acknowledge with due recognitions in varied ways from law to the sciences by employing them in different administrative posts.

All these expressions of the Indian minds outmoded the British personalities of highest orders. The Indian intelligentsia also understood the contribution of the British and the Europeans, including the Germans, Russians and Americans, in general.

India's was a triumphant expression later in arguing for Indian independence through a linguistic sphere along with efficiencies in other spheres from philosophy and politics to sciences and technologies. India's independent position was recognized in all aspects of experiential processes from literature to science and politics.

Simultaneously another process ensued in the country by some members of the Committee on Public Instruction who had studied Sanskrit, Persian and Arabic, and discovered the wealth of varied cultural processes contained in their literatures.

India is undoubtedly rich in entertaining mankind of entire near East and South Asian countries, providing enough materials for aesthetic enjoyments through songs, music and varied cinema products, with a mastery reaching an accuracy in a parallel level with science and technology.

These have added to the business-prosperity of the competent Indians, in these varied fields. India's engineers and doctors, scientists and technologists get employments in many advanced countries. India is a pointer to new spheres of cultural processes, undreamt of before. Its interest for ancient civilization of the new Hemisphere, of Mayan and other civilizations stretching from the central to South American countries can provide an Indian scholar an opportunity to study and write from a comparative perspective of different civilizations of the ancient world. (Cf. *Mysteries of the Ancient Americans*, and *The World Before Columbus*, cf., The Reader Digest Association, Inc., Pleasanton, New York, 1986). It is wrong to be exclusively interested in one's own country or culture, when there are similar processes in other parts of the world, now drawn to one's door-step. No longer as an Indian one should extol one's civilization as earliest in the world, without studying the ancient civilizations of the world at large from the old to the new Hemisphere.

These are also comparable with the ancient civilizations of the old Hemispheres, viz., Mesopotamian and Babylonian civilizations of Iraq and other places of the Middle East countries, as associated with India's Harappa and Mahenjodaro civilizations.

These finds led Sir John Marshall, Director General of Archaeology in India, to think that this excavation of both cities is a record of the discovery of ancient India's first civilization. Dr. Possehl, at the University of Pennsylvania in Philadelphia, in his further investigation, finds out that some old bricks at Mahenjodaro, were used to build a Buddhist shrine in the second century A.D. At Harappa very few ruins are still in tact. In the late 1800's the city's ancient bricks were carted away and used to build a railroad between Lahore and Multan. (Cf. *Mysteries of the Ancient World*, National Geographic Society,

Washington D.C., 1991). These are further studies in historical process. But one has to see and examine other things also.

Equally interesting is also the civilization of the artic regions, in this context, for progressive extension of the scope of civilized processes, specially of the Inuits of North Canada, their lives in sub-zero weather conditions, and their language (Cf. *The Inuit: Life As It Was*, by Richard Harrington, Hurtig Publishers Ltd., Edmonton, Alberta, Canada, 1981).

There are possible extensions to the study of ancient civilizations. They provide a contrasting background for emerging advanced currents in Americas, as the Aztec civilization is an emergence out of the early civilization of the Toltec, and the Inca civilization is from the early Peruvian.

This is a liberal expression through an evolution in the direction of knowledge through different types of civilizations of the New Hemisphere, and that the civilizations of the New Hemisphere of the pre-Columbian periods have to be referred to different civilizations of the Old Hemisphere.

The dominant supposition now is that the cultural processes of the two hemispheres were somehow interconnected, not only from the pre-Columbus period, but also from pre-historic period from very early periods of the geological history of the Earth itself.

This is also an international need. Civilization, as a world-wide process is a subject of international interest, and is located in many parts of the world. It also provides food for wide comparative studies, to get rid of the bias for a particular region of the world or its people in a particular period of history. It is a wrong pride to claim one's own civilization as the oldest in the world, as all are interlinked processes.

The progressive interest in the study of ancient civilizations is good for anthropological studies, providing fresh areas of unknown languages of the world.

This kind of linguistic future was initiated into India after the arrival in India of the several British and foreign orientalists, like Sir William Jones (1746-1794) and others as Arnold, awakening many German, French and Italian specialists in all aspects of Indian thought from the myths to the sciences.

Jones was a pioneer of Asian studies, his arrival in Calcutta in 1784, gave a decisive impulse to the founding of Asiatic Society of Bengal. His Grammar of the Persian language, and his words of praise for the wonderful structure of Sanskrit surpassing Greek, Latin and other languages of the world – raised the prestige of Indo-Persian-Sanskrit cultural process, of India, as an elevating civilized process. Though really the continuation of the linguistic process is not a requisite process for this book, my consideration of it is an illustration, to indicate that all aspects of experiences are interrelated processes.

With this new linguistic revival and rejuvination – the pride of knowing the other oriental and foreign languages with a scientific zest – has become a pride of the really learned intelligentsia of India in different parts of the world as a whole. Interested people can understand that human beings by such discoveries can consider themselves as belonging to one family, which needs constant extension. Every experiential process touches other processes. One has to extend this knowledge-sphere to the different areas of the New World from the period when the avenue through the 'Bering Strait' was opened up.

This again raised the prestige of India from the perspectives of the foreign eyes in general (cf. Will Durant, *Our Oriental Heritage*, Simon and Schuster, New York, 1954).

In this connection one should read also David Diringer's *The Alphabet: A Key to the History of Mankind* (Hutchinson's Scientific and Technological Publications, London, New York, Sydney, Cape Town, 1947, specially, Ch.VII Ancient Central America and Mexico, and Their Scripts of Pre-Columbian America).

The two Hemispheres were once geologically connected, and they lost links, but that can be regained by progressive insightful researches by applications of advanced research skills. This enhances one's knowledge of geology and oceanography. Life in a comprehensive sense is not possible without this general knowledge of knowing all subjects, in their entirety, and in interrelation.

The study of comparative civilization also awakens man to realize that the 'ancient achievements' – of the old or the new hemisphere – are not merely comparable with the achievements of the present – but sometimes one has to wait and think for a while to evaluate the engineering skills of the Peruvian and the inca periods in contrast to the civilization of the Greeks and Romans as stretching to the scientific and technological skills demonstrated in South America – whether in road-making or in agriculture. (For details vide, *America's Ancient Civilizations* by A. Hyatt Verill & Ruth Verill, Capricorn Books, New York, 1967, Chs.20-23).

Rammohun's knowledge of the several of oriental languages was a real surprise when the Indian government in 1823 decided to find a support for New College for Sanskrit Studies. Rammohun with his superb knowledge of Persian, Sanskrit, Bengali (his mother tongue) and also the knowledge of Greek, Latin and other foreign languages – earned for him the necessary credit for all times to come. His was not in the least – a spirit of sound and fury signifying nothing, but he symbolized a precious gem of his time, revolutionizing India, and the linguistic world in general.

His many-sided linguistic efficiency was indeed an inspiration for oriental studies along with English.

This mellowed the force of Macauley's over-emphasis for education through English alone. The new move added to the strength of a cultivation of traditional or antiquated cultural process, from a multi-language perspective.

The place of English language is one of these socially-oriented linguistic family of languages. Hence its importance is from a different perspective, not just as a language, but in the aspect of an interrelated language-situation. Knowledge in interrelatedness is an emporium of power vibrating all centres of powers. Hence it has an intercultural value.

It gave a progressive open-minded attitude of Indians in general, during Rammohun's time and in the present times, when misdirected suggestions and experiences cast polluting dusts in the minds of those who have wrong emotions or uncontrollable dispositions to take some advantage over the less powerful persons.

The fight for exclusive religion, language, science or politics is a mistake. It is a far cry from a desert land where there is nobody to hear one's call for any help.

Just as the Muslim-conquest of India injected a fresh stream of religious thought and fervour into the veins of the Hindu society, so the British conquest brought with it new views of the world, man and religious experience, through a language. It made India presentable to the whole nations of the world, as a scholarly nation. It gave a new elevating spirit to the Indian minds, to be operative with independence and strength with the English speaking people, as could be found in the courageous gestures expressed by the Indian government to the United States decision against India's sole control over the Indian Ocean, where no foreign power has any access, as a superior power, without India government's permission.

Indian Ambassador, Abid Hussain, on 30 April 1992, sharply protested, taking strong exceptions and supporting P. Chidambaram's aptly worded linguistic gesture. As a Union Minister of State for Commerce, Chidambaram said that 'I feel the action or the decision of the United States was unjustified and unwarranted' (*India News*, 16-30 April 1992). This enforced linguistic flare, is often a need in a critical political situation.

There is always a possibility of an improved religious and linguistic consciousness, to directly cultivate the best elements of all religious and linguistic elements of the world, in all their subtleties, as scientific principles – and apply their criticisms with a cogency and appropriateness. This serves as an appeal to cultivate an improved spirit, in togetherness for rising to a common self-improvement.

In the present day advanced aspects of scientific researches, to exclude or neglect a language, a religion or civilization is an 'islandish attitude.'

India has been enriched by the foreign connections and interventions throughout its history, hence its civilization is a collective product and process, enriched by various aspects of world-civilization in continuous developments with a process of give and take.

It will be more so, if the present day Indians become progressively interested in the studies of the American ancient civilizations both of some places and of the different tribal and polynesian people of different islands, from a comparative perspective.

This project might be in relation to the civilization of the tribal people of different parts of India and other adjacent islands to India and in Indian Ocean at length from pre-historic periods, stretching to geological, anthropological and archeological history of mankind in general.

Viewed from all these perspectives, there was a definite improvement in Hindu religious process, which was initiated by Rammohun with an improvement and modification of social practices. In this aspect, the British or Western education was a boon for India from all aspects of experiential processes.

Some further ideas in this direction, initiated by Rammohun's new religious process beyond early Hinduism, with studies of Bible and the life of Jesus and his moral precepts, added further improved elements into India's religious current in general. His new advanced religion of 'Brahma Social Process and Movement,' not only updated India's religious process in 'inter-religious' spheres, but it also opened up a new possibility for getting beyond any specific religion.

All these became evident through the followers of Rammuhan, from Debendranath Tagore and Keshab Chandra Sen to M.G. Ranade, and from these to G.K. Gokale, and Mahatma Gandhi, in progressive ways.

All these liberal minds linked up the religious process with Nationalist Movements of India, in different ways, as weapons to eliminate progressively only the British influence or the rule of India. Rammohun's activities have been specially chosen for his many-sided influence.

This fruitfully establishes the thesis of this volume that the transition is from the stage of views and their restrictedness to the stage of effective actualization of these processes, in togetherness, which opened up the Indian minds to deal with, in a progressive sense of equality, generating a nationalist consciousness, with an internationalist attitude, which was not so dominantly present in India's history, except in certain pockets.

Now by a contrast it was a unified nationalist consciousness in making the country free from the British Rule, a unique achievement of India's transcendentally oriented objective. Now it became open to international perspective with a broad humanistic drive.

India, in this adventure, retained the useful, discarding the useless. Therefore, if this volume is prepared for utilizing mostly the ideas of my fourth volume of *Dynamic Facets of Indian Thought*, and supplementing them by several additions—one can be made conscious through my further progressive works, specially in the new aspects of philosophical deliberations, from many aspects of intercultural perspectives.

Here, in this volume, a stress will be to scientific and technological thinking, and to politics in the advanced aspects, without being troubled by the policies of many other advancing or developing countries of the world—restricted as they are to some specific philosophy, religion, politics or sciences in general. By a contrast India is undoubtedly an advanced country.

To my mind, all countries are developing countries. The real so-called superpowers are not superpowers from all possible cultural perspectives.

They will always remain at the level of developing powers if wrongly they cherish an ultimate in different orders of experiential processes.

They should have the courage to declare that though there is no ultimate principle as God – they are still religious in the advanced orders of religious experience. Similarly they must be capable of saying that they are fully scientific, even transcending their exclusive theories.

India courageously declares that it is religious even without postulating God as an ultimate, God, as an experienced process, comes in the stage of cultivation of religious experience, but it is not an end-process, it can be transcended by a further meditative possibility.

The meditative possibilities are not just 'religious', it is widely variant, along multifarious experiential processes, as man is capable of thinking of, and now with further experiential processes, it is on a progressive increase, in progressive transcendent areas, crossing all traditionally held conceptual and empirical processes and boundaries. A living personality is not an exclusive personality; his personality is a contrasting process in the milieu of other experiential processes and possibilities.

Similarly scientific pursuit, proceeds through some specific or general theories – but this can also be transcended by further penetrative modes of scientific deliberations – as indicated in the Einstein's reflection through special and general theories of relativity, then this process was transcended by a field theory beyond both. At this stage it is no longer a theory, but a 'transcendent possibility'.

All experiential processes from the myths to the sciences – are continuous, transcending their respective restrictive experiential processes, and they can run into one another, retaining their separate specialities, to introduce, varieties into a prospective experiential continuum or possibility.

So none of the experiential processes are exclusively important, they as confronted experiences – have a possibility of being transcended losing their ultimateness.

They emerge and vanish to be replaced by another alternative order with all freshness – with no restricted categorical pastness. No experience – whether as past or as future – is categorically represented. They are alternative possibilities – with 'if-then' characters. A 'personality', likewise, is such an interrelated possibility'.

The 'transcendent order' and even the 'confronted orders' – are not experienced in 'any entitative' way. They are just possibles – but for some reasons;

they are not imaginary they are concrete confronted processes, fully realistic, so long as they are individually or commonly experienced. But from the transcendent insightful perspectives they are replaceable till they are confronted.

This deficient behavioural attitude prevents it from reaching the insightful process of the transcendent order, and hence it returns by a kind of frustration, on account of its deficiencies in some aspects.

These are India's Upanisadic expressions in a different and advanced choice of new dress or fashion.

India does not want to be rich only in economic aspects, as a jackdaw trying to be adorned with peacock's feathers with some borrowings, without trying to know what really is a peacock, abandoning the interest in the best aspects of all experiential processes.

In this interrelated process lies the real wealth of the country. If a country, for example is fabulously rich in its economic prosperity, but there are homeless men, women, and children – that country is really poor and not culturally perfect, or really insightfully affluent. It is a forceful external effort without the inner ingredient.

To illustrate this long cultural process, I may simply refer to my books without further illustrating this process.

Without any direct help of the teachers the students are not capable of reaching the higher orders of the meditative processes. I followed Yajnavalkya's method of the ancient Upanisadic period, throughout my teaching career, as a symbol of a compassionate attitude towards students.

In contrast, in the Buddhist meditative disciplines, the transcendent experience, was not intellectualized by any entitative ways, it was symbolically oriented, with a total detachment from all intellectual forms. It was compassionate in a further progressive sense.

But this detachment is not losing touch with the continuous transcendent feeling-experience. Hence in a sense, the Buddhist scheme was a compassionate process in a further direction.

In contrast to these and other pathways in pure meditative directions the present day intellectual pursuits are directed to sciences and technology, with the same meditative insight and discipline. But here also the same sort of compassionate process has to be introduced.

The many-sided alternative philosophically oriented social and political movements awakened India to feel the importance of the concrete benefits that can be acquired through scientific and technological training of the mind, which are process for acquisition of efficiencies in different areas of experiential process. Modern India needs such powers, to be matched with *concentration*, *meditation* and *compassion*. Here *compassion* is not a feeling but a comprehensive post-intellectual process and possibility – an insightful possibility.

The advanced philosophical meditative trends are directed to scientific ideas in their progressive advanced aspects even over-reaching the post-Einsteinian physics, and later to prospective astrophysicist possibilities, through mathematical and algebraic symbolical ways, with a progressive skill through computers.

If all powers by all nations were given to the United Nations, by all the powers of the world with a detached meditative attitude – then, the prospects of the world as a whole could have been quickly focussed on to a new humanistic policy, in the political sphere, with a balanced business oriented economic prosperity of the world.

There is no 'poverty' as an objective phenomenon, it is primarily a human creation, when through progressive stratification it becomes objectively realized as a confronted phenomenon.

Buddhism in its depths, denounces the place of any so-called absolutely existing phenomenon, as the trees or mountains, oceans or valleys. Hence Marx's theory of economics as a resultant process of class-complex has a place, but it needs to be transcended as all other specific processes.

They come as a process in the confronted existential sphere in course of time, for underlying all these aspects of experiential processes – there is an intervening *time* as a transforming agent, and a total vanishing later.

During this time period something inconceivable happens, which needs to be confronted when such crisis-situation emerges. Even so, all experiential processes are modified by mental interventions of a person or personalities.

India had to learn to face all these varied types after independence. It had to learn to tackle any critical-situation through the scientific-technological efficiencies – for self-defence, specially from West Pakistan.

There is no need to mention about West Pakistan's activities as I have already referred to it. But the problem remains – and India's vigilance has to be maintained, with insight as a political policy.

But mankind is not thinking in terms of such depth-ways, except as a specialized study. But such deliberative ways are continuous processes for a civilized mind. Only in critical or crisis-moments, man begins to think, philosophically, scientifically and psychologically, to pass into the depth processes, with some detached attitude, but they are constantly missing to realize this possibility, coming back once more to the already experienced historical process of mere repetition.

The development of civilized process is a continuously emergent process, possible when man in his reflective mode, tries to tap the fundamental core of the civilized process that flows over the minds of nations which really aim at peaceful 'coexistence' with mutual appreciation, as a compassionate teacher transmits the process to one's dear student.

My transition is from relative restricted views to effective actual prospects in alternative spheres of experiential processes.

To realize this continuously progressive spheres of knowledge, I am stepping into the next Chapter of this work, to create more grounds for further improvements. I may as well point out that the first chapter, of the previous part was directed towards my discourse on the contemporary philosophical trends, in an advanced aspect, but this chapter casts further more of realist elements through its deliberations in consequent directions of progressive democratizations with no exclusive ideologies or exclusive proneness for empiricism.

It is a search for continuous freedom, within and without, with an eye to a progressive transcendent meditative possibility. The move is now to science and technology – with India's progressive advancements.

REFERENCES

In this reference for Ch.IV, I have chosen to refer to books by Barbara Ward, *Spaceship Earth*, K.M. Panikkar, *A Survey of Indian History*, John H. Herz, *International Politics in the Atomic Age*, and also some books, *Peaceful Change in Modern Society*, edited with an Introduction by E. Berkeley Tompkins, and *Iqbal: Poet-Philosopher of Pakistan*, edited by Hafeez Malik.

From each of these books I became interested in some novel ideas as Barbara Ward's reference to the present day life within a 'Spaceship Earth' with its vulnerability. K.M. Panikkar's representation of India's history from the ancient times to the period of its independence and thereafter, and his role, in India's political situation before and after independence, John H. Harz's 'International Politics in the Atomic Age' with some apprehensions in international political situation. I have taken note of the other two books in relevant contexts.

I have taken their ideas for consideration with some insightfulness but I have made a departure from their positions, with the changing times, presenting 'history' with considerable changes – and indicating that since their historical period, the world has advanced to a stage where mere idealist supposition which fitted with ideological politics, or mere theory oriented philosophy or science has totally vanished or is in progressive vanishing stages.

In these possible phases – each of the confronted aspects of experiential processes – from myths, philosophies and religions to sciences, technologies and politics – has lost its ultimateness.

With the loss of the boundaries among the different experiential processes, one is entering into the processes of the others, mutually impregnating one with the other in multifariuous ways so that a repetitive historical process has become an antiquated fossil, or obsolete. This world is heading towards a *New History.*

With this insigntful understanding the core of different orders and aspects of experiential processes has changed and each is taking different dynamic shapes of processes and possibilities losing its categorical exclusive character, as influencing the world-process with a current and outlook, moving towards a forced order ofr 'Peace', which was not experienced in the past historical socio-political process. But peace has become a possibility, because history today is no longer ideological or soft, it is like a steel, hammered repeatedly, with hard facts, offering no more a soft alternative. Hence the goal ultimately is for world-wide Peace, facing hard facts.

Hence from this perspective this chapter and the book as a whole has a claim for a *New History*, as mentioned.

This new historical process, this writer feels, will be a process towards a disciplined psychological attitude, which was not clearly seen earlier in any period of human history.

In this chapter the path for a prospective international attitude has been presented from all aspects of experiential processes – with no exclusive preference for an aspect of considering that aspect as *Ultimate*. We are living in an age of *no Ultimates* as all aspects of experience interpenetrate one another to prepare the ground for progressive unifying visions, with no end, and this prospect is created not in any 'categorical way', but always in an 'if-then' way, as the so-called confronted phenomena as found, may or may not be found, as one would normally expect, but it is accepted as when it happens.

Man, in day to day life, does not think this way, but one who follows the path of advanced ways, interculturally known how, from all aspects of experiential processes has to accept this mode of life with a spirit of acceptance.

This chapter is addressed to this prospect.

This chapter admits of the alternative spellings – Rammohun or Rammohan.

CHAPTER V

India's Scientific and Technological Developments in Tune with Nuclear and Space Age

This chapter considers briefly India's scientific and technological developments after India's independence from 1947 to August of 1991, taking note of President Venkataraman's broadcast to the nation on the eve of Independence day, as published in *India News*, 1-15 August 1991. India's scientific and technological developments in continuity are stepping towards the eighth five year Planning Programme from 1992 to 1997.

This consideration also opens up the later processes to the present with a lead towards the future.

This chapter also utilizes many specialist's papers in that issue of *India News*, giving some general ideas of the varied and many-sided responsibilities of a freshly earned independence from a powerful colonial master.

In this resume, there is no claim of a specialist's version on any scientific or technological area, but it is only an indication of a new direction from a philosophic perspective, which is no longer locked by any *abstract speculation.* It is a departure from an *abstract* philosophic attitude, and a stepping into the *concrete* realistic programme of science and technology to bring about fast social and political change in India with progressive self-defence.

Independent India has to maintain itself with a relative strength in the international community of mankind as a whole, with some demonstrable achievements, to indicate India's progress, in contrast to its progress after India's independence, from year to year, or from time to time, following some definite schedule of insightful five year plannings. These plannings are many-dimensional with visions of choices for priorities with a confrontation of a dual vision of the processes within India, and in relation to changing external milieu.

Addressing all, and duly respecting the past freedom fighters, President Venkataraman of the Indian Republic draws attention to India's present advantages of independent position, as freed from the colonial sovereignty.

India now stands as one of the developing nations, with a freedom to choose its own destiny by effecting a progress in all aspects of science and technology, for defence and as well for progressive self-development, according to its economic assets, and also in accordance with its cultural heritage.

In course of its forty-four years' of independence India must be proud of its great strides in fields like computers, oceanography, space research, genetic engineering and biotechnology. Drawing attention to another aspect, the President further says that we are also one of a handful countries which can maintain the full nuclear fuel cycle – from the refining of the ore to the reprocessing of the waste. This is a balanced attitude for all developing countries.

India's scientific knowledge and technological skill is for building a new society based on social justice, hence each individual of India, whether living in India, or as a resident of any foreign country, must be reflectively operative in the spirit of a 'peace-loving nation, with democratic ideals,' confronting its multifarious problems, of diversity of geographical position, culture, religion et al., often spoilt by short-sighted terrorist attitudes, population explosion and proliferation of ignorance.

India's progressive attitude will always remain linked with the Indians abroad gone to foreign countries for their different qualities and for different purposes, where the different types of intelligentsia from students to specialists in all experiential spheres can prosper, and the Indian people as a whole will appreciate them all, in different ways, with the changing times, giving up the past notions that such migrations are 'brain-drain processes.' On the other hand, one can see the new prospects of these migrants, and their contributions to the countries wherever they have gone and their roles. This reciprocal relationship has enriched India and foreign countries, as a perennial process. Each Indian in a foreign land, is a messenger of India's culture with an outlook of appreciation of the foreign cultural processes.

But all these processes, if entailed with some emergent problems, arise—those have to be understood, with patience and foresight—have to be faced, tackled and dealt with, by insightful educational schemes, making the country enlightened progressively with further insightful prospects obtained from Indian heritage.

Here the objective avenues for progress, both outside and within India—the resident Indians of India should be determined, with a sense of distinction between the education of the masses and the selective higher forms of education, in the higher aspects of scientific and technological knowledge, or vocational education, for a select few, depending on one's aptitude, capability, and income et al.

The primary asset of a country within India lies in its own people, men and women, boys and girls, increased number of intelligentsia, with a reasonable standard of the scientific and technological efficiencies, and, awareness of country's natural resources.

One should not be scared by India's soaring population, for when there is a development of a uniform programme for education mixed with a simultaneous programme for vocational possibilities – there is obviously no fresh problems emerging from population explosion created, but could be solved through the intellectual and insightful educated personalities, with vocational capabilities.

The booming population is not then a menace but a boon. One's own people can always be preferable to the experts from other countries, but this statement does not exclude taking necessary help and advice from foreign specialists and adepts. Foreign help is also a reasonable need, for foreign aid meets the demands of the needed coffers to launch a project which a country cannot afford to meet just at the moment.

A country can meet the money required for paying the interests for the loan amount, but not the entire amount for a new project, and in such arrangements, with the loan amount, the immediate strain of a country is removed, and the country proceeds ahead speedily, not merely by foreign aids from affluent countries, but also through commercial methods. With such an attitude, a country can maintain its independent status with some prestige and strength, and with a speed. Any developing country maintains a steady interest for improvement within, with a prestigious position within an international milieu. These suggestions are from an international perspective, with progressive experience and insight. India will always acknowledge that India is one of the developing countries, and hence dependent on many powers superior to it, in the international milieu. Moreover, India's attitude will remain firmly fixed to its cultural attitude, which is non-violent in spirit, and absolutely detached from any craving for stepping into a 'superior power,' in any restricted intellectualist sense.

But more than anything, a country needs, is its discovery of the most advanced methods and skills for self-defence, by a judicious understanding of any antagonistic motives of its neighbouring countries through satellites and most efficient telegraphic equipments. A country like India, has also to build up a strong group-allies with the closely linked places around the seas, and also the mountainous sides.

Vikram Sarabhai (1919-1971) – a Tripos and Ph.D. from Cambridge University – worked with C.V. Raman, at the Indian Institute of Science at Bangalore.

Raman, Nobel Prize winner, known for his non-linear scattering theory – experimentally indicated the connection between physics and the aesthetic experience through the sounds of *Mridanga* (a musical instrument).

Taking account of these vast possibilities in socio-cultural direction Sarabhai set up the Physical Research Laboratory at Ahmedabad. He was also instrumental in setting up the space research organization, and helped establish a number of space complexes in the country. In 1966, he became Chairman of the department of atomic energy, which post he held till his death in 1971 India lost the further scientific possibilities for his early death. But there are other emergent eminent scientists to continue his programmes.

Sarabhai, through his various writings, expresses a dream for India's manifold scientific possibilities, which have stepped up India's large nuclear power stations, serving agro-industrial complexes of synchronous satellites for direct broadcast television and of the many peaceful uses of nuclear energy. This is sufficiently indicated by India's broadcasting house of All India Radio, which links India with the countries to which the various language broadcasts are directed. His long life would have benefitted more. But the stark reality has always to be faced, with sorrows and confidence.

V.N. Tripathi, a civil servant, in a paper, entitled, 'External Services of All India Radio' in *India News* (1-15 September 1991), states the first external services was started in 1942. The reference is here from a pro-British stage to the present period of Indian independence. India, since independence has brought about a sea of change in the aims of external broadcasting – with a noble objective to reflect the ideas and achievements of an open society to present the Indian perception of the world affairs.

After Indian independence it immediately became known to all – home and abroad – that free India would follow an independent line on all international matters. Its attitude will always remain culturally oriented.

Enmity towards none and friendship for all is India's avowed policy. India is going hand in hand with its 'nonaligned' foreign policy, of peace, goodwill, amity, coexistence and non-interference with a country's internal affairs (*Panchaseel*). This noble Buddhist principle was initiated by India's first Prime Minister and it was steadfastly continued with further improvements of international relations by subsequent prime ministers and presidents – from Rajendra Prasad, Radhakrishnan to others. My object is to indicate the emergence of such personalities, detached from all notions of restricted 'ultimates'.

India's voice is heard in various countries of the world from Fiji Islands to West Indies, and from Africa to the European countries, including Russia, and is stretched to the Americas, and through the East and South East Asian countries to Australia and New Zealand—in various Indian languages, and in English specially, as over seas services. The reference to all these does not merely suggest that in those places, there are Indians, but also the foreigners who can be aware of the India's subtle cultural processes from varied aesthetic perspectives.

India's broadcast services are composite in nature – from news, programmes on sports, literature, films, socio-economic and to political issues, including India's scientific and technological and business advancements. This speciality India will maintain, as its cultural ideal for all times to come.

1990 was a watershed year for the United Nations. It was a period of growing superpower cooperation, the emergence of unified Germany, European integration, Japan's economic growth, and the Gulf Crisis, war and its aftermath. All these were reflected with an indication of the role of UN, its functions through General Assembly and Security Council, where India's permanent Representative to the United Nations played some important roles.

Here there was a special reference to the environmental issues by the 45th UN General assembly, in consequence of the Gulf War, with the resultant increase in the oil price, its effects on all affecting countries, including the South Asian countries, and extra-budgetary expenditure of UN. Here India's role of 'nonalignment policy' was reshuffled to confront the emergent international problems.

In this change of the world situation and complexity, as stated I am reminded of Radhakrishnan's valuable statements about the working of democracy in India, according to the constitution of India. The reference is here to a paper entitled, 'Ideal of Democracy: Scope for Expression and Development of Individual Human Being' in *India News* (1-15 September 1991).

With his depth experience of the India's constitution, and of his understanding of India's democracy as issuing out of man's 'solitary side' of his nature, Radhakrishnan developed Indian Democracy, by pointing out that 'democracy' is the expression of a Man when he is truly human and 'democratic in spirit.' This kind of insightful expression reveals his mind as 'culturally oriented.'

Both in his behavioural aspect and in his cognitive sphere – Man, in his contemplative solitude – manifests his appreciation of the great work of art, literature, philosophy, religion, science and technology through which Man impregnates his experience by the intensity of his own vision. Here interconnectedness of experiential processes from philosophy to other reflective activities are comprehensively indicated with a subtlty and insight. In this exalted aspect man maintains a dignity of his own, and sets 'value' to his activity-process.

If one has understood properly Marx's capitalist industrialism, from this exalted contemplative attitude one will have found that it has dehumanized man, and has gone against a 'socialist society.' Man has missed his 'spiritual freedom' to be expressive in the free exchange and clash of ideas. Herein lies the development of one's personality. Radhakrishnan brings in that evaluative process to eliminate the inadequacy of the mere capitalist industrialism.

This right of a human attitude has to be respected and not encroached upon in the exercise of one's individual freedom.

The parliamentary democracy – Radhakrishnan contends – which India has adopted is perhaps the best form of Government, with a happy blend of dignity and stability. It is the government of the people by the people through their elected representatives, and not by fortuitous leaders through an abstract intellectualist principle.

Political democracy gives one the right of lifting up one's voice and uttering one's opinions from an exalted insightfully oriented cultural perspective. This implies education, and periodic elections to contact with the people to determine their wishes, and to find out what level of education they have risen to. So there is a need of training people's mind with requisite training to understand the implication of education in political democracy which is effected through individual education and economic opportunity for all to understand one's own nationality as a part of international community. One's nationality, though one has to maintain, should not in any way lose touch with 'international obligations.'

To achieve this goal the international bodies need to remove several of the deficiencies that are still present in the UN, which in wholly controlled by the superpowers through their exclusive ideologies, that control their dependent satellites. This organization, as Radhakrishnan thought, must be represented by all nations of the world, and not controlled by 'cold-war strategy.'

The UN has not devised a proper machinery for effecting changes in the present order. They are operative through their own bias and prejudices.

So Radhakrishnan suggested, with some criticism, that the United Nations should adopt a more positive kind of internationalism, so that a nation, small or great, will be able to play an important part in international conduct. No nation, however great, has any authority to police the world and interfere in the affairs of another nation.

The United Nations is a symbol and a hope, so it has to be an organization in the real sense, with an effective authority and power, with laws, and with a vision. It is a comprehensive political arrangement – a social and economic technique – and an international attitude and approach, with a sense of obligation, appreciation and detachment.

Its prospect is towards a single or One World. This is a possibility for a 'Civilized Process' to be cultivated by international bodies. It should be operative with a common acceptable language, to be operative in terms of advanced science and technology. Man's technology must be addressed to man's sense of values and mutual appreciation, adopting a right standard of behaviour for growth in human nature.

All these insightful statements of Radhakrishnan indicate that his was not just speculation – but also some insightful suggestions for the change of political ideas as reflected in mere communist frame of reference, which by itself is not perfect, but suffers from one-sidedness. His reference to the United Nations, with some suggested improvements indicates how from an international perspective – some possible improvements in the world-situation are possible, but only when its members become conscious of the drawbacks of some restricted ideologies. One has therefore to take note of the Buddha's continuous transcendence of intellectualforms, and Whitehead's transcendent post-intellectual wisdom order beyond the preceding restricted order of restless intellectuality.

Both stand for a transcendent post-intellectual continously developing insightful order, detaching oneself from restricted orders of their early cultural bases.

Reviewing other areas of this brochure in this connection, I find some very interesting discourses on India's advances in high-tech areas, as if making India aware as one of the steadily emerging 'technological superpowers,' in the fields of nuclear energy, space, biotechnology, and deep-sea mining, for, simple thinking is not effective till its potentiality is applied to the scientific and technological spheres. "Thinking" therefore has always to be determined not by an ideal wish but always strengthened and balanced by scientific and technological insights.

India has scientifically learned the capability of development of superconductors to save energy, anti-fertility vaccine to check population growth, use of biotechnology to solve many problems. India needs to improve upon its medicine, agriculture and industry. At the same time India has to attend to the process of development of rocket launches and energy, to extend scientific and technological advancing processes for peaceful purposes. All these different suggestions are meant for knowing that all these pursuits should not just be pursuits in their own accounts but they should be riveted to all human interests, elevating the concrete realist demands of a situation.

The development of superconductors could lead to incredible savings in energy: as necessary for the 'trains' that speed across the country-side, at hundreds of miles per hour on a cushion of magnetism; practical electric cars, powerful, yet conducted by smaller computers. In these statements, the suggestion is of a development in the different scientific and technological areas, with a balance, foresight, and language – always taking into consideration, country's economic capabilities. These are many-sided programmes to make the country both internally and externally strong, from all conceivable perspective, progressively.

Besides all these operations, for the country, one should help not only the neighbouring countries, but also distant middle Eastern countries and many places of African countries. With this kind of broad and liberal outlooks India also is a country which welcomes foreigners to visit the country for various purposes, and also a country of tourist interest.

India is a great country of tourist interest. India's tourism has to be developed, with advertisements, for welcoming the foreigners for visiting the country for becoming aware of its past heritage and present developments in science and technological spheres. But this invitation has to be enriched by meeting the tourist facilities by improving upon the sanitary facilities, at the tourist centres, as of the United States, Japan or Singapore.

Tourism has to be developed in two ways, as (i) a great country of religious pilgrimage, of Hinduism, Buddhism and Jainism, and of other religions of India, in different places from Zoroastrianism, Judaism, Christianity and Islam of different mystic forms, stretching to other minor religions, with a tradition of a hoary past, for the many Asian and other adjacent countries are culturally connected with India. India encourages all religions and cultures to thrive, and derive benefits by this confluence and exchange of ideas. Every religion has some novelty, and socio-cultural implication.

It is not a situation of competing with the different religions prevailing in India, but elevating all, with each one's assets, to build up a unified emporium of India's religious wealth. India should demonstrate its speciality in this great religious sphere, with a pursuit beyond all religions, as modern advanced scientists are pursuing beyond all past theories, or politics beyond all restricted ideologies.

It should be noted that these statements are not meant for mere statements but for interrelating the different aspects of experience with a togetherness, in that process alone, there is a prospect in interrelatedness, not separating one from the other.

Russia is not very far from certain points of India, viz. the centre central Asia, serves as the corridor for intercontinental connections, not merely as a trade route but also as different culture centres connected with Buddhism, and art of India in general. India as an important centre for the world civilization and culture is a place of a great interest for Russia, also, for building up 'common heritage,' in many of its adjacent places.

Russia in its hoary past, opened out the passage to the new world through the Bering Strait. The study of the early history of migration of Asia from Russia will extend the path of 'culture and civilization' that spread into the New Hemisphere, before its discovery by the Europeans, at large, in the historical periods. Some recent publications are pointers in this direction. They provide materials for enriching one's 'cultural wealths' further with fresh 'cultural elements,' as a new scientific discovery leads to prospective scientific thinking.

India is a mystery-land as a centre of 'innovative thinking' from the myths to the sciences, from mathematics to astrophysics. India is also an interesting place for the study of the scholars in various scientific and technological spheres, for its discovery of insightful background for '0' (zero) as operative behind and after any positive number from '1 to 9,' the symbolic suggestions by letters through the processes of algebra, and differential calculus, as indicating the progressive aspects of 'Symbolical Thinking' with unimaginable speed.

India's impact through zero, touching the infinity from many-dimensional aspects is an undoubted fact – an eye opener for interstellar connection, which bridges up all geographical and cultural gulfs. The world presently is placed in the milieu of infinity, if viewed from 'astrophysicist's perspectives', in contrast to the perspectives, of the vastly variant 'Micro-cosmic Universe.'

The zero, present, on both sides of the confronted positive numbers extends infinitely the scope of knowledge – both positively and negatively – in which zero retains its symbolic suggestions, without changing its nature as a 'number' as but a *symbol.* It is a ground for numbers to be operative and not restricted to a positive number. It opens up the avenue for establishing a corridor between the finite and the infinite.

Interconnectedness of confronted positive experiences are placed between the infinites on both sides of the confronted numbers. Some advanced ideas on *Zero* can be found in two books by Constance Reids, entitled, *From Zero to Infinity and Introduction to Higher Mathematics* (Apollo Edition, 1962, USA).

How India is stepping into the developmental processes of all contrasting orders of experiential processes, and of one country with another – with differences – how at the same time India can indicate further possibilities, by pointing out that besides its more or less ignorant masses, it has an intelligentsia which is numerically the largest intelligentsia in the world, when compared with France, England or America, taken separately.

But India is devising a scheme of progressive education for enlightening the India's mass, by a general sort of essential education. 'General understanding' sometimes is better than 'specialized learning,' where often man loses touch with common sense.

India knows how to enlighten the country's ignorant masses, with a speed, in a general way, giving some general education with a planned programme, by character-training and insightful yet relevant scientific-philosophic training, within a shorter time than other nations of the world, and with an economy.

India's method of teaching follows a path of encouragement in some essential and practical directions, by encouraging voluntary group-service. The spirit of learning by mutual help, spreads from village to village with a willing heart, utilizing one's people, by encouraging the masses, with the offer of some privileges only, without taking recourse to the modern method of expensive services. This is a spread of 'friendship' and indebtedness' that prevails between 'teachers' and 'students' that lasts through lives. India avoids the use of too much expensive technological gadgets as practised in affluent countries, which go by self-oriented form of education, with not much help from others.

Here in India, there are at least some places for non-technical ordinary good-hearted persons who are willing to serve others with a smiling face. Hence teaching and learning are interlinked cultural processes as human relationship. It excludes the path of the use of money with a considerable moderation, letting higher processes become operative in man. India demonstrates that 'learning' is possible with least expenditure of money.

In contrast, India teaches a way of 'self-sacrifice' or 'giving' whole-heartedly, as a human qualification. In these different aspects, this particular issue of the *India News*, is a very useful guide for enlightening the masses, through some of India's specialists, in science and technology, in matters of defence and trade, in aspects of insightful education, insightful foreign policy, and foreign relation, and through India's already cultural assets.

Knowledge extends from mathematics to varied aesthetic orders of experiential processes, where sciences are not cut off from vibrating cosmic-psychic aesthetic expressions, of ordinary persons, but become insightful guides, turning out adepts in art, architecture, dance, music, and in the writing of poetry. This kind of general education is for the growth of home industry, and of home education for different types of crafts and arts, pottery, sculpture et al. The ideal involved in these different aspects is for making man and woman more human and appreciative of other's qualities.

I have, therefore, studied some of the papers in this *News* edition, beginning with the paper entitled, 'India's Advances in High-Tech Areas' by Bal Krishna. His thesis lies in showing how India is stepping towards a steadily emerging technological superpower in an area of the world,' breaking big powers' monopoly in the fields of nuclear energy, space, biotechnology, and deep-sea mining.

Within 44 years of India's independence, with five-year plans India has outmoded the achievements of its 2000 years of past. This is of course due to present day advancements of the world as whole, in all spheres of science and technology.

But the credit goes to India's intelligentsia, and their competitive sense and capacity with *insight*. There is also some great contributions from some insightful mass-minds. India never thinks that knowledge emanates only from the 'intelligentsia' and not from the so-called ignorant masses, no one knows wherefrom real insightful knowledge emerges. India is a land that equally respects its teeming masses.

Through scientific disciplines, India is making rapid strides in the acquisition of mastering the spheres of atomic and nuclear energies, space, aeronautics, biotechnology, electronics, oceanography, and Antarctic expeditions, through its manpower as emerging from colleges, universities, centres or institutions of advanced studies, and as well from the far flung villages that surround emerging cities. Some evidence in both these directions will be given in this chapter progressively.

But one should also remain aware, learning in the highest form, in any area is not possible for all to acquire, but a general knowledge is always possible. General knowledge lies in common sense, and opening oneself to hearing about different experiences. This is a balanced attitude, avoiding one-sidedness. It is worth knowing in all educational processes.

These sets of people are mastering agriculture, as to transform it to the varied processes of commercial agriculture, and when risen to that level, can lead India to the process of a rapid progress in the production of steel, non-ferrous metals, cement, paper, chemicals, fertilizers, as well as heavy engineering and manufactured goods, including ship-building.

The construction of off-shore oil-drilling platforms and equipment, helps the growth of locomotive industries. In the context of ship building, Narottam Morarjee's initiative during the British period, was an achievement, and it records India's tradition of establishing connection with distant placed through trade and commerce through oceanic routes.

It is interesting to see that India, in the field of biotechnology, has become the first developing country to make use of science to solve many problems faced in medicine, agriculture and industry. Here the reference is certainly to the specialists, but some ordinary persons also, with moderate understanding, can pick up some essential elements of each of the subjects, and can volunteer services in different ways, with a willingness.

Using instruments of biotechnology, India has made significant advances in development of anti-fertility vaccine which is undergoing trials in the country and abroad. India's first biotechnology product – 'a pregnancy kit' – has already entered US market.

This I feel will stealthily stem the tide of injudicious population explosion, creating conflict, controversy, and Supreme court's ethical-religious problems, as in the United States, resulting from Anti-Abortion laws prevalent in different parts of the United States. It is interesting to note, that India with the presence of many religions, with some understanding of the depth-processes is not creating a moot problem for discussion, so that India can accept scientific means as the best avenue for self and social advancements. It is a kind of 'operation' with a 'silence,' and not through 'much of hot discussions.' India cherishes this 'exalted attitude', that often higher possibilities are opened up by following the path of insightful *silence* while pursuing the investigation in various pursuits from philosophical to scientific deliberations.

When the mind knows to open itself to the empty sphere where no specific knowledge has any ultimate place, there to block the continuity of onward process of innovative ideas of all areas – 'arise' or 'emerge' suddenly, from the myths to the sciences, and from the 'intelligentsia' and the 'masses' which are insightful. They have a place to be operative, as we shall see, is well indicated in this chapter. Indian culture has that open-out possibility. This book is broadly addressed to these processes.

India's accomplishments in the field of oceanography, include nine expeditions to Antarctica, and establishment of two permanent bases there and development of know-how for extraction of metals from polymetalic nodules.

India has been credited by the United Nations as the 'pioneer investors' for deep seabed mining which gives it exclusive right of exploring upto 1.5 lakh square km area in the central Indian ocean. In response to which the department of ocean development has a programme for carrying out deep-sea mining and extraction of metals from polymetalic nodules by 1995.

In still another field of space research, India is now set to take complex missions like building polar 'satellite launch vehicle' for remote sensing satellites et al.

The various scientific departments will progressively develop the process of interlinked possibilities, and will be operative in various fields of medicine, food grain production, drinking water, dairy farming, broadcasting, and improvement of communication specially in the aspect of spread of education.

Improving the sense of quality of life, India will also be ecologically conscious, and will pay adequate attention for retarding the process of degradation of natural resources, and water pollution from industrial effluents.

With an improved technological skill with new directions—the specialists will help the country for achieving the new goals and prepare India for the next century.

National Geographic Society monthly magazine for example (May 1992) also draws attention to 'India's Wildlife Dilemma,' for its possible loss of its unique natural heritage, due to the growing numbers of poverty stricken farmers competing for land.

While in another part of the world, in Tanzania's Ngorongoro Crater, a sizable lion population appears healthy, but is, in fact, severely inbred. This gives a chance to the researchers in animal biology, how this process of gradual extinction of the lion population can be prevented. This observation of the researchers will as well open the eyes of many researchers in the human sphere, where this phenomenon occurs in human situation, in India, among the Parsee community, or among the Eskimos in the artic regions of the different parts of the world (Cf. *National Geographic Society* Magazine, April 1992).

Knowledge, in its growth in many spheres closes the gaps between the human and the animal worlds, and between the poor farmers and the affluent businessmen of the different parts of the civilized world, raising biological, ecological and economic problems.

India is luckily open to many-sided learning in contrast to exclusive specialized learning of Western tradition. Very few Western thinkers are aware of this openness of attitude towards 'knowledge-process.'

A.N. Whitehead well-known Western mathematician and philosopher – in his *Adventures of Ideas* admits the place of 'togetherness' of knowing-process accepting the place of many 'logics.' This involves the admission of many-psychologies – and for that matter many-sided attitudes towards all spheres of knowing. This book, following this principle tries to suggest, and develop, in different ways how philosophical thinking, by cancelling the 'ultimateness,' of each of the aspects of experiential processes can help the process of 'interrelatedness' of all experiential processes – and how 'Sciences from Physics,' by abandoning any 'theory', as 'ultimate' – can merge into prospective – 'fields' beyond the contrasting characters 'particles' or 'waves.' It can run into 'fields' – and how again finally the politics, abandoning ideologies, can come together to solve all human problems in a concrete sense, looking into common human interests.

India can understand the possibilities for many 'logics,' 'philosophies', 'religions', 'psychologies,' 'physics' et al., stretching to many-dimensional experiential processes and possibilities. I have demonstrated this inter-linked process in many of my writings. My mission in life is linking up various experiential processes as confronted with the transcendent, which is nothing but an open field lying beyond all confronted understanding. I have indicated in my recent book, *The Mysteries of Vajrayana Buddhism from Atisha to Dalai Lama* that a lift to a transcendent *psychological process* can be possible by a detachment from, the *confronted psychological experiences*.

I may choose another paper by Anil K. Joseph, *Indigenous Infrastructure in Science and Technology* to indicate the country's heavy investments to build up an indigenous infrastructure and capabilities.

To improve progressively, many-sided capabilities in the area of science and technology – to obtain the high returns, in terms of three-million strong force of scientists and related manpower – over 1000 research and development laboratories in the public and private sectors, and over 150 institutions India is engaged in research under various scientific departments.

I am not just enumerating the numbers, but drawing attention to their achievements in revolutionizing the country's self-sufficiency in foodgrains, and with this other areas and aspects. India's primary attention is for sufficiency of food, and then concern for other things.

To facilitate increased crop production the Department of Agriculture has located several thrust areas with programmes for special foodgrain production of rice, wheat, maize, millets, pulses and oilseeds. The department, further has also instituted National Watershed Development Programme for Rainfed Agriculture. India's consumption of fertilizers has increased from 1950 to 1990 with a need for new fertilizer units, and the farmers are given assistance to purchase these, at economical prices, locally after a due test in laboratories spread all over the country.

The tractors have become popular as in the advanced countries, in the different states of India from the South to the North, or from the West to the East. The Institute of Budni is the main testing station for conducting performing test on tractors and power tillers. There is another Institute at Hissar, for testing smaller machines.

India accepts in general the advanced technological ways of the West in speeding up the growth in agricultural products, but recently there is a deviation towards an innovative process towards 'Organic Farming.'

The farming practices can be traditional, chemical or natural. These methods differ mainly on their effects on soil. In contrast to the first two methods, the third goes beyond organic farming in advocating a 'do-nothing' approach.

It is a kind of watching nature as doing all the hard work. It is contrary to modern approach, which may not be beneficial for human beings. This is an evidence from a small farm near Bangalore, where the use of ecology-supportive farming techniques are yielding amazing results.

The reference is here to L. Narayan Reddy's *farming technique* and his discovery after repeated experiments on *fragility of the soil system.* He knows through his experiments that 'the more you plough the soil, the more you break up soil particles and disturb microbial activity.' This is not taught at the schools. His is a discovery of a new method, without going to any school, but learning from his own experience in a village milieu, and family experience.

Instead of using chemicals and repeated ploughing, he increased the application of 'organic matter' (compost prepared with cowdung slur, farm wastes, even roadside leaf litter) while simultaneously cutting down on chemical fertilizers. It cuts down also costs in agricultural projects. India remains always alert to prevent calamities emerging from affluent countries and famine stricken countries of Africa.

M.V. Nadkarni, head of the ecology-economics unit of the Institute of Social & Economic Change, Bangalore, studied Reddy's case, for ten years, when Reddy completely stopped the use of the chemicals, but found the growth in the yield, with a rise in income by more than three times. Reddy also pointed out the reduction of water by taking recourse to 'drip irrigation,' cutting the need 30 hours of pumping water to only 6 hours covering the same area.

This case also indicates that a new innovative possibility may emerge in different areas of experience both from the insightful 'intelligentsia' and 'teeming mass minds.'

Readers may refer here to Ramakrishna Paramhamsa, Rabindranath Tagore, M.N. Roy, in the different areas – religion, literature, politics and others – to point out even without going to schools one is capable of opening up new avenues of insightful thinking.

Reddy's innovative contribution to the sphere of agriculture opens up some similar possibilities, in an important experiential aspect, and this innovative process can be stretched to the insightful uses of chemically treated fertilizers. Here Reddy also pointed out, by his direct experience for some years, that too much use of chemicals is not necessary, but they can be reduced reasonably with the knowledge of the soil conditions. But this optimist prospect depends on continuous activity in the direction of a chosen experiential direction.

Experiential processes are interconnected, possible innovative discoveries are always possible from inconceivable aspects. Possible emergence is a mysterious phenomenon – modern science can not eliminate it. This is also true of all other aspects of experience, where also there is a place for mysterious possibilities. The world is not totally determinate or determinable. But idleness is not fostered.

There is no doubt that India is taking varied modern shapes, and it is also pleasing to see, that in some busy streets of Varanasi there are still cows in front of one's house giving an ancient grace and recognition of animal life, and the tolerance for this animal on the streets where the car drivers show their concern for these roaming animals. Sometimes too quick changes are not tolerated by the masses or even by the intelligentsia.

India does not mind the progress of the world in a particular direction, but does not become intolerant of preserving something of ancient civilization. More than speed, there is some kind of reasonable understanding giving some space between processes. This may help cut the process of unnecessary restlessness or over-activity.

Reddy's triumph lies in giving also a new direction to 'ecology-supportive farming,' with a plan for reduction of the use of fertilizers and water. His achievement further was his practical experience, without schooling (cf. *India News*, 1-15 July 1992).

I am also tempted to mention about India's 'Novel Railway Project' for solving twin problems of space scarcity and lack of resources by the use of the air space over the stations and exploit the railway buildings and land commercially. Here the reference is to the 'rail-plan' to take some burden off Bombay, India's business capital, by developing a rapid mass transfort link with the satellite town of New Bombay.

This railway link is a part of a grand plan which, apart from the above project, have five other corridors, including one ring railway to serve commuters within Belapur area. The six corridors will cover a total of 157 km. Of the five stations Vashi and Belapur will have a commercial complex, and three others, will have a shopping arcade, to cater to the agricultural producer's requirements. This is the first time in the history of the Indian railways that air space above the station building is being exploited for commercial purposes.

There is also in this newspaper a reference to another achievement of India as it develops digital high definition receivers.

Bangalore's Indian scientists have achieved a breakthrough in the development of the world's first digital high definition (HD) TV receivers that can also be applied to home information systems such as videophones.

J.P. Raina coordinator of a Rs.180 million national project of the Department of Electronics to develop digital TV, home-information systems and HD-TV suggests that this technology will enable the Indian consumer electronic industry to bring out TV sets that will meet most of the world's TV standards through vastly improved audio and video quality. Raina further foresees the prospects of the future of the world's electronic community, in appraising this technology for bringing out a multi-standard digital TV. The demand for such marketable materials has already attracted attention of Germany and Japan.

This project was started in September 1990, and involves nine premier institutions – the IITs at Madras, Bombay, Delhi and Kharagpur, Bangalore, Trivandrum, and other places. Money-wise this project in Europe and America will cost more, so naturally India will be first to capture the world market in this field (with some changes, from *India News*, 1-15 July 1992).

All these references to India's innovative skills and also avoiding some kind of unnecessary *restlessness* for too quick changes suggests that India wants some changes but not too quick changes.

Coming back to my travels through Sarnath area I had this novel experience while proceeding towards Sarnath, which has become a place of pilgrimage, and a coveted place for the tourists from foreign countries and from different places of India. The Indians, who often want to revive their inspiring historical past, by frequent visits – perhaps have the same kindred feeling. In this tolerant attitude of matching new advancement with some thinking for the past India retains its ancient look without any show of primitiveness.

Animal husbandry received the much needed attention of the Government and other several sectors for increased milk production, and for cattle-breeding farms, supplying high-quality breeding bulls.

Scientists at the National Institute of Immunology in New Delhi have developed a single shot sterization injection to sterilize scrap bulls, without resulting in any loss of libido that would affect their work quality.

As agriculture cannot develop without an increase in a number of fields, the country is heading towards establishing a large number of centres of science and technology, in the areas of nuclear science, space technology, defence and health.

Besides these, the Indian Meteorological Department has established a large number of data collection centres, to help them forecast with reasonable accuracy the behaviour of Monsoon at least a month before the onset of the Monsoon season.

In the field of biotechnology, the research is on the development of the new vaccines for immunization programmes, with novel and simple approaches to birth-control.

The Indian scientists are also forging ahead in earth sciences and oceanography. At the National Institute of Oceanography in Goa in South-West India, facing Arabian Ocean – biologists are processing marine resources as possible of pharmaceutical drugs. Scientists also are developing techniques to help mine polymetalic nodules from the sea-bed.

Travelling to Goa, on my way to attend an international conference at

Osmania University, in 1984, I found that Goa has a future, even when compared with San Francisco. I think India retains its contrasting looks for centuries.

Along with these achievements, the Indian scientists, as mentioned earlier, have developed indigenous nuclear reactors, and have mastered all operations from the extraction and processing of uranium ore to the fabrication of fuel rods.

Radiation technology is used for development of radioisotopes for use in medical technology diagnosis and treatment, and also for the development of high-yielding crops. India is waiting for solar energy as a project in near future, to get beyond the limitation of water resources and shortage of gas and electricity.

In space technology India's capability to build and launch satellites has been developed, looking into the country's monetary capability and priority. The satellites are now used for telecommunications, radio and television broadcasting, and for meteorological purposes. India's remote sensing satellite is used for the management of the country's dwindling natural resources like forests. India's satellite missile project of Agni has created a concern in the mind of West Pakistan diplomats, and Russia to insist whether it was only meant for peaceful purposes, and not for enhancing war potential. In this project India believes in tasting a little of all experiences without draining itself out for one project as Russia had to learn by spending to the brim for proving itself a 'superpower' nation, and then finally losing its entitative importance.

India's operational schemes, as narrated above, may seem too much repetitive, but one must as well note that there is a basic life-process of the people in general that goes on – so men in general – have to observe patiently and then devise the way how to tackle a situation, and give necessary insightful suggestions. Viewed from this perspective, the mention of these innovative ways has some merits – it is pleasing to the Indian people in general, and as well to the international bodies – that India is not far behind in looking into its future progress maintaining its cultural outlook all-through. India's is a cautious move.

The electronic industry is now exporting several key components and computers. Indian software has grown to a position of strength and status in the international market of computer technology. So man has risen to progressive order of detached attitude.

All these are given to show that previously the primary attention was given to the subtlety in abstract philosophical thinking, but now the emphasis is in the concrete realistic processes of scientific and technological aspects of experiential order, as confronted processes. This is a detachment with progressive insight.

All these improved attitudes are still dependent on the *transcendental bases* of experiential process, but after India's independence, without ignoring the transcendental bases Indian mind has learned to proceed towards scientific and technological training of the mind, as a prior need before every other consideration, and therefore now the progress lies in the scientific and technological spheres, but often the Indian scientists turn to show that these advanced attitudes are pursuing the transcendental modes as well. The scientific pursuits have a relative value, as all confronted experiential modes.

This comprehensive realistic attitude, which is many-sided and non-exclusive of any experiential possibilities – is well evidenced in the researches of the renowned scientists from C.V. Raman to S. Chandrasekhar and others. The universe is continuous from physics and astrophysics, to aesthetics in its different aspects. Raman has indicated this process, in his many-dimensional scattering rays – from physics to musical vibrations – and Chandrasekhar has indicated this process from astrophysics to aesthetics, scientifically, and by a further extension as a continuous experiential phenomenon in his philosophical and literary writings. Life is not merely a pursuit in the advanced aspects – it seeks also a descent downwards for giving some opportunities for the enjoyment of life in varied aesthetic directions – with a descent to the masses.

In his paper 'Space Research and Satellite Communication Systems', Biman Basu mentions about the satellite network as set up pervades all over India. He here also speaks of the varied conveniences, its reachability through telecommunication system to any part of India and the world at large. This kind of prospect is satisfying from many aspects of giving inspirations to life-process in many directions.

He also refers to additional conveniences through public facsimile (FAX) transmission facilities. Here, in this context, the satellite network has revolutionized business communication, and at the last election time it has speeded up parliamentary election process, linking up different parts of India, and relaying the news all over the world, in a moment.

India, through such facilities can live with an awareness mechanism, as of 'one unit' operative with a control over the masses, with no threat of a process of mass-disconnection.

Previously India solely depended on Yogic and Tantric visions of some limited sections, for foreknowledge, or for knowing things at a distance, depending on persons who rightly or wrongly claimed that kind of foreknowledge.

But now this kind of knowledge through scientific media has become a common social property to be shared evenly without having to rest on some faith in those select persons.

Here there is no place for the non-believers to argue for nothing, it has become a unversally distributable commodity of knowledge, as one finds, in conformity with this scientific and technological age, where distance can be conquered within a short time through telephone, air travels, or other means. Switching on to another kind of experiential process is always judicious, rather than trying to extract all kinds of experiential processes – from only one form of experience – philosophy, religion, science or politics. None of these pursuits are ultimate, nor are they reservoirs of everything. They have values in a relativist sense – the ground for which is provided by the transcendent process and possibility, that runs commonly in all spheres of experiential processes, when none of the experiential processes are mistakenly taken as *ultimate*.

The talk of philosophy is one thing, and actually realizing this process – in concrete experiential situation – is another.

India has now stepped into this new age of scientific and technological advancements, but as well with insights for all other aspects of experiential processes in interrelation.

A new class of scientific Yogins have also emerged in the present century, applying India's meditative skills to many scientific studies, so that 'Yogic exercises' have become a piece of scientific experience – in some kind of concrete vibration in the cerebral processes. India is not also lagging behind in indicating some similar demonstrable programmes. Practice of detachment is now a progressive science and technology.

These novel inventive capabilities are practised in America, fostering some scientific possibilities involved in Yogic discoveries, as one may find popularized by Mahesh Yogi. He has some followers among the scientists, from the mathematicians to the physicists, biologists and psychologists.

Yoga and Tantra with powers – to have an access to all nerve centres – are not merely helpful for mastering one's body from within for keeping it healthy and powerful, but in certain periods of India's history, they are both ancient and modern.

Yoga, with Tantric meditational processes can even have control over the elements of Nature, more than science. One may here refer to my book *The Mysteries of Vajrayana Buddhism from Atisha to Dalai Lạma*. Here the reference is to the Sakyapa schools and other adepts of Yogic and Tantric disciplines in controlling the elements of Nature, and human organism. A new creative science is soon to emerge, besides the usual science with the knowledge of the elements and principles – now they can be controlled in yogic-Tantric ways.

In these days of scientific developments one may as well take note of Sakuntala Devi – an Indian lady who has a speed in calculation faster than a calculator. She visited Hong Kong and also several universities in America and demonstrated her ability which is a challenge to science, or some insightful scientist might discover certain avenues for some future research in some aspect of science, in minutely observing the insightful activities of some extraordinary personalities. Perhaps this capability is due to one's intensive concentration, which generates *power.*

From my detailed study of India's cultural history I have given some reasons to suggest that population, as such, is no longer a curse, but a boon, if people are scientifically trained with insight, maintaining in tact India's cultural process, to serve mankind with the best of ideas, and with intense concentration.

Though it might sound as injudicious to speak for population growth, but no one can ignore the future possibilities of Tagore-family of India which provided a galaxy of geniuses in different fields of learning, in aesthetic and literary spheres. It excelled in concentration.

Rabindranath Tagore became a Nobel laureate in the sphere of poetry, with displays of insightful writings in many literary and artistic directions. He did not take advantage of school or college education, but followed his own ways in acquiring knowledge in different areas of literature, drama and art – with deep concentration.

It is at least a good example to stop arguments against population growth, or regular ways of learning, from a *categorical perspective.* Such arguments are no longer absolute statements, but stays as a *probable argument*, with some value, but lacking the spirit of concentration.

Human mind has some specialities of its own, as often found among many scholars of economics, sociology, politics, or any scientists, opening up new avenues of inventions, creating an atmosphere not totally against population as such. Worthy persons of any sort are always appreciated.

Pursuing through a previous copy of *India News* dated 27 July 1987 – I found a review of Deviprasad Chattopadhyaya's book, *History of Science and Technology in Ancient India: The Beginnings*, Foreword by Joseph Needham (Firma KLM Ltd, Calcutta), where there is a reference to his early work entitled, *Lokayat*, or Indian Materialism, published in Bengali and English during the years 1957 and 1959, from West-East perspective, and other subsequent works.

In this connection, in his recent work with the title, *Studies in the History of Science in India* in two volumes (Editorial Enterprises, New Delhi, 1982), he rightly adds to the value of India's transcendentally based subject-matters, with a materialist stress as the need of the present day scientifically and technologically oriented world. Here one can see a balanced attitude displayed even if one is 'transcendentally oriented.' It is not an offence to be transcendentally oriented, for otherwise, how one can be disciplined, without becoming dogmatic? Transcendental outlook opens up fresh avenues for reflection, detaching oneself from abstract formalist thinking, and also from the confusions between the inspiring and basic transcendental possibility – and the confronted presented experiential process of varied types.

In every sphere of experiences there is an aspect of dogmatic reasoning, whenever a person fosters too much of arguments from a presented restricted conceptual or empirical order or from transcendental aspect, which is not presented but insightfully experienced.

This work tries to develop a balanced attitude – indicating interests for all aspects of experiential processes in interrelation, keeping apart the transcendent vision which is a different order other than the presented.

Here one can refer to several of the Indian scientists like Jagadish Bose, P.C. Ray, Bajendralal Seal, M.N. Saha, C.V. Raman, Homi Bhabha and philosophers of the type S.N. Dasgupta, S. Radhakrishnan, and others, of India before and after India's independence, to show how they were operative *confronting* the concrete experiential processes, with a transcendent outlook and possibility.

All these thinkers understood that all confronted experiential processes – philosophic or scientific, are in togetherness – a scientist intruding into philosophy, and a philosopher at least appreciating scientific thinking as not exclusive.

There is always a need for science and technology for speeding up the social and economic status of man. Their's were pro-philosophic and pro-scientific attitudes in togetherness. They are detached in their pursuits.

I would like to add more to this list to refer to the emergent younger generations, who follow the path of these predecessors, whenever necessary.

The reference is specially to India's innovative aspects of deliberations in the spheres of astronomy and mathematics. In both of these scientific areas there is a place for 'Zero', decimals, algebric symbols, trignonometry, differential calculus, et al., as always operative from a transcendent sphere in different ways demonstrating how to connect the confronted finites, with the possible infinites, in varied symbolic ways. 'Incas' of South America developed thinking through the strands of knotted yarns, called 'Quipus', which helped symbolical thinking, and recording, like zero used in India.

This novel two-dimension mode of confrontation of experiential processes outmoded the early Greeco-Roman geometry, art, architecture, pottery, textiles, seals, metal technology, chemistry, and progressive urbanization – by a reference to extensive archaeological excavations. But here the range of discovery or invention is not limited to any special order of people – intelligentsia or mass. They emerge from all orders of mankind. The gulfs which existed among the sciences and men are now gradually vanishing.

The excavation process in archeology, in different parts of the world can be extended to probing into geological investigation in the different regions of India in connection with its several rivers – Ganges, Mahanadi, Godavari, Narmada, Cauvery and Krishna valleys, – providing clues, to the various stages of progressive urbanization processes of the past in India. There is here also an attention paid to the importance of the teamwork in research.

Chattopadhyaya's work in scientific directions was, also a recognition of the contributions of this group, as a group-ethics, necessary to carry forward the spirit of cooperative works.

Through my recent book, with the title, *Zero: Its Role and Prospects* as mentioned earlier I have fully demonstrated the possibility of the advanced processes of the development of sciences and technologies through the different periods of India's history from the basis of a transcendental attitude.

In the confronted presented universe, there is always a growing need for researches and utilizing the results through scientific and technological means to change India in varied aspects as to maintain its independent status among the nations of the world.

Here what is in demand for man is progressive freedom from economic strains, which create tension and strain of the mind and body, creating unhealthy condition of both body and mind.

In the case of the superpowers they are always concerned for maintaining their international status, through progressive military potentials which ultimately create another kind of tension between the superpowers.

So both the developing nations and the superior nations remain always with some pressures or apprehensions. Their fallacy lies in remaining confined to the presented universe, they do not direct their attention to the transcendent insightful aspect of experiential process.

India's frequent assurances – of pursuing an ideal of nonalignment have a general effect, but this transcendental gesture of India, does not have any appeal to its neighbouring border nations like China and Pakistan – they often create disturbing situations by invasion or by creation of some tension. To guard against a possible attack from these foreign nations, India, has to make its self-defence – stronger by progressive scientific and technological ways.

Evidently, India's political strategy is of a reasonable precautionary measure for self-defence, though its core-principle continues to hold fast to its objective of non-violence.

India's choice for nuclear efficiently is for maintaining its civilization, with a potentiality for progressive social change of India and its people through insightful education – uniting intelligence with wisdom.

Continuing the above progressive attitudes in the scientific and technological directions, further, I shall be referring to India's 'Battle for Protection and Preservation of Environment,' by an article of D.R. Rajagopal, with the above title.

In this connection I shall draw attention to its several gigantic programmes, connected with the projects of (a) cleansing the waters of the Ganges, flowing from the Himalayas to the Bay of Bengal, and (b) construction of the two dams across the Narmada river.

The Narmada project was in the name of (i) Sardar Vallabhai Patel, India's first Deputy Prime Minister. It will be executed in two regions with the titles: (i)Sardar Sarovar and(ii) Narmada Sagar. Both were colossal reservoirs, to be built upstream in the state of Madhya Pradesh, with a further extended project to the river Cauvery, in the South Western region of India.

Besides these two colossal dams, 30 large dams, 135 medium-sized dams and 3000 minor dams are to be built across the Narmada river, guiding the water into thousands of kilometre of irrigation canals. In this project there is also a display of scientific-technological skill of controlling the water discharges to only 16 per cent of the river water into the Arabian sea. What a colossal scientific-engineering project with restraint and understanding the value of water, the effort to use it to its utmost, and not letting out water to be drained into the sea.

Here service to the growing population is also implied, with a feeling for the population of different provinces, living on the banks, including the revolutionary changes of the tribal people, and their civilization, directing them to the modernizing process. In these advanced scientific and technological processes – India's main ideas are for creating Indian mind as one United Whole, living with happiness and peace.

The third project is with regard to the Tehri dam project in Uttar Pradesh, this has been opposed for the short life span of only 62 years because of rapid siltation. Here one can see India's progressive future vision.

The people of the Himalayan villages are agitating non-violently in the peaceful persuasive Gandhian style for a drastic revision of Forest Settlement, leading to the Chipko movement, to protect the forests against the axe-men destroying them. It is also for preventing deadly forest fire.

The speciality of this scientific-technological scheme is for changing the economic prospect of India, but it runs into social problem of convenience and inconvenience, or for or against some people dwelling in those regions displacing to a new region involving migration.

Here there would emerge the leadership of different regional people, tribal people, youth movement or women movements causing problems for the local ministers in different ways.

These socially oriented scientific-technological schemes will invite the scientist and engineers for debates.

The foreigners watching India's improvements will either praise the Indian Government for its achievements or the jealous foreign powers will try to disturb the prospective future prospects of India. There will be others occupied with the problems of India's Nature-oriented maintenance of environment.

Human existence is a complex social affair – a man exists for others, and also if others allow one to exist, in a particular way. A man's existence also depends on Nature's generous feeling for man or any of the biological species.

Another modernizing process lies in progressive 'March Towards Total Literacy' as indicated earlier in a different context. Now this subject will be faced in the context of a scheme emerging from R.C. Rajamani's paper, with the above title. It has a claim of an innovative method.

This or other kinds of alternative projects have been mentioned before. Here is another alternative project, to spread literacy among the totally ignorant masses, from village to village and from state to state, by teaching people to reach a certain standard of learning.

Here a variety of themes like the small family norm, health and hygiene, immunization, women equality, and civic duties, according to age-groups, through voluntary teachers, students, scouts, guides, voluntary agencies, trade unions, political parties, private and public sector agencies or club houses (*Jana Sikhsan Nilayams*) have been emphasized to undertake literary promotion and teaching, to eliminate huge expenditure for teaching.

These voluntary teachers, with a freedom were inspired by ideals or obligations suggesting that nothing is more important than education, but this ideal has to be understood from a perspective of a compassionate attitude. Educating others is also a *gift* to people, which can never be paid back.

It is the chief of the developmental processes for independent India, hence these volunteers do not receive any financial assistance, but only appreciation, whereas the learners are provided with the materials for learning, as slates, pencils, notebooks, blackboards, and that also through local sponsorship and voluntary contributors. It is a psychological project – on both sides from the authorities and the teachers, and also with the students, with their assumed willingness to learn, and giving back their learning to others when they become teachers.

A reference to all these are mentioned to clarify the speciality of India's progressive social changes, with a peaceful revolution, suiting the Indian mind and the limited economical capability of the Indian Government and its people – rich and poor.

At present, mass campaigns for total literacy are underway in 36 districts in the states of Andhra Pradesh, Bihar, Karnataka, Madhya Pradesh, Maharashtra, Uttar Pradesh and West Bengal.

Earlier, Kerala served as an ideal as having a total success, and Tamil Nadu coming next to this. This scheme of a tolerable social life is a great demand for mankind in general otherwise, if the masses are left to themselves – there is a possibility of increasing number of persisting with ignorance which will foster the process of criminals.

This is also a method of keeping men occupied. It is also a kind of 'give and take' process, where simple way of life, with a consideration for others continues.

India is fostering this new idealist project to carry out as everyone's life treated as a 'mission.' Without some 'idealist-compassionate attitude' a culture cannot be fostered not will it have any 'value.'

This kind of generation-process of values stems out from developing transcendental consciousness, which is a development of human psychology, in post-intellectual order. It is riveted to a disciplined meditative direction without advocating or fostering any process towards an *ism* or *cult*.

This process of educating the masses was first initiated by Gandhi and his followers, and later also from Nehru to the present Prime Ministers.

Vinoba Bhave had also a policy of receiving the unused land from the landowners to be distributed to the landless needy cultivators to cultivate, and thereby changing the social and economic shape of India's masses, and solving the political problems emerging out of social clashes between the affluent and the 'have-nots,' with a new kind of socio-political movement, of giving one's surplus to the needy or the deprived, with a good heart – between the giver and the taker.

India is aware of Pakistan's continuous development of weapon-oriented nuclear programmes. For its ballistic missiles and technology – from Chinese assistance – India's Defence Ministry has to remain always on the alert, and is compelled to take necessary measures to safeguard the country's security, and also making efforts to improve relations with Pakistan. Though India nourishes a wish for an improved socio-economic political process, it remains with a concern.

India's present Prime Minister, P.V. Narasimha Rao, welcomed on 2 August 1992, the Treaty on Strategic Arms Reduction singned by the United States and the Soviet Union, which raised the prospects of 'a world free from nuclear weapons,' and hopes for further arms reduction, preparing the path of progressive aspects of world, free from wars.

India and the Soviet Union on 8 August 1991, announced extension of historic Indo-Soviet Treaty for Peace, Friendship and Cooperation for a further period of 20 years. The two countries will take up publications of all Indo-Soviet documents of historic importance since Indian independence, and also decided to invite dance groups, and present exhibitions of royal Indian costumes in the Soviet Union.

This kind of development of political relationship between countries, has to be progressively developed for cultural relations, which will foster real humanist attitude.

In this complex international background, this writer will develop India's comprehensive and changing parliamentary democracy from different prime ministers in succession, after India's independence in 1947, from Nehru to later periods, in Part III of this Volume, by pursuing some principal ideas of the book entitled, *The Government and Politics of India* by W.H. Morris-Jones. He is concerned with the development of India's capability of framing its special Constitution, in transition from the British Constitution, as a non-violent process of transfer, in its dual-dynamic processes, between the Government and the National Movements, both before and after Indian independence.

As the formation of the Constitution of a country, after colonial control and sovereignty, presupposes a lot of changes fitting the requirements of a newly emergent independent country, with provisions for international developmental process through judicious plannings and skillful foreign policies with the changing international processes – India had to start with a prior importance given to its defence to preserve its independence. Hence with these preliminary statements, the writer will wait for further detailed statements to the Part III of this volume.

It should be mentioned in passing, that India will always remain awakened to Barbara Ward's economic progress of any country – as all scientifically and technologically oriented – and has an unimaginable speed, of progress, in contrast to the past histories of mankind, in general.

This is also true of India, as a developing country operative with science and technology in progress. But this possibility of progress will emerge, with a stability, in a real world-wide pattern, only if man develops a non-violent attitude, and appreciates the contributions of all countries in general – small or big, minor or super, with a proper use of nuclear powers or potentialities.

India has accepted that peaceful life of co-existence with the people of the world as a whole, in the spirit of give and take way. The attitude of non-violence with a precautionary scientific-technological defence is a judicious scheme with an economic control of money, and saving human interest from multifarious perspectives.

In this chapter it has been clarified that both philosophy and science have been completely freed from the clutches of all intellectual forms, which often create situations of conflicts, with no insightful solutions.

This transcendent attitude – in both philosophical and scientific deliberations – has freed itself from any abstract speculation in these spheres; hence the rise is in both to a transcendent order of infinite possibilities.

The writer, therefore, claims that India's advanced technological developments – in both philosophic and scientitic spheres, so freed – have paved the path of technological developments, in a special direction, absolutely freed from abstractions to meet the concrete needs of the people at large.

The political process, too, in the aspects of process and possibility, therefore, in India, will be in continuous *insightful directions* serving the people of India for which they need proper education in these advanced aspects. The chapters of the Part III of this book have been harnessed in these directions.

[illegible] convenient.

[illegible]

[illegible]

[illegible]

[illegible]

[illegible]

[illegible]

[illegible]

PART III

Politics as a Dynamic Phoenix

CHAPTER VI

India's Democratic Constitution from an Intercultural Perspective

To shape my 'Triadic Avenues of India's Cultural Prospects—Philosophy, Physics and Politics'—I would very much like to use some ideas of John H. Herz's long Introduction to his book, with the title, *International Politics in the Atomic Age*[1] (Columbia University Press, New York, 1959).

The speciality of this volume lies in presenting Herz's ideas as he wants to indicate that as a 'free scholar,' he was an independence, in contrast to a team-work or of a collective study. It is a shaping of the author's own individual view representing the 'indefiniteness and the newness' of the atomic and post-atomic age, when military and technological powers in international politics have become a suicidal adventure.

It has given a tremendous power to a superpower – a nation or a person, who has an extraordinary capacity to deal with the human affairs, with a policy or with an inconceivable challenge. Such a personality often takes advantage of a success. He often lacks the spirit of consideration or some kind of compassionate disposition and hence the failure of the entire process.

In this connection, I would like to bring in a further contrasting experiential process to test the truth of some relevant quotations from Richard Attenborough's work, entitled, *The Words of Gandhi*[2] (Newmarket Press, Navajivan Trust, Bombay, 1982). I have found out that this post-atomic world-situation leads one to a powerful psychological era, in a fresh *transcendental order*. The adherence to mere conceptual and empirical processes prolongs the repetitive historical process. The world is crying for a new order.

There is a need for changing the present confronted order by *transcending* it, so that a new consciousness will emerge. India has already a well-established transcendent consciousness of giving a new basis to the presented universe, by saying that what is confronted or seen is not a real or direct vision, but a wrong vision as emergent from an ever-perishing renascent process and possibility.

Hence what is needed is a detachment from the confronted order, and an adherence to an ever-fresh self-shining transcendent background sheltering infinitely variant processes and possibilities – perishing and emerging. We live in a symbolical universe of possibilities, and not a causally oriented perishing universe. So there is no place for causality in philosophic or scientific sphere. The world is symbolically oriented from no specific subjective or objective perspective. All confronted perspectives are truncated phenomena deficiently actual seeking for completion in a progressive possible sphere with no beginning or end. India's transcendental presupposition is dually expressed as below.

This has been symbolized in one way in the Vedas, and in another way in the Upanisads, if one chooses to interpret the confronted experiential process from India's transcendent perspective and background. There are other possible ways as one traverses through India's history of the continuous development of transcendent possibilities. The transcendent possibility observes but is never observed. It reverses the confronted, but its 'selfshiningness' is never 'reversed.' All these confronted experiential possibilities are well indicated in the progressive cancellation of the confronted intellectual forms in Buddhist thought, with another order of self-shining transcendent processes of Vedantic expressions from Sankara onwards to many-sided pathways, leading to the modern era.

The adventure is neither for a superpowered nation, nor for a superpowered person, but one has to proceed with a non-egoist attitude, and remain operative with a total silence, before which there is nothing to confront, not even one's self – either as transcendent or existent. Without this kind of pursuit there will be repeated fallacies as confusing the transcendent with some thing as presented. The transcendent is the ever-enlightening spirit of all which are confronted and replaced by another confrontation. The world as confronted is not categorically present, but presented as an ever-fresh possibility – with an 'if-then' in a symbolical sense.

This kind of attitude as presented here can give some possible answer to Herz's problems in present-day international situation.

In the Western World, too, there is a similar movement, under the advanced philosophic outlook of Whitehead. He wants to bring in the *trans-intellectual* 'Wisdom Mode,' interpreting intellectuality as an emergence, in the background of a non-sensuous process of Nature, which is a vast sphere of connectedness or a feeling process, but it carries forward the intellectual possibility by a further transcendence. This post-intellectual transcendent process and possibility is towards a continuous insightful process of perishing and emergence, with no specific presented character, as subject or object. It is a progressive non-dual continuity or intensification of the insightful process.

This is Whitehead's post-intellectual 'Wisdom Mode,' where there is neither intellectual restlessness nor intellectual restrictedness or abstraction, but a perennial possibility, with a freedom from intellectuality. It is a transcendent situation, where adventure of abstractive' process has ceased to exist constantly replacing the presented order by an order of perennial peace. This advanced Western attitude, in confronting the dual confronted situation – confronting the intellectually patterned situation with a possibility of transcendence to a post-intellectual Wisdom-Process is a coming in line with the many-dimensional possible spheres of experiential possibilities. This also can present some alternative answers to Herz's problems in international situation as presented earlier.

This is nothing but feeling a need for a transcendental attitude – non-different from the Buddhist search for a 'continuous Wisdom Mode' or 'meditative attitude of Silence or Peace' – a continuous energetic continuum, akin to Buddhist Asvaghosa's series of meditative possibilities, or alternatively Nagarjuna's non-intellectual possibilities, eliminating all intellectual presented forms by the weapon of 'neither-nor' process, offering no conceivable choice.

This entire Indian transcendental development of psychology, is accepted by Whitehead, by a reference to Buddhism, without diving deep into Buddhist process. But this is certainly a core of Buddhist meditative attitude – cessation of *suffering*, resulting out of abstractive ratiocination.

This Buddhist attitude along with the Whiteheadian attitude, noted above, as 'non-different' – I am showing through my writings on both, since 1968.

The same attitude is conspicuously present in India's Jaina mode of life, which dominated the life and mind of Gandhi of the present century – though Gandhi was not a Jaina, but wholly influenced by the Jaina spirit of progressive detachment, and a life of transcendental 'aloneness.'

In simple, it will be implicitly and explicitly shown, that now the advanced world-process is a study in contrast – in the advanced aspects of Indian and Western thought.

It has come to a stage where the intellectuals both of the philosophic and scientific spheres have realized a transcendent freedom from all abstract intellectual forms.

They have welcomed a new period of human adventure freed from all ideologies and are trying to reflect about this universe from the *transcendent perspective* of process and possibility, fully convinced that nothing is determinable categorically as the confronted universe – subjectively or objectively. It is an 'if-then' process and possibility.

Insightful persons of the world as a whole are no longer categorically oriented. They have stepped into a new historical order of infinite alternative possibilities with indefiniteness. Here advanced philosophical deliberations are on par with the scientific researches from physics to astrophysics (cf. *Socrates to Sartre* by Samuel Enoch Slumpf, McGraw-Hill Book Company, New York, 1966 – is illustrative of this historical process.

The emergence of anything confronted may or may not emerge, or might not have emerged, in any expectant definite form.

In the context of Gandhi and Whitehead the process towards the transcendent pursuit is applied to tap the complexities of international politics, in the atomic and post-atomic age, as will be variously and symbolically presented in select ways by Herz, in his above book.

Placing some of the expressions towards the end of the Introduction of the above work of Herz, I would like to quote the following: 'Thus in view of the various trends and developments in technology and in the power sphere, concepts, approaches, and policies are bound to be bifurcated or trifurcated. This accounts for the indefiniteness of the age and the uncertainties of the future. The uncertainties are of two different kinds: they may be connected with our lack of knowing what turns develop themselves will take, for instance, what kind of intercontinental missile or space weapon will have what kind of military effect; or they are caused by the fact that even, on the basis of certain conditions and the availability of such-and-such military equipment, objectives, intentions, and policies – something may differ and thus are not clearly predictable. No body knows, for instance, whether certain 'all-out' weapons – will be used or not. It is above these latter uncertainty which renders all calculations and policy planning so enormously complicated. Even if one knows his own mind, he cannot fully know the mind of his opponent. If we take the present (or impending) nuclear stalemate between the two superpowers as an illustration, there are so many possibilities for anticipating events that what actually will happen depends largely on one or the other sides 'ideas of what is and will be.'

In simple, if there is no certainty in the presented observational field of experiential process – then what is certain is the direct unchangeable dimension of transcendent process and possibility, where there is only a place for confrontation and observing its reversibility. By a contrast, in the direct transcendent experience, there is multidimensional process and possibility, without any modification, or change of dimension. It is a continuous concentration or intensity of experiential continuity.

This kind of thinking can be fostered in this world-process today, if still a place is given to a superpower to be operative, for, with the dawn of the two transcendent trends of Whitehead and Gandhi – a process of post-intellectual transcendent mode in terms of indefiniteness and possibilities – is becoming conspicuous, going beyond the restrictive or dogmatic claims of the superpowers, who are wrongly indulging in mere restricted 'intellectually oriented thinking'.

The previous superpowers, lived with a simplicist world of their making, now it has gone beyond that stage to *an age of indefiniteness*, with multifarious approximations. Instead of a useless infinity of calculations – it is always judicious to wait for the possible ways of confrontations as in man's waiting for any earthquake possibilities, in earthquake areas.

Reviewing the entire historical processes as referred to by Herz, in this Introduction he observes a series of events. After the victory of the First World

War, through the gunpower of the victors, and of the Second World War, through the discharge of atom bombs, there is a period leading to consequent periods of cold war between the two superpowers – Russia and America – the world had faced a gradual proliferation of nuclear materials, in China and other countries.

One could see with insight by way of contrast—some concrete values in the statements of Gandhi, so well-shown in the flaps by Attenborough, and also within the body of his work *The Works of Gandhi*, mentioned before.

The drift of the flaps runs as follows: 'In this age of wonders no one will say that a thing or an idea is worthless because it is new. To say it impossible because it is difficult, is again not in consonance with the spirit of the age. Things undreamt of are daily being seen, the impossible is ever becoming possible. We are constantly astonished these days at the amazing discoveries in the field of violence. But I maintain that far more undreamt of and seemingly impossible discoveries will be made in the field of non-violence.' Our life is helplessly placed between two contrasts – violence and non-violence.

In this incalculable, uncertain world of today perhaps the alternative possibility of non-violence as an international policy can be tried. This non-violent attitude has to be improved upon by a continuous contemplation with an aloneness, to a new unrestricted discovery of a comprehensive experiential possibility – in the dual ways of the Buddhist savants stretched to the present and Whitehead, replacing the past Western-possibility and science by the flashes of his insightful possibilities as mentioned.

With this internationally oriented philosophic-scientific attitude, one will have to devise possible precautionary measures with scientific-technological protective skills.

Here, in this choice – one will have a chance to find the insightful *transcendental psychology* of Gandhi, unfolded in dynamic ways in five perspectives of life – Daily Life, Cooperation, Non-violence, Faith and Peace. This implies an educational scheme for mankind as a whole.

All these meditative ways of life and expectations, of the future world are dynamically interrelated processes of pursuits of an individual's life towards Truth, inextricably bound with something beyond the reach of calculated ways of intellectuality.

This is an ascent to a post-intellectual order of peace and non-violence – in the line of Whitehead also – to seek a religious experience from one's inner self (in one's depths of solitude), to a God of one's heart, faithfully nourished with faith and placidity, ready to turn an opponent, to one's dear, alter or complementary Self.

This humanist attitude, as emergent – can be exercised – as a process of activity, in the post-superpower stage of the world. Its transcendent vibration will be akin to the dancing songs of Tagore, expressed in his Nobel prize winning poetic work – *Gitanjali.*

But all these idealist possibilities are realizable, if mankind chooses to accept the *peaceful* or *non-violent* attitude, otherwise such expectations are futile assertions and expectations – a wild goose chase.

But if one thinks back to Radhakrishnan's understanding of democracy in politics one must know that real democratic political process emerges in man when one passes into the depths of his silence, and can dwell on the fundamental reflective aspects of the core of all experiential processes – from philosophy and religion, to science and aesthetics to the refined aspects of the human soul, when all are transformed to appreciate each other's good and need.

This philosophic-scientific experience, as a possibility has been realized not just by one person, but by a series of personalities, religiously and philosophically oriented, mentioned above – Gandhi, Whitehead, Tagore, Radhakrishnan, and others—and among the scientists – as Einstein, who wondered how could there be a Gandhi in this nuclear age, and other post-Einsteinian scientists also both Western and Indian – Heisenberg, Schrodinger and Dirac and Raman, J.C. Bose, P.C. Roy, S. Chandrasekhar and others got rid of restricted theoretical modes.

India's Constitution with a deep insight from all these changing perspectives, ranging from philosophy to the sciences, expresses a difference from the British and American forms and tries to realize a complex process and possibility – by checks and balances – and stepping into post-intellectual insightful process of distributing *Powers* to the President, Prime Minister, Two Houses, Speaker, Provincial Governors, Chief Ministers and down to the *people* at last, in a continuous relativity, from one to the other, as if between related pairs. The 'life' as lifted to the order of politics, from all these dynamic indefinitely dispersed political process is not a matter of 'simplified situation' to be dealt with by a single-personality, it has to be continuously transferred to a series of 'insightful' dedicated and detached personalities. India symbolically represents such a possibility through its symbolic Constitution.

We are in a different age of human history of uncertainties and possibilities where 'might' had been tried, nuclear destructive power had been experienced – so the world is somewhat inclined to see once again some possibility of a benign attitude through people's government, in international contexts. India seems to provide that possibility – in a different direction.

Gandhi's above words, in their inner senses expressed in varied situations, were not only theoretically true, but also were practically true, when the British, with the strength of more than hundred years of powerful rule of India had chosen to quit India with a friendly gesture, for, India chose a new pathway, not challenging 'violence with violence' but with 'non-violence', which puzzled the British mind.

The British, were made progressively civilized, coming in contact with India's extraordinary personalities including Gandhi, Tagore, Nehru and Radhakrishnan among others.

The British administrators felt powerless to be operative, violently, so they became operative in the political sphere with a new diplomacy, for a kind of psychological change, thus establishing an unforeseeable historical revolution in terms of a 'peace movement.' Power is no longer seen in 'might,' but in terms of some expressions of amicable settlement of problems, from a broad humanist perspective. Man has been raised to a height of experiential sphere, with some flickers of helplessly oriented hopes.

Exhibition of *might* is a demonstration of defeat and frustation – with no solution, but smeared with nightmares.

This change of attitude of a great nation of the world, is a hallmark of World History, to remain, so long as the civilized world will persist. I have shown in a later portion of this work how the process of a Constitutional transference of power to some extraordinary personalities of India, viz, Nehru, Patel and others was possible, for Lord Mountbatten, though he came with a suggestion of a 'Union of India,' in which the alternative plan for partition of India, was not excluded, so that the latter could be offered, in critical situations. The possible emergent political situations, for their infinitude of indefinite uncertainties cannot be solved so easily. So events happened that way.

Confronting the complexity of India's situation Mountbatten became operative placidly, by negotiating with the leaders. He pointed out during the interim process, with the best of his intentions, and with a placidity, several alternatives to choose—which ranged from the offer of a 'dominion status' to a Commonwealth Process, under the British Crown, and finally also to a full independence of the Indian Republic with a similar operational capability of its other part—Pakistan.

The processes as mentioned were transitional stages, which both the British administrators and the Indian parliamentary oriented personalities could visualize as peaceful transitional processes that could be entertained for a historical change of India's political history, under the circumstances, as then emerged.

Transition through 'parliamentary process of succession,' in countries like Great Britain, France and West Germany was insightfully studied, and this author found that the transference of power by Great Britain to India was comparatively smooth – when compared with the succession-processes of the democratic varieties of France and West Germany or the different 'Communist forms' of Government as prevalent in Russia, and in other parts, of the world – China, North Korea or North Vietnam.

The reference is here to the book *Cases in Comparative Politics*, edited by James B. Christoph, to understand the speciality of India's Constitution by a contrast after the transfer process. But none of these writers on Constitution has drawn attention to India's disciplined transcendental attitude of experience as a 'confronted process' in the background of a transcendental order as prior or even posterior to the confronted order – as a 'continuous process'.

Before the transfer of 'parliamentary form' of Government by the British to India – in India, there were two radical controversies going on among the followers of: (i) M.N. Roy, Jayaprakash Narayan and Subash Chandra Bose and (ii) Gandhi and (iii) Gandhi of the his later days when he agreed with Jayaprakash Narayan.

M.N. Roy after his close experience with Lenin, Trotsky, and Stalin was no more inclined towards a solution in the communist ways, but, for his own safety, his choice was for a mysterious escape from Russia to Mexico, and a return to India finally with a Muslim name.

After all these varied experiences of Russian communism Roy held the view that Russia's communist movement was full of uncertainties, and risky for any person, or for, any kind of dedicated service of a person for another person.

So, from his point of view, this form of Government, with uncertainties as mentioned will not be effective for India, and for Russia also, in the long run.

It will be a transitory process, as Roy thought, so long as Russia remained dependent on 'European capitalist system,' without taking note of the colonial people and their prospective change of attitudes.

Roy had a foresight of a possible market economics of the present day world situation, if there were a collapse of 'Russian Union.' Needless to mention it has become a possibility.

S.N. Roy, a supporter of M.N. Roy, pointed out that just as Russian revolution was a departure from what Marx and Engels envisaged, so had the communist seizure of power in China, by a departure from Russia.

North Korea and North Vietnam took the same lead from their respective cultural differences, in different forms of revolutionary attitudes.

Each was a different type of emergent communism, in terms of its respective cultural basis. The rise of a new social force depends on the rise of the new aspirations of the peoples of emerging nations. But the communism of Marx and Engels was restricted, to an *abstract* 'ism' – it was 'Eurocentric' and concept-oriented.

Now the world has to tackle a new emergence of economic and political phase, possibly there will be a total disappearance of any of the past ideological positions, and will be directed to each individual's interest, as will emerge from one's cultural milieu. The world is destined towards some form of *people's government*, when people will so *transform* themselves through progressive education in insightful directions of progressive histories.

It is a confrontion to the hard world of *people* with varied interests to tackle which, one has to seek for a tolerant and compassionate attitude and not some kind of *abstract ratiocinative process*. It is a post-ideological age, where different politically oriented forms – democracy or communism – take multiple forms of one's cultural background, but that also of whatever shape is not an end-process it is a ground for infinite other types of emergent possibilities. Hence there is no freedom so long as man adheres to a form. Hence this author suggests an ascent to a transcendent attitude beyond all forms to a stable transcendent *psychic condition* of 'Peace' or 'Balanced Attitude,' with progressive defence schemes, with a non-violent attitude. This is a stage of a new vision of progressive possibilities in advanced Buddhist-Whiteheadian post-intellectual direction.

But the deferred process was envisaged by a series of wars – from Vietnam to Korea. By a contrast, the new move of Gorbachev was a *peaceful* process, but one has still to see the future international process and possibility or crisis.

Hegel, who inspired the dialectic process – philosophically and *abstractly* – if re-examined with insight – has now to be rescued from his unchangeable abstract position and one has to step out of his castle to breathe a fresh air, to enable the world-process to continue its life-span – in a new atmosphere or garb, with a dignity – if possible by the transcendent suggestions from India's Buddhism, in progressive pathways of India's alternative Vedantic expressions, in association with the Whiteheadian suggestion towards a prospect of transcendent post-intellectual modes towards a pursuit of "Peace" or "Wisdom Mode", as repeatedly suggested.

Hegel and his indirect followers Marx et al., had a very low opinion about any Asian country, all these, again, were narrow attitudes, leading finally for 'Germanization' of the Europeans. It was a shift from one's narrowness to a vast portion of it ended up with racism, and massacre of a vast portion of humanity. A fresh rise of this process has to be discouraged.

In support of his *transformable attitude*, and readiness for change in many possible ways Roy could think of alternative possibilities, quickly, he could develop an attitude towards a non-capitalist pathway for development.

This forms the new economic basis of communism in Asian countries. It is a step into some form of 'international relationship.'

This is Roy's dynamic communism, in application, touching finally the upliftment of the village community, providing it with opportunities for a variety of economic resources and possibilities for people at large. It was a kind of shift towards village economy, with an aspiration for 'people's rule, with alternative aspirations – a voyage towards a progressive democracy.

This new turn towards the economic choice of village people, from a compassionate humanist attitude – Roy touched Jayaprakash Narayan, in an aspect, and later Gandhi, who also gradually turned towards Jayaprakash Narayan. But this transition was a later possibility, which India can realize in its post-independence period in process and progress, but it was a premature demand when the British were not ready for accepting such a process, at the time of its decision for quitting Indian Rule.

Implicitly it was an exercise with Nagarjuna's 'neither-nor' scheme, with no offer of any specific choice, but leaving the world to an openness – where everyone will be offered a freedom from a transcendent perspective with depths of insights. A difficult path too sharp to tread on, but the world yet proceeds, though so hard to traverse. Human instinct or normal disposition is always an attachment to confronted-conceptual-empirical process, but one has to be detached from this proneness, for the confronted universe does not categorically exist – it emerges and perishes with no remnants – and apparent givenness is no 'givenness' – from the post-conceptual-empirical perspectives of both the Buddha and Whitehead.

All these contrary movements were not supportive of the transference of British parliamentary powers to India. But according to my judgement Nehru's decision to accept the *transition* by parliamentary form of Government was pragmatic and timely – for this helped a *smooth transition.* Otherwise the British would have become puzzled by the varieties of people's choice. In this post-independence period – the various other possible forms will gradually emerge. But really there is no possible solution in any form. What is suggested is mankind's choice for a transcendent *balanced* attitude, not as an end-process, but as a progressive pathway of possibilities. I have stressed this point beyond post-Hegelian, Marxian Dialectic.

Now, with this choice for a smooth transition India's stepping into its later Constitutional difference was nothing but a consequent possibility. It was a novel emergence with functional changes, with a continuity, suiting India's culture and people's wishes, at that time – in Buddhist-Vedantic 'neither-nor' trans-dialectic transcendent condition of real freedom.

India, after a *transition* continued with progressive modifications from an initial frame.

Any demand for a change at the beginning would have been worse for India, for, the British would not have understood this, and India would have missed *independence*. The change in *history* takes time, till the world-process as a whole comes to a point when people in general are left with no other alternative that becomes the right time for change.

The feeling for the consequent possibility as indicated above with a possible waiting period, was not patent before, and it would have been difficult for a quick communication to the British, who then had only an understanding of their own parliamentary form of government, and no other alternative.

For Roy also, there was not an acceptance of Communism. He wanted to utilize its form to be stretched to some kind of market economics, from a different cultural perspective, with a criticism. It is just a kind of offer to some types of people who are business-minded.

His was a glimpse of many alternatives, by a suggestion that there is a further possibility, from the 'Russian form' also, by a prospective emergence into 'Asian market economy.'

The world-process is neither Eurocentric nor Asian or Afrocentric – it is widely open even to incorporate in its frame of reference of the New Hemisphere – in its present phase with a possibility of taking in the best of the past cultural processes of the pre-Indian masses entering into the processes of progressive Mayan civilization from pre-Olmec and Toltec bases, or of Incan process from the early Peruvian processes, as bases. The later European impact on the American Indians was a later historical impact, with the forceful elimination of the past traditions. But who knows that there will not be a new emergence like that of the Phoenix of old?

If there were so many variations in Communism – as Roy pointed out, so there are similar variations also in the democratic forms of Government, as of France or West Germany.

After a perusal in some depths, it seems to me that, in contrast to the British form of democracy – the French and the German forms have more complexity.

Behind any 'ism,' there is a cultural background or impact. So transition from a somewhat kindred form of democracy to another slightly modified form of democracy is smoother. 'Indian form' of Constitution was that kind of smooth transition. If politics is progressively culture-centric – it is deeply psychological, but in a progressive depth-aspect – and not in restricted conceptual-empirical direction, or simply not ideologically oriented. Hence this kind of a psychic attitude can be symbolically expressed, but there is no such thing as a categorical solution – though there is within a form there is a change beyond it.

Martin Luther King, Jr.'s statement, at the outer cover of the above work, so appropriately collected by Attenborough, has been added as if, as a further psychic force in support of Gandhi's wonderfully expressed power, which is a kind of change in constitution, in the form of a compassionate attitude.

Compassionate attitude is the core of the major religions of India, it is a process, so special to Indian mind in continuous development. This process can be a possibility in the political sphere, in India's cultural background. It is not a prescription for other cultural processes. Be it noted that 'compassionate attitude' emerges beyond all restricted conceptual-empirical processes.

But neither Gandhi nor Martin Luther King clearly spelt out explicitly any political or Constitutional frame of reference. They were restricted to their respective Constitutions in their developments of non-violence, as indicated by me when considering about Indian Constitution. Only there was a general reference to a compassionate attitude.

But my reference in the case of M.N. Roy was from the Russian Communist frame of reference, in the case of Gandhi in the context of India, and in the case of Martin Luther King, from American frame of reference.

A compassionate or an open door attitude is also prescribed for international politics, which is becoming indirectly a new force, after the Gulf War, and after the collapse of the Russian Union. Its implication varies from culture to culture and place to place.

Gandhi's was a transference of a liberal religious principle into international politics in the same sort of compassionate spirit during the British period. Now it is a cognizance of a transcendent process, which is found in all aspects of experience, not merely in the religious order of experiential process.

This new historical process of Open Market attitude is becoming progressive democratic choice for many emergent nations of the later post-Gulf War and the post-collapse period of Russian Union. The world is with a many-sided activity-process from play to the business. Undoubtedly the entire mankind is within a milieu of a progressive change towards a new 'Psychic Possibility,' of which needs to be detached from the basic conceptual-empirical process, and has to be transcendentally transplanted.

But behind all political phases there is a place for a transcendent compassionate process. I am indicating all these different political varieties and possibilities in this process of the changing universe offering the places for many-sided activities, repeating the Bhagavad-gita's truth that 'Arjuna' (or any 'Man') cannot remain without some activity-process, as emerging from one's basic disposition. Arjuna was given a chance to be operative with a freedom and detachment.

Mere abstract form, in any sphere of experience – from religion to politics – is an abstraction, and hence I am for the activity-process from one's dispositional basis, but that needs to be disciplined from a transcendent perspective. One can find some detailed consideration, regarding different types of discipline suiting one's life and profession from the kings to the shepherds – in my recently published book, with the title, *The Mysteries of Vajrayana Buddhism from Atisha to Dalai Lama*. It is a book dealing with a tolerant attitude for any dispositional process with a transcendent attitude, so that it fosters an activity-process without any strain, suiting one's nature.

I have placed below, some of my deliberations, to give some relevance to an appropriateness, for any effective advanced form of democratic movement or its further possibilities.

Attenborough's statements, were relevant in some other contexts, but suiting my purpose, also, so with a little change of my mind, I will let him express his mind, as they were used by him: 'Gandhi was inevitable. If humanity is to progress, Gandhi is inescapable. He lived, thought and acted, inspired by the vision of humanity evolving towards a world of peace and harmony. We may ignore him at our own risk.' Gandhi was also aware of various political forms other than democratic, but he had a choice for the democratic, because of his cultural background and for the prevalence of the British form of democracy in India.

But from my complex philosophical disposition, Attenborough's simplicist statements have some value, as an alternative. This was good for the period of a previous historical change. Now, with several other historical changes the international situation has to take note of my other concrete processes and suggestions for possible peace-movements.

My suggestion will be for a further possibility beyond this Gandhian or any type of progressive democratic possibility.

But all these statements, including mine, so far, are only insightful views, from the perspectives of confronted emergent phenomena, but they also lack many other alternative perspectives. Hence I have not found any merit in any concrete analysis of different concrete situations as emerging within the country, and also in outside world, creating further emergent crisis-situations.

These emergent phenomena will be taken up, with relevant directions of 'if-then' process, and not in any 'categorical form' in this and other chapters of this volume by reference to the writing of some specialists to be mentioned from time to time, as may be found earlier in this volume, and also by the development of its later processes, below. But I am not presenting any solution, I am indicating all symbolical and alternative possibilities.

But the world-process in any stage normally takes the way of violence, enforcing one's ideas on others; in any frame of political ideology, the mighty dominates over the weak and the fragile. The world with its varied human types with chosen ideals and ideologists, takes time for change and transformation. Hence the emergences of some revolutionary or even dogmatic attitudes in recurrence, can never be excluded as a possibility till the world as a whole understands the true character of the transcendent possibilities which confront many historical situations as changing or perishing confrontations. There are a series of emergent histories with *sufferings*, with no final solution, but they have a merit, only as a suggestion for an alternative process.

Here I find that Herz also declares through subsequent chapters of his above work, a similar or not so similar such statements. His is also a reference to changing confronted processes of several alternatives with 'indefiniteness.'

The different parts of this present volume, progressively give dynamic shapes to my newly framed avenues of philosophy and science in the context of astrophysics – and these are stretched to international politics in contrasting ways with no exhibition of power. But the world-situation after all remains as one of indefiniteness and uncertainties. I do not suggest any solution, but I would prefer to say that it is one alternative possibility among other infinite possibilities.

I choose to elaborate my position by concrete ideas, now enriched by the new psychology of Gandhi, and also by the political thinker like Herz. The world is heading towards a pathway of varied openness.

In each of these reflective areas of human activities, and also by some specialists in different experiential fields from philosophy to politics my method of thinking is not in the past abstract ways, but in 'concrete lively psychological ways,' with an insightful use of the scientific powers for a common benefit to share and develop. But all these activity-processes are patterned to suit the confronted universe in 'if-then process.'

But in this period of scientific and technological advancements mankind has to pave the path for a transcendental psychology in progressive developments through confronted experiential and historical processes, and with a sense of perennial transcendent openness in varied symbolic ways.

This path is one of peace, and a movement in the direction of a compassionate form of democracy, no more restricted to mere conceptuality, or a form of empiricism. The terms *transcendent* and *compassionate* are used in a general non-specific sense.

If I think of any other form of government my understanding of them will not be abstractly ideological, but softened through concrete experience and with a cultural difference, to progressive benign aspects.

In this sphere of one's experience, the role of a personality, also has to be understood, from an advanced Buddhist-Whiteheadian sense of personality, where it is not also something ultimate, but one who travels in that sphere as progressively disciplined and insightful. This pathway can contribute towards the change of the socio-cultural process, by relieving human mind from the obstructive processes of one's instinctive propensities, and retarding thinking from 'intellectual abstractions'.

Such a personality has to cultivate a progressive compassionate attitude, in progressive meditational direction, by a detachment from all confronted conceptual-empirical forms to uphold a position with a tolerant and compassionate attitude, knowing fully that the confronted personality also is not finally experiential in any categorical form. It is touched with the symbolical ways of 'neither-nor' process followed by continuous insightful experiences.

Such a transformed personality is so important as capable of bringing in a total change or revolution in the history of mankind in general.

Gandhi could eliminate the British rule of India, because he could be operative with a detached attitude. Martin Luther King, similarly could be operative with a detachment to bring some social and political change in the American social milieu and constitution in favour of the Black Afro-American community.

Gorbachev also could be operative in a different ideological sphere with a detachment to bring about similar socio-political change in the Russian Communist ideas, by introducing some kind of democratic or liberal process by his *perestroika*.

These are expressions of non-different processes in operation, in different ideological contexts, and at different international situations in different historical contexts and processes by releasing human mind from mere abstract thinking. But often, in historical process, there is a return to wrong gestures of the past, preventing advancements. There are insightful visions, by a freedom from *abstract intellectual forms of arguments* but such followers are not many.

Historical process is not merely a naturalist process, it is a resultant possibility through naturalist-humanist process, towards further developmental process, where some kind of extraordinary personalities have important places and roles. Here one has also to study deeply one's life-history, how one formulates or shapes it, in a socio-cultural environment. Besides all these roles of personality-processes, one must also give some attention to the kind of ruling powers that are operative in a country in a historical period.

To communicate all these insightful varieties one has to choose appropriate language for adequate expression.

What is suggested in this context is some kind of reciprocal operation between two contending parties. The British, in India, responded appropriately to Gandhi's movements along with others, facilitating Gandhi's movement. That is true as well of the United States, where some kind of cultural appreciation was the characteristic speciality of the constitution-makers. That is as well true of Russia, where also, though an 'ism' was pursued in all its ruthless stringency, still there was an underlying understanding of some form of human culture and appreciation, on account of which change was possible.

In this context, without very much of details, I shall refer to some specialities of Gandhi's formulation of a psychology through the varied experiences of life, his early life as a child, his educational period in England, and his return to India, with a degree as a barrister, his luck or ill-luck in going to Africa, and becoming transformed through his activity, facing the Governing British, with anti-racial feelings and attitude, in Africa.

Gandhi, returning to India, tried hard to find a place in the India's Congress movement. His activities as a lawyer with his novel psychic spirit helped mankind. His life was a pursuit towards 'Truth,' avoiding lying to one's inner Self, which is nothing but cultivating violence. But this new way of thinking, came to his mind, not when he was in India, but when he went to Africa to help the Indian merchants in a foreign land, where race discrimination was predominantly present, while in India it was not so pronounced. So his simple business responsibility became complicated with the problems that prevailed in that country. It was a place of coloured people, they were placed along with the dominantly placed European 'Whites.' What prevailed there was an 'apartheid process' – a suffering awaiting 'transcendence'.

But the pursuit to Truth as an insightful experience in all its transcendental purity by a contrast – in its true sense is an operative process of love, even against one's opponent, and hence in this superior order of political activity, Gandhi, with a transformed mental attitude, challenged the legal system that prevailed in India, during the British period, after his return from Africa, understanding the core of African experience.

Gandhi discerned a proneness for a discriminatory attitude, in the British rule of India. The word justice, that was administered to the Indians, in contrast to the British rulers, or to the Europeans in general, was different.

Even some judges observing Gandhi's attitude – because of the cultural environment of India – frankly admitted that Gandhi was a different type of personality, so he had to be treated differently.

They used to salute Gandhi in the open court, but still passed some judgements with a reluctance. But they saw behind such actions, some element of injustice. But this feeling in them was because of India's cultural environment. This 'feeling' progressively developed in the minds of some of the ruling officers sent from London's India office.

Gandhi tried to change the British mind towards a really just attitude. In this context, he shifted his attention to a different field, demonstrating how he would realize a dynamic God in intercultural context, by a pursuit of Truth, with love for humanity, and Love within one's self, getting rid of any attitude of discrimination against the British rulers. He became operative from two experiential centres confronted and transcendent, and went beyond the confronted.

With this as a background, Gandhi proceeded forward to reform the British minds as rulers, but being moved by Gandhi, the British also gradually gave up their propensities of exploiting Indian minds and also the material resources of India, by fostering its own business, in varied ways and fields, but still depriving Indians of their natural powers and material resources. To Gandhi, it was a kind of violation of a moral and religious principle, which he wanted to correct by entering into the political sphere. His descent to the political sphere, by an extension of his religious experience, was a novel experiential process catered to mankind.

In an aspect, it was an understanding of a core of all religions of the world, eliminating their scriptural differences, so that in lifting up of the Indian masses, he suggested them to follow one's religion with a depth-study of all types of religions, of India and also of the West or other countries, to develop an internationalist consciousness, by progresive 'transcendence'.

India's understanding of religion was a continuous pursuit, and not a restriction to or an adherence to any scripture. It was a supra-cultural living process and activity, as of scientific pursuit and activity. Hence religious life and experience, cannot be separated from any other experiential processes.

Gandhi's was an experiment with religious experience, in the changing political environment of India. It was a concrete practice of religious consciousness from an international perspective a real platform for 'independence,' in a 'transformed political situation.'

In India, Gandhi was operative in a different cultural environment, so he could successively proceed, whereas Africa needed the emergence of its own people to fight against racism of its own country.

The violent operations of the British were to suppress the Indian movements of varied types prevailing in India. Later a process of promise with a delaying tactics prevailed in India, in the transference of the political power to India.

The British or European policy in Africa was different. Africa was a place of many European Whites confronting the black Africans. The problems were more complicated.

The British were somewhat differently operative in India. The British had the insight to be operative differently in an effective way in India by a through study of the Indian mind, its history and culture, and so they were operative through its *India House* in England. It was highly an intellectual orientation on par with India's growing intelligentsia. In India this process was well displayed.

Gandhi, through his association with the Indian masses, was deeply aware of the mass-psychology. According to Gandhi, mass has no mind or premeditation, the *mass* acts in a frenzy, as its mind is normally a wandering mind. In this wavering and wandering disposition, of the Indian mass, Gandhi suggests, that the mass mind could be evolved, and developed and the emergence of a disciplined mass-mind would be a possibility – out of a mob disorder. This new emergence was a threatening phenomenon for the British to confront.

So Gandhi had a faith in this type of the emergence of the people of India, and he had a boundless faith in such a powerful Indian mass, and it was for this faith in people that Gandhi could recruit volunteers for carrying out his non-cooperation movement against the British rule. Gandhi's was a kind of marching with a new type of regiment, pitted against the British rulers. It was an inauguration of a tremendously powerful mass-phenomenon of a disciplined sort – a challenging threat for an opponent. This mass was not intellectually oriented, but *transcendentally inspired.*

In this active emerging possibility, Gandhi could eliminate from the *mass minds* the elements of hatred (an instinctive disposition) for the British people or British rulers, but making them operative, from a transcendent sphere (Upanisadic-Buddhist insightful potentiality), so that the Indian masses were all ready for self-sacrifice with no expression of power but with expressions of a courage to confront any situation—even death. This was Upanisadic-Buddhist spirit.

The adamantine *energetic attitude* was sufficient to puzzle the British mind, and this touched the British moral core, so the British rulers became totally incapable to tackle such critical situations.

Empirical psychology was no match for *a transcendental psychology*. India has a hoary past in the cultural development of a transcendental psychology. This process was insightfully developed in a post-intellectual direction from the Upanisads to present movements.

Gandhi utilized it in the aspect of a political sphere, generating a new force from the mass minds by utilizing and transforming them. The British were compelled to take note of this emergent mass movement.

Gandhi became operative with the British, from his transcendent psychic perspective, with another programme of raising the masses to an elevated consciousness with a new demand.

After recruiting the volunteers, Gandhi would arrange for their general training for acting appropriately in a critical situation. His was a harnessing of the mass mind against the injustice of the British rulers.

Faith in the *mass*, and *faith* in the development of the confrontation of the political sphere in the garb of a religious psyche was a continuous dynamic phenomenon, or a creative-experimental process or scheme, suiting the *mass-mind*. Gandhi, became operative from this new field – thus challenging the British with a novel political weapon.

So, in the case of recruiting the volunteers from the masses, he cherished a faith in the *masses*; his faith or belief in God, was not a theory, but as a *fact*, more real than that of life itself. It is finding God in the masses. Gandhi was a creative artist in a new direction, in the line of Swami Vivekananda.

In the present political situation India's public should be given a general education, as to understand with insight the prospects of a commonly shared interest, forgetting one's exclusive self-interest, or restricted attitudes towards one's relations with people living close to one another.

Country and its people had to be elevated to a height – and that attitude towards life should be the primary motive of all. This method of controlling the *mass-mind* was more than the mere ordinary educational process which is a 'delaying process.'

To Gandhi, or to any one who understands religion as a process of progressive self-perfection, in an aspect – as any science or art, perfects a person, in an aspect, it is the tie that binds one to one's creator – from that transcendent level, one can find the truth in the following statement: Whilst the body perishes, as it has to – religion – so viewed – enters into other experiences in relation and contrast – elevating them all. Such an elevating process, as I view, persists even after death.

To my mind, not *religion* alone, but it is a *pursuit* in the direction of a knowledge in any aspect or art of life, and this outlives one's life, or one's individuality.

Gandhi's creation of a *mass-movement* involved all these elevating experiential processes, so it became a unique forceful-phenomenon. Here the gulf between the rich and the poor, intelligent and non-intelligent sank into nothingness. Humanity became one over-all power.

My reference is here to the transcendent aspect behind all forms of experiential processes from the myths to the sciences, not excluding the science of politics. Here something persists beyond the present confronted existence.

Such a refined mind thinks that the human mind and the human society are not divided into watertight compartments of social, political, and religious frames of references. They act and react upon one another, perfecting one by the other. It is a continuous milieu.

Hence to him, spiritual law, has an exclusive field of operation, it expresses itself in ordinary activities of life, affecting economic, social and political fields, in togetherness.

My interest in Buddhism is along this aspect of experiential process, involved in Buddhism, and this is indicated through the political processes in the present advanced aspects.

Religion as converging upon the same point, and for that matter any refined pursuit takes different roads, to reach the same goal. It is also an efficient principle to be operative with.

It is interesting to see at this moment, while writing this chapter this writer is faced with a social problem at the United States, where women are fighting for establishing their strength in every aspect of life, at home, and in offices, where they are employed.

Often the problem arises between the chief male officer and the female secretaries. There is a general feeling among the females for losing their jobs, if the officer in charge is not kept in good humour, this leads often to a clash between the officer and secretaries from a psychological perspective.

In course of always trying to please the higher officials – often some situation emerges, when one does not understand the other. This is graphically demonstrated in the case between the officer, Judge of the High Court, Clarence Thomas – nominated by President Bush, for one of the Associate Judges of the Supreme Court – and Prof. Anita Hill – a tenured Professor of Law of Oklahoma University.

Anita previously, worked as a Secretary under Judge Thomas. Though the case was a clumsy one from an aesthetic and moral perspective, it had an important role for American social process. It is after-all a non-violent protest.

I am not comparing the American social process, with any other social process, between European, Asian or Middle-Asian or African social processes, but I am considering this emergent social process, where there is a dominance of Freudian psychological current.

But from India's transcendent disciplined form of psychology, human Nature is viewed as belonging to two orders: (i) a basic psychic condition, where jealousy and covetousness predominate – there is always an illusory or wrong vision, and (ii) a transcendent disciplined process, as prescribed – in Buddhist triadic process – which is a cultivation of a 'transcendent process', beyond all intellectual arguments. A person, in this disciplined condition, is silently meditative, which frees him/her from wrong illusory visions. A person so disciplined is more in the direction of right perspective of things. But here also I am not standing for anything absolutely final, as there is no ultimate.

When a person, rises to that disciplined contemplative or meditative order – he/she is freed from the egoistic propensities of jealousy and covetousness, and hence freed from the touches of Freudian psychology, with its presupposition of sex as the most dominant of the instincts.

Freed from the aforementioned instinctive level to a detached condition of *freedom* – man or woman becomes a new potentiality.

In India, also there might be some similar cases, wherever there is a contact between a male and a female officer – but such cases will be few as there are other psychological and social laws operative in India, from the past to the present. The cultivation of disciplined ways elevates the social process, and lifts it to a height worthy of human appreciation.

Hence the process is not a simple psychological phenomenon of one instinct in operation, with another, but it is a bit different, and hence people in general develop a culturally oriented disciplined behavioural process, in a different way, which is a medley of complex operations of many other instincts besides sex, with a socio-cultural elevating process, in togetherness. A disciplined man/woman, with a detached vision can improve upon the social relationship between persons in general, and also between men and females. Any situation that has emerged in American social process is not really a serious problem, but can be improved with insight.

It is already there in social process, but it has to be simply transcended, by cultivating a disciplined path, which transcends both ordinary levels of morality, and, fosters a development of religious experience beyond theistic disposition, as progressively indicated in Indian thought, from the Upanisadic to Buddhist and Vedandtic periods.

This process is continuously developed to the contemporary stage of India's varied movements, including Gandhi's.

Religion, in development – is not just theistic – it is post-theistic, yogic and tantric, in further evolution and development. It is in a disciplined line. Similar disciplined prospect is also possible in all other experiential processes, with which religious experience, in disciplined aspect is interconnected.

If there is a meditative possibility beyond the restricted sphere of intellectuality the theistic aspect of religious experience can be transcended by post-theistic meditative possibility.

In America, the above judicial case has set up a future problem, how henceforth, male or the female will behave in any normal social situation. The Senate members were in a fix, and the public were waiting for an emergence of a novel social process, out of this situation. I have indicated a possibility, from the advanced Buddhist meditative disciplined processes, as also in the Whiteheadian suggestion for a transcendent process beyond restrictive intellectual abstractions.

The case however was decided in favour of Thomas, but the case had cast hidden pollutions to human social process, if the American masses chose to stick to Freudian psychology. To face this concrete, social, moral, psychological and religious experience or situation, in togetherness, devotion to Truth and Love is essential. The prayer that one has to cultivate in religious situation, is not just theistic, but transcends that process as it has to be stretched to all religions, and finally to continuous transcendent orders of all aspects of experiential processes, not alone in one exclusive aspect.

Experiential processes are interrelated. Religious experience is not dissociated from other experiential processes. Each experiential process crosses one another to fertilize man's entire psychic bodily and social structure, and one, with such exercises can rise to progressive international and intercultural consciousness.

But all these onward marches suggest simply a transcendence of restricted abstractive intellectuality and a choice for a goal beyond all 'intellectual forms,' to progressive insightful understanding.

Religious life, as an aspect of human experiential process, is an important aspect of social life. Religious life can be lived in one's depths of solitude, where there is no place for any theory, any specific God of a religion, or belief in theism. If any one cherishes some form of atheism, it does not mean that one is irreligious. Religious experience is not intellectually expressed, it is an experience by itself – transcendentally ever-present whether one follows or not – just as any person cannot violate any law of nature, or even moral and social laws.

I am making all these statements from my own experiential perspectives as placed in different parts of the world with the major responsibilities for teaching and research with a kind of dedication.

Social life does not depend on any one's opinion. It is present continuously, with its speciality as the river flows or the wind blows, or the lightening flashes, or any other currents of life, flow, et al.

Life is physical, physicochemical, vital, mental, social et al. and is governed by each one's special laws but it is also a process of life with that of others – with love, compassion and tolerance – hence religious process of life flows with other experiential processes in relation and prospect. Symbolically, it goes or rises to transcendental orders with 'Prayer as the key of the morning and bolt of the evening.'

In my case, religious life is a further process, as operative, where the notion of prayer is nothing but a practice of continuous concentration. Concentration can be directed to the cultivation of religious experience, and to all other experiential processes, including science. It is a powerful prayerful attitude, which is evenly shared in cultivation of any experiential process in depth.

It is only possible when there is an admission of the Indian mode of life and discipline. But it is difficult in non-Indian socio-political situation, where priority or prominence is given to analysis of the particular case of a socio-moral process, with not much of an appeal to each one's religious frame of experiential process. Experience, in a continuous insightful direction, cherishes no impassable barrier, it runs into all experiential processes. Life lies in continuous concentration and not analysis.

Gandhi's reduction of Truth and Love, with 'God' was understood in a transcendent sense of continuous concentration, or meditative possibility. Hence it was not taken in any exclusive religious sense, of Hinduism, alone it was taken as a purpose and a possibility in the meditative sense. Religious experience runs into the core-processes of all religions. Any religious process is pervasively present in all religious processes of the world. Hence this religious process is 'Humanistically oriented.' It is a supreme 'concentration' – a prayer.

While Gandhi said that, in the spirit of his time, but I am adding my extended view of religious experience, symbolically, in a mathematical way – treating it as a positive number, with a sequence between any two positive numbers, for there is an ever-present '0', as a remnant process needed in counting and joining with the *unfathomable infinite* on both sides of the *confronted numbers.*

Any positive religion, like any positive number runs into the infinity, through '0' that prepares the ground for its *emergence* with a consequent possibility towards *nothingness* with *no end.*

Religion – if Gandhi viewed it from a depth-aspect of the Sermon on the Mount – becomes the core of all religious processes. It is an understanding of a creative process of energizing oneself to a purified force so that it conquers all, from a progressive transcendental perspective of depth of one's silence – when all are present as non-different from one's own self.

All religions run into the depths of *silence*, which is the fountain of emergence of all religions.

The life-blood of this energetic religious continuum is non-violent activity – a path towards 'Peace', a 'transcendence' of one's egoistic motives and aspirations, a dedication, a surrender – and if it is operative through nationalism it also steps into the process of internationalism.

The transcendent experiential process is one of progressive continuity insightfully getting rid of discursive forms. It lifts all experiential processes to a transcendent level wherefrom one becomes operative with a *potency* which is not available from the confronted level which is restricted to its form. Man needs such a transcendent process of lifting potentiality.

Here the roles of both males and females are equally important, or sometimes one has to admit more or less power and capacity, in '*either or*' way, between men and women or between different races, or different classes, grades, and age.

This turn to an *exalted mind* is also viewed among the great scientists of today, when I found in the March issue of *Science* magazine of American Association for Advancement of Science – general editor, Daniel E. Koshland, Jr. in his appropriately headed title, 'The Race for the Gold in Research' – was pointing out that in the present-day of advanced scientific sphere – there is a competitive race of priorities, when a nation or an individual of a country has to face a scientific world of 'first-to-file' approach, with a 'grace period.'

It is a case of the international scientific community, whether to allow the same privileges to all nationalities, and also it is a recognition for filling their basics in any part of the world, or evaluate their priorities by an arbitrary decision of a superpower scientific community of a country, which has a claim of a superiority over others.

In every sphere of experiential process, the real potentiality emerges only when an individual lifts himself to the transcendent level with continuous possibilities. There the possibilities are no more restricted. Hence it has infinite potentiality.

But this refined individual has also to point out that every scientist, and every country can formulate its basics to join finally to the international status. This is a broad scientific process, touching the best of moral and religious processes.

American Association will recognize their research achievements without any partiality. Since any scientific contribution is a kind of gold, any nation or any individual contributing the most funds to basic research has to be recognized with a sense of equality, for, the international community is the ultimate beneficiary. This is a gesture of appreciation from a truly international scientific perspective. It is indeed praiseworthy.

In the international and interpolitical milieu, India has declared its use of nuclear power or programme for peaceful ends only, but it is reluctant to sign the Nuclear Non-Proliferation Treaty (NPT) or attend any five Nation Conference, for maintaining the fundamentals of its political policy of nonalignment, which fosters a discursive attitude.

With regard to the control of the Indian Ocean, it is not willing to part with its right over Indian Ocean, and is ready for parlance with the United States, in a bilateral capacity, requesting the United States to concretize the obligations. Washington was willing to ensure nuclear non-proliferation by a treaty with Russia, China, Pakistan and Israel, but did not extend to India the same privileges, or it is placing India on the watch list, expressing a dominating attitude? India's House of People (Lok Sabha) protested through its Union Minister of State P. Chidambaram, and by Indian ambassador Abid Hussain – as unjustified and unwarranted or even unfortunate (*India News*, 16-30 Aoril 1992, Washington, D.C.).

This courageous attitude and protest of India had been accepted verbally earlier by the Deputy Secretary of State, Lawrence Eagleburger. He told the

Indian Foreign Secretary, J.N.Dixitnot to take note of as a serious input (India News, 1-15 March 1992). But this assurance was not carried out, hence India's protest against United States' 'Presidential Proclamation' as mentioned.

To Gandhi, in his time, this was also the Message of Asia, to the West, against the display of violence and lying, and taking revenge upon mankind by discharging 'atom or bydrogen bombs, or proliferation of chemical and biologicalgerms all over the world, ocean or atmosphere – violating religious and moral laws.

The arms race at any time is an anti-human process, going contrary to civilization, a wiping out of humanity finally, and destroying the normal processes of nature, disrupting it by the insertion of counter elemental forces or elements, infecting human body, life and mind.

The atom bom brought empty victory, and for the time-being destroyed the soul of Japan. It is not just a day to day life of conflict – between person and person, male and female, class and class, or races or nations – but going beyond all conceivable boundaries. it touches the international boundaries, infecting the whole world.

Gandhi's ideal statements were not idealistic, viewed from this hard realistic perspective. It was a realistic adventure and experiment through his life and experience beyond tragedies and sufferings. Gandhi's was a *movement* to a path of Truth, Non-violence, Non-cooperation, Love and Peace, and 'an aloneness', without flinching, repeating nothing new, but repeating still, the Truth, in a symbolical sense, as long as there are men who disbelieved it (cf. the life-history of Gandhi in Africa and India, challenging the British masters).

These were also the essence of Max Muller's appreciation of Indian thought of this century to remain vibrant for ever.

But all these statements of Gandhi, were energetic continua, when mankind as a whole would respond appropriately to the processes and would also try to spread into the inner recesses of such enlightened orders. These were not enforcements but willing acceptance of progressive self-improvement.

But if there was a change in this frame of reference, then all the above statements would lose all their forces.

The next phase of political history of India changed during the life-time of Gandhi, when counter-processes emerged, and violence became rampant with the partition of India. Man lost faith in man, and the world-process took a chaotic turn, Gandhi met with an untimely tragic end, may be due to some excess in his (Gandhi's mode) activity-process, but the tragic event happened, terminating the age of Gandhi, stepping into the process of violence and chaos, and uncertainty. The world-process is too slow to rise to the exalted 'transcendent level'.

India's democratic secular process – from different regimes of Prime Ministers to the present, after Nehru – is facing such tragic uncertain processes.

But this is all due to mass mind over which no one, or any government has any control, but both man and government all over the world should cooperate to create a good atmosphere. Here man at heart should be willing to 'transform himself' to raise himself to the *transcendent* heights.

In this critical situation of India, I feel like considering in details the book by Morris-Jones entitlted, *The Government and Politics of India* (Anchor Book, Doubleday & Co., Inc., New York, 1967).

Morris-Jones, a Professor of Political Theory at the University of Durham (England) since 1955, served Indian Army from 1941 to 1946, and was Constitutional Advisor to the Viceroy of India in 1947. After his service at the Indian School of International Studies in New Delhi, he was a Visiting Professor at the University of Chicago, since 1962. On leave from the University of Chicago, he came to India for a few months, on a grant from Rockefeller Foundation. His purpose was for collection of some materials for formulating a graduate course on India's changing political scenes of the different emergent states of India, after Indian independence.

But to understand these processes – changes and possibilities – between one state to another one has to think of India's pre-independent and post-independent processes, by a contrast, to understand the relative possibilities after Indian independence.

They are complex psychological-political emergents from operative processes as legacies, under four heads: (i) Government, (ii) Movement, (iii) Intermediates, and (iv) Problems and Promises.

The objective in common of these processes are for realizing the objectives finally for the preservation of the India's Union through federal processes of the Provinces and Territories (some special places or island countries surrounding India).

Government before Indian independence was 'British' – after Indian independence 'Indian' – in progressive formation.

Before Indian independence the 'Movements' of many parties, were contending for dominance and after independence the Congress party became dominant, but later opposition parties – emerged, with majorities in some of the States. If some of such parties cooperate to form opposition groups to be operative together – they will threaten the dominant majority party controlling the Government.

The intermediates – the intelligentsia of lawyers, teachers, doctors and judiciaries in varied aspects – before independence fed both the streams of Government and the emergent movements, with manpower, and after independence the feeding process through the first two legacies, has to be construed differently.

Problems and promises, before independence, were pro-British or anti-British and after independence similarly, though in a different sense – pro-or anti-Governmental – fostering some movements of the different parties, sometimes confusing and sometimes conflicting.

Morris-Jones puts his thoughts, in his conceptual-empirical frames of reference, but this writer, by way of contrast, observing so many changes after India's independence, places them all – as only confronted processes – to be evaluated from India's changing transcendental experiential processes and possibilities, to safeguard the Governmental possibilities, with a progressive responsibility, and as a Republic, to maintain its solidarity, with a self-sufficiency, as its own concern.

But the mission finally will be for the good of the people of India as a whole. All these different possibilities have been stated from the prime-ministry of V.P. Singh to Chandrasekhar and to P.V. Narasimha Rao.

Congress Government had to stabilize its position by a majority of people's vote, taking note of the roles of the different opposite parties. Soon after Nehru, Shastri and Indira Gandhi (in the later stages of her Ministry) had to face the relative strength of Opposition Parties, which disturbed Indira Gandhi's position, by an interval with the emergence of short time Ministries from Morarji Desai and others. Indira Gandhi regained her ministry and exhibited her power in helping the emergent possibilities of 'Bangladesh' (East Pakistan) from the clutches of West Pakistan. She as well, saved India from the invasion of West Pakistan, but internal problems, ended her Ministry with her tragic assassination.

Her son, Rajiv Gandhi, again revived the majority rule of the Congress, but soon that was disturbed by the rise of the dominant opposition party with a transition, and after his tragic murder through a foreign intervention, the Congress majority came back with some progressive powers of the present Prime Minister, P.V. Narasimha Rao, of the Congress Party. But one can see here clearly that this process of the change of ministers, from the Congress majority to other minor powers indicates that the Constitutional process is not always a smooth run. The role of people ar large is an ever-present dominating force, controlling the process of the different parties. But to settle this conflict between the majority and minority parties – the mass-minds or people need proper education and constitution change.

India's new Constitution in formation through its dynamic Structure, or Houses, and through the functional relation between the President and the Prime Minister and his cabinet ministers, and also their relation between the Prime Minister and the Chief Ministers of the different States – is a complicated process. The solution in this sphere, needs proper schooling of the people in general.

This relational process is further complicated by its executive and the judiciary powers, as inherited from the British period. There were some changes after the independence, hence all need an understanding in some details, not only in the context of India, but also in the context of the democratic political structures and functions of other countries, specially of Great Britain and America.

In the Indian frame of Parliamentary democracy between representatives of the two Parliamentary Houses – Rajya and Lok Sabhas – there is a good deal of complex emergent situations, which are not present in either Great Britain or America.

Hence decision, in any critical situation, does not lie in a leadership of any single head, but is a determination in a relative process, where a personality has a relative place, but still a place nevertheless, offering others with a suggestion, not violating the process-aspect of the constitution as well.

To settle all these anomalies there is a further process of going finally to the mandate of the people at large, with some qualification or restraint. It must be remembered that to hear the voices of such a large number of population is not easy. India certainly deserves to be considered from this perspective, as representing the largest democracy in the world. But people in general have to be trained in this specialized direction by a progressive plan or scheme. Here the Election Commissioner has an important role to play.

If India's political structure is from the democratic perspective, and people-oriented process, finally, it has to be contrasted with Constitutions of the democratic countries – British and American – and non-democratic countries – communist Russia and China, with distinct cultural backgrounds.

This is an interesting specialized study from a comparative perspective. Here the comparative study is not a positivist version but a statement by continuous contrasts.

Hence Indian intelligentsia and masses, need some basic and also specialized training in Constitutional matters for establishing a superior order of democratic-secular processes to eliminate some critical problems in a peaceful way.

The consideration however will be in terms of India's cultural history, and immediate background of the British rule of India, or some influences from such countries as America or Russia. India's Constitution has been framed by a contrast to British and American Constitutions. People should be trained to understand this process.

India's speciality lies in a consideration from a comparative international political frame of reference, which is not usually found in democratic countries of Great Britain or America, they are self-oriented in terms of their respective Constitutions. There is no place for a transcendent mode of understanding all aspects of confronted experiential processes from philosophy, religion and sciences to politics, as yet, with some impacts of Buddhist and Whiteheadian process-philosophy in the present century. More of an application of Buddhist-Whiteheadian mode in all aspects of experiential process, is a need. Hence

Indian Constitution in the political sphere, expresses a basic philosophic attitude, which flouts any specific positivist perspective.

India needed to eliminate the restrictiveness of confronted processes by the reflective and post-reflective insightful fighters, for achieving the stages of progressive forms of Independence (movements chiefly of the Congress and of some contrasting parties), with progressive secular outlooks during the British period, and now after independence in its own frame of reference, not losing sight of the international political perspectives. Symbolically India's politics pursues a dynamic frame of reference, fostering continuous possibilities to respond appropriately to people's demands for progressive improvements.

I will not like to foster any more self-centred narrowly oriented personages fighting for their restricted groups or religions – after India's independence.

Previously, during the British period – the extremely efficient diplomatic British rulers – backed up by their Parliamentary form of Government – with a select group of British administrative officials – Viceroy, Governors, and ICS Officers et al. – used to govern India with a force.

All these formative processes, with a comparative insight – had been operative for some time, after Indian independence in its early stages in contrast to the Constitutions of the different parts of the world, and with special reference to India.

But soon after these, little departures did not attract the other advanced types of ideas that prevailed in India.

Hence in India, some able-minded persons, of the Constituent Assembly emerged to reform the Constitution with the help of a drafting committee. This newly energized Constitutional bodies under the very able Chairmanship of Dr. Ambedkar prepared the detailed draft Constitution. Ample debates took place on the committee reports on the draft for 165 days, and later that process was stretched through for more than two years to reach the final stage. The advanced processes were beyond caste-oriented classes, and stretched to India's Buddhists, outcastes and tribal peoples scattered through the lengths and breadths of India.

A comprehensive perspective of the Indian Constitution, after Indian independence was built up for a progressive 'dynamic democracy' through the stages of different prime ministers from Nehru onwards.

These advanced processes have been clearly represented by Karan Singh an Ambassador of India to the United States, in the January 1990 issue of *India News* with one of its titles – 'Triumph of Democracy' – in the context of India's ninth General Elections. This kind of programme is also a need from time to time.

Karan Singh's deliberation was in contrast to the democracies of other democratic and other forms of Government prevalent in different parts of the world, bringing out the speciality of India's democracy.

There are other contributors to this above journal, as of Narayan Keshavan a special correspondent for the *New York City Tribune* et al. he, with a title, of a paper 'Indian Democracy Promises Hope for Others' clarifies his presentation of the subtleties of India's changing Constitution. This process also has a need to be emphasized.

I am using them together. The drive is in defence of Indian Parliamentary System, with its hope to preserve this process it is not in a static categorical aspect, but with an 'if and then' *possibility* to indicate its ever-changing improved character.

Here my intention is to indicate that with the advancing ideas of Buddhism, and Whitehead, directed to philosophical deliberations one can look upon politics also from an advanced progressive line of 'if-then' process.

This is my interpretation of philosophy from an advanced philosophic-scientific perspective from an intercultural study for which I trained myself through education, research and teaching.

This, if imbibed, by the politicians will help them realize with a speed, giving up the proneness for thinking in categorical ways. My progressive publications are in this direction.

I would like to suggest further that India's intelligentsia and the *masses* should be given proper progressive training for understanding the special character of India's democracy by way of a contrast and not just arguing from a general perspective of the political processes of India, with a kind of pride or sense of prestige as belonging to India. All heated discussions often must be mellowed down with an insightful calmness.

Specially referring to the elected Prime Minister of India, in the context of India's above General Elections, when V.P. Singh, as an elected Prime Minister the leader of the second largest party, he was allowed to form the Government by the President of India, even though he prepresented a minority party.

The new minister confronted the former Prime Minister, Rajiv Gandhi, with a dignity – and Rajiv Gandhi in turn, with a propound placidity and show of honour warmly congratulated his successor. He remained as the leader of the opposition party.

It was a demonstration of a superb democratic gesture. It was an acceptance of a change with a gesture of smile, for after all, one has to respect peoples wish as provided by the Constitution.

After the election, there was no more any animosity of pre-election scenes. Once the verdict of the voters was in, there was a smooth transition of power to the new government.

But this shaping of a government, cannot prevent problems of a contrary sort to emerge in near future, and this happened, for the human psychology is variously operative. It is a changing process for progress. And this state of things happens in all form of Government, and in all countries, whether democratic or not; as no forms of Government is perfect or an expression of an ultimate version.

Indian democracy follows this common naturalist pattern and possibility. But the stream flows in multi-dimensional ways where people remain dynamically operative.

The same sort of multifarious problems have emerged in Soviet Union and Eastern Europe, by Gorbachev's Revolutionary scheme and changes through *Perestroika*, introducing some democratic process. And there is emergence of further possibilities.

A consideration of progressive democratic process, therefore, in the Indian frame of reference is a realistic possibility, in contrast to British frame of reference, befitting India's cultural and historical processes.

This mobility in Indian Constitution leads one to understand some inner principles of the India's Constitution that emerged in the background of the dual processes of the-then British Government in change and transference, and the movements that the Indians took recourse to, to make the transference of the British power to the post-colonial power and process.

India emerged through with the commotion of a partition but still who knows, it prevented civil war in India? Hence one has to construe that the transition was comparatively smooth – by a contrast. The sole responsibility fell into the hands of the Congress to negotiate with other contending movements, or party formations, as they expressed different groups or attitudes of different orders of people of India.

These two processes were operative together, with the departure of the British, at the end of struggle for independence of the Congress and other parties and the British.

But this transitional process and party politics will go on, taking advantage of India's weakness, or efficiency, in its different historical stages as an apparent phase, and, not often smoothly, with a dignity, as when the operative power goes to the dominant Congress party, or when there was no alternative, there is a shift of power towards other parties, grouping together to out-number the Congress party. A country has to remain operative to continue its governmental process. India is operative in all these dynamic and alternative ways since Independence.

A reflective and insightful understanding of India's Constitution from an advanced psychological perspective can be developed if there is a detailed curriculum for the study of Constitution from a comparative perspective in all schools and colleges so that India in future can turn out an arena of specialists in this study.

In this training-process India will be able to bring out experts in Constitutional matters, to turn out Presidents and Prime Ministers, and other officials – with progressive efficiency – in a fully competitive spirit, making the country fully democratic from a secular perspective in a right spirit.

No longer India will have to continue the special privileges to the minorities, or to the depressed classes. The country can be placed in the level of genuine competitive basis, with an even prestige. It should be raised to the competitive level giving equal attention to all religions, sexes and persons of different ages and positions. As India includes several tribes, within and in its surrounding borders – and also different types of peoples of the islands of the Indian ocean – India should proudly recognize them with their cultural privileges – and with the spread of education in progressive insightful directions India should avoid too much stress on abstract intellectual forms. An education in an abstract single subject with no relation to other subjects is no real education.

During the early process-period of transition from the British to the Indian a similar situation emerged so the India's sober politicians chose the best of the alternatives, for the delay was dangerous, and the majority of the Congress personalities became disgusted with the further monotonous continuity of the British rule.

Hence independence was chosen with partition. Net result of this historical gesture of India's admission of whatever was to take advantage of a British offer became a unique phenomenon that emerged between the British and the Indian politician specials.

After Indian independence, with a self-acquired power India became free from any foreign pressure, and flowed with a gradual process of departure from the colonial to the post-colonial process of India's novel constitutional democratic transference of power. This led to a 'Sovereign Democratic Socialist Secular Republic' structure or power, of India, after India's independence.

This process of India's progressive advancement I have indicated with concrete illustrations.

But now India's responsibility has been multiplied as people's demands have become *more* than before, with a progressive and changing consciousness.

I am not blaming the Indians for their progressive demands, but my suggestion will be for a progressive insightful process, as I am trying to communicate through my ideas of the triadic processes of this present text – from philosophy to politics through science and technologies.

The people have become more powerful than before, once becoming fully aware of their future prospects, in the Indian Constitution, as mentioned. But India has to step into its progressive realistic phases whatever more might be, in course of its evolution and developments through progressive education in practical directions.

In India, the same sorts of social classes with different religious bodies will remain as before, so the Partition was not a solution, it had given only an inevitable turn, at the period of Indian independence and transference of power. Now a departure is needed.

To be fully successful in carrying forward this constitutional power, India's intelligentsia and masses should maintain 'Indian Union' with some continuity as primary concern. In this advancement India will prosper progressively, if a corresponding educational scheme is improvised in insightful directions with a progressive detachment.

It is difficult to imagine, that a Constitution will remain without any problems. So, one has to know the basic frames of reference of the operational principles of India's special Constitution of the dynamic democratic process. But through education – it is possible in 'if-then' aspects with detachments in insightful aspects avoiding vain discussions.

Indian people need to be educated from the school-stage to higher stages to imbibe the real spirit, behind these skeletons of Constitutional frameworks. This is undoubtedly a primary programme to be enforced into the minds of the people from the masses to the intelligentsia in different levels with insights.

Indian people will have to learn to look into the capabilities of understanding of the different orders of people. Indian independence, with all its drawbacks – was, in an aspect 'good' in the Leibnization mode of philosophic deliberation, that this world, manifests choice for the best of the possible choices, with a transcendent placidity rejecting all ultimates – from any experiential spheres.

One has to learn, to accept with tolerance, whatever happens only one has to offer some insightful suggestions, with no hatred or jealousy declaring the interrelatedness of all experiences and discarding all ultimates as religions or politics, abolishing all notions of majority and minority problems.

To be successful in all these efficient directions an expert education is needed. A Constitution in formation is a perennial process and it is never a fixed entity. Therefore, learning about it more and more is like the study of a subject, with a seriousness. Only the meritorious has to be given a superior place with no discrimination.

India is a Union of 25 States and 7 centrally administered Union Territories, in different locations of India from New Delhi to Manipur and far Eastern parts of India and the Andaman and Nichobar Islands.

These, areas are centrally-administered, through the President as the Constitutional Head of the Union, and Commander-in-Chief of the armed forces for some special reasons.

The President is elected for a term of five years by an electoral college, consisting of elected members of both the Houses (Sabhas): (i) Rajya Sabha (the Council of States) and (ii) Lok Sabha (House of the People).

The Rajya Sabha is presided over by the Vice President, consisting of not more than 250 members, including 12 members nominated by the President.

The Lok Sabha consists of 544 members of which 530 are directly elected from the 25 States, and 12 from the 7 Union Territories. Two members are nominated by the President to represent the Anglo-Indian community. The Lok Sabha elects the Speaker who is the presiding officer, and has a five year term.

With these two basic forms of Presidential government, between the President and the Vice President, two Houses and the Speaker, the Indian Government is dynamically operative where no one remains absolutely static. Leadership if any, is a sharing of a power in a process.

In the case of the Vice President at the United States, his functions are almost nothing, as if waiting as a static entity, to crop up only to act as President to be or serving the President in critical situations.

India's two Presidents are operative in two different functional ways, handing over their relative executive powers through a Council of Ministers, headed by the Prime Minister, whose mandate is to aid and advise the Presidents in the exercise of their respective functions.

In this functional phase India's Presidents, unlike the Presidents of the United States are operative only generally not as 'executive leaders' but in emergency in crisis situations.

Their powers are shared with a major transference of power to the elected Prime Minister whose power as a leader is not absolute but is within the frame of a process – so long as he carries the confidence of the people at large.

The Council is collectively responsible to the Lok Sabha. All these statements are assumed to be known by people, as in all types of Governments in different parts of the world, in the context of their respective governments, but if India wants to be some what special India needs a progressive programme for interested persons to know the system, and its operational process.

My suggestion in this context is an appropriate teaching programme at the schools, colleges and universities for learning about the Constitution and its subtle operations with a special test.

This is a long-drawn lesson to be transmitted to all classes and types of people, through leading institutions as a continuing process for life, so that the people of India will gradually understand India's Constitution and its complexities with relevant insights avoiding useless arguments.

Its complexities, may emerge through population explosion, through internal turmoils over the execution of some technological projects, et al., and also for some sudden international crises, and so on.

The other occasional problems, other than socio-political problems – may arise through Nature's whimsical behaviours, through earthquakes, flood, epidemic or railway or plane accidents or riots et al. Or it may emerge from parliamentary discussions on socio-cultural problems or some complex relations with foreign countries, too many to mention, and too many to deal with at once, so these are left out as general events and occurrences – waiting for events to happen.

Indian Constitution is open to all these, with its own specialities. In bringing about India-wide changes through its progressive five-year plans, India solves the problems progressively with a principle, that events happen – as a possibility and not with an absolute certainty.

In the above statements with regard to the Constitutional structure and its functions, there were some mechanical processes but a specialist who is interested in this subject has to learn with patience, through school and college courses many subtle points, so that one will know that to pass an abrupt judgment over some political affairs has not much value.

People, with proper education will not be over-critical of everything without diving deep into any complexities of an emergent situation.

India's insightful sober mind, can be made an asset, when manpower can be properly channeled as efficient when it will be able to demonstrate its special character symbolizing its specialities, as ensuing from its early builders of Indian Constitutional processes from all-sided disciplined processes following the broad ideals of a series of persons who thought of Indian nation as serving humanity as a whole.

These personalities were Vivekananda, Rabindranath Tagore, Aurobindo, Gandhi, Nehru, Patel, and others in association with them, who understood India, its people, and confronted different situations during their life-time.

Their's was a contribution for the present generation of Indian people, and hence they are always to be recognized with proper respect. India should advance beyond its past with progressive insights.

The process that is referred to here is India's faith and acceptance of a trans-empirical process – transcending both confronted conceptual and empirical modes – with a progressive detachment. Detachment does not mean disconnection with confronted facts but properly understanding them.

Here one can compare the Presidential power of India as relative and dependent – yet finally, in an aspect, supreme. In Indian cultural process – President's operational process can be compared with India's transcendent power, which becomes operative in critical moments. It is a reminder of the Vedic sages suggesting that the confronted forces have a meaning or strength only when transcendentally based.

In America, the President as an executive Head, can be operative with the Congress as the Leader of two democratic parties – Democrats and Republicans – with proper checks and balances.

But the President has the power to veto both, and pursue his arbitrary option, by a kind of persistence till that is realized. In a sense, he is still a monarch with a democratic mandate.

This is dominantly operative as a special prerogative, for the President of the United States, for some historical reasons of the past of the United States.

But, India's Presidents or Prime Ministers know that the people's will or mandate is primary, and that is expressed through votes. Explicit statement is better than the implicit process. India is for the explicit process to realize its objectives in its adventure towards a dynamic democracy of India.

In extreme critical situations the President or the Prime Minister takes a responsibility to save the situation by some prompt decision, but otherwise, normally all decisions are made by a fresh election.

When a Ministry fails to solve any problem, whether at the Centre or in the states, the President takes his Constitutional power to exercise. The Council of Ministers, in the Indian Constitution comprises ministers of State, and is operative through the ministers of States responsible for various departments, and deputy ministers.

Since Indian independence with the disappearance of the rule by the Maharajas of the different States the people belonging to the States have chosen the former Maharajas as Governors, with executive powers, operative through officials responsible for various departments.

But finally, the Chief Minister of each State, or finally the Prime Minister of India, or still later the President of the entire Constitution decides for the People.

These frames of references also need proper elucidations with appropriate educational methods in all aforesaid stages of educational fabric. These processes involve progressive insightful procedures and learning with experiences.

As Indian Constitution has denounced the ultimate from all experiential processes from the myths to the sciences – India's Constitution places its faith in process and possibility, as already mentioned.

I am not referring here to the Chinese type of present Communist preference, in the Chinese sense of experiment with the Western Forms of Communism, from another type of transcendent perspective – as Chinese sociopolitical process is an allowance, or concession, from its early bases on Taoism. This alternative process, rests on the notion of Nature's Way and Power. 'Nature's Way and Power' is a pre-conceptual and pre-intellectual process, so the Chinese, from a Taoist transcendent perspective, advocates a detachment from thought and language and adheres to a non-attached transcendental attitude.

But this is again a wrong absolute attitude amounting to a dogmatism. India avoids this attitude of primary and secondary process. They are together in India's cultural frame of reference.

Confucianism of China admitting, a little of intellectualism and linguistic freedom in its socio-political evolution could last only for a few centuries, when it reached the stage of its legalist extreme phase, it collapsed.

At this critical moment Buddhism from India went to China and probed into its depths, and rescued it from some kind of its dogmatic slumber on an abstract nature-oriented transcendentalism.

Buddhism's suggestion was by a continuous meditative possibility, which presents experience in dual ways of confrontation and transcendence. There was no longer a need of total detachment from the confronted present, nor was there any need for the allowance of a Confucianist mystic form of Government on an ultimate moral basis of an ideal king's rule, or the control of the head of the family, as the supreme power controlling the members of the family, by an unbreakable family obligations.

I do not find, however, any such continuous development of life-process, or a continuous development of political ideas to this century as in India's Constitutional frames of reference.

It had a later development of a philosophic or a religious process in an aspect of 'Chan Buddhism,' which denounced all confronted order as a secondary order, missing the transcendent prior background of experiential process. It was only a conformation to an extreme Taoist attitude, in development through its cultural backgrounds, or in its another experiment at present in association with Russian Communism, within its cultural background, in a political frame of reference, with not much of reference to international frames.

But in present century, China cannot avoid the straight way of thinking confronting the Universe of Nature or socio-cultural processes as they normally are, with an understanding as prevailed or current in intercultural spheres. Hence continuous exercise in meditative practice confronting the concrete situations is a realistic attitude.

The Chinese politicians must take note of the foreign impacts of the different powers which are operative in international milieu – of the Russians, British or American, with which it may be involved in historical process. This is learning in concrete ways, not merely adhering to an 'abstract' as practised in China, in the past.

By contrast India's political and religious experiences – like all other experiential processes – are dynamic and fluent in the sphere of process and possibilities. It is an international process – both from the confronted naturalist and socio-cultural milieu. India, as shown by this writer, has changed, and is in change with progressive insights. But people in general are not so trained. Hence training is a need.

India's religious consciousness, as also political consciousness, as already indicated, are interrelated processes and are *insightfully* discerned, from a dual perspective – *transcendental* and the *confronted*. India throughout the centuries, whether in the development of its own cultural processes, or in influencing the major Asian countries, drew special attention to prepare the grounds for thinking in progressive transcendental ways – freeing oneself from abstract conceptual or empirical modes. (For details of the change of the basic cultural attitudes of the major Asian countries from the Tibet and China to Korea and Japan, vide my *Dynamic Facets of Indian Thought*, Vol.2, Ch.7.)

What I have presented through my books, belongs to the intelligentsia, but it can as well be fitted to the ordinary people's minds. The *mass minds* move around mostly by their instincts, and by common courtesy, if they are simple or considerate, but there are persons of varied natures, over which no body has any control. They are just admitted or tolerated in society, unless they create problems.

My main contention, in the case of political matters, is to indicate, that in the Indian frame of reference, – since no Constitution is ultimate but is only experientially oriented – it like all spheres of experience has to be considered as a process and a possibility. Hence training in this sphere is a need.

I will not therefore any more try to distinguish Indian politics from other kinds of political frames of reference as expressed through their Constitutions, for, India's political experience is one of continuous change and possibility, and never ultimate. Hence a continuous study and training is suggested with an impartial attitude.

It is a *departure* from any notion of 'Ultimates' either from any transcendental perspectives or from any *Ultimates* in confronted conceptual and empirical spheres of experience, in terms of any theories, ideologies or *isms*. Even Karl Marx felt the need of an ultimate communist society with a calmness and placidity.

India pursues an attitude of confronting any experiential process in a 'continuous frame of if-then process' denouncing anything as continuously present. The Universe as presented is with a 'if-then' frame of reference.

All these statements of mine, in the context of politics by way of a contrast, with other cultural areas of the world need a thorough course-study of comparative philosophy, religion, psychology, politics and sciences, in general from, varied and changed perspectives of the different cultural processes of the world. But in every case my point will remain, that none of their political attitudes can remain static. If they remain static they will not carry historical importance or value.

I do not think, that any of the experiential processes, is an exclusive process. They are all interrelated processes – they are experiences as process and possibilities in their respective frames of references, and in those frames of references they are not *ultimates* but just processes and possibilities in international contexts.

Now, in the present historical process, all such 'Constitutions' have lost their absolute ideological frames – they are no more ultimates. Hence none of their political ideas can be treated as exclusives or imperishables.

My works are riveted towards this comprehensive direction of change and possibility. It offers them a place for argumentation, but does not admit of their eternality, they are perishing processes, with possibilities of further improvements. Hence arguments are not ultimates, they are entertained to be transcended from various alternative perspectives and possibilities.

It is for this reason, that I would like to preseve English to remain as language for cultivation, or for deliberation, not merely for subtle and critical understanding but also for 'communication on a world-wide basis' with strength and courage.

It is a language, though a foreign language that had an important historical role, which recorded the end of the British rule of India, and was replaced by India's Independent process – so that further possibilities of Indian politics with progressive changes, could be envisaged from that Indo-British linguistic perspective.

English ceased to be a foreign language, but it got a new spiritual outlook by a process of Indianization, both linguistically and politically, and as well historically – with an international spirit.

English language became progressively more perfect, with the assimilative elements of India's cultural processes, than what it was before. It became enriched with elements of transcendental possibilities, which the non-Indian traditions lack. These processes and possibilities are continuously present – with both ends open.

With the pride and privilege, of an Indian as pursuing deeply its cultural heritage and process I would like to preserve, its linguistic efficiency in the communicative sphere, without any element of dogmatism, so that I will not be lost in the environmental forces where English is applicable for the advancement of knowledge in the spheres of science and technology.

The same nationalist-internationalist process permeates through the local languages, of various geographical regions, in all directions. These processes, from an international perspective can be perfected with the subtle communicable possibilities of English as a means of communication.

The other nations of the world, European, American, Asian and African, in general are slowly trying to realize, this international possibility of a political liberation through English.

In stating this I am never disparaging the process of intensive study of each one's mother language, which shapes man's inner self with a spontaneity and naturality. That can be transmitted to English, to develop the multifarious prospects and possibilities of English as a dynamic language.

In my long teaching experience, and through my publications I have helped a lot, not only the students of Sri Lanka, where I taught for twenty years at the University of Ceylon, but the students of the different parts of Asian countries, used to come to the University, and when I started teaching in America I had students from China, Japan, Burma, Vietnam, as also the American, Canadanian and British students. My function was to shape their minds in progressive international directions, by a detachment from their exclusive cultural perspectives, through a training in transcendent insightful meditative directions.

My main theme of teaching was to communicate to them the need to understand the insightful transcendental ways of perfecting the human mind through psychological training of the human mind through Buddhist-Vedantist-Jainist ways of India and of Whitehead's process-philosophy, by way of contrast and relation.

This process I could always communicate to my students and colleagues without any strain or difficulty through English. I am operative here with a common language, transmitting to all the importance of the transcendent spirit in all of their thinking activities, announcing to them that there is no ultimateness in every experiential sphere – all run into one another enriching one by the other. It is a complex fertilizing process.

I have also realized that English as a language is not restricted to a country of origin, but it has an international and intercultural possibility and development. I am speaking of a language, that has a possibility of change and development through varied environmental forces.

It becomes an asset for many countries to be utilized effectively, when it can be extensively utilized for communicating with a speed through computer, as an astrophysicist gets from the telescopic, and microscopic instruments to extend one's vision to the galactic universe. I am here reminded of the novel researches of C.V. Raman, in the scattering of radiation process in multilinear directions, from physics to aesthetics, from the sphere of the physical universe, to subtle vibrations of the musical instruments.

There is use in specialized learning of one's regional language, its value is considerably enhanced if the author can communicate himself as in English or through any other foreign language, if he happens to be so efficient in that language. This makes an access into the whole world, crossing the boundaries of one's country.

Coming back to the Indian Constitution – I would like to draw my attention to the judiciary headed by the Supreme Court Judge and his cabinet.

It is kept separate or independent of the Constitutional affairs, though the Chief Justice and other officials need the final sanction from the President of India, for the selection or nomination process. This is a continuation of the British administrative process. India does not give any exclusive power to any elected member from the President to the Provincial Governors, they are cooperative officials of a federal unified Republic.

Before independence the British government, used to function from the Viceroy to Governors of the different Provinces, and to the subordinate officers from the district magistrates to the village Tashildars, or revenue officers as a regular office procedure.

According to the present India's Constitution the Prime Minister acts for the President, and President acts in consultation with the Prime Minister, and both serving the will of the people, exhibiting a reciprocal process with an acknowledgement of each one's indebtedness to the other official and the people at large. Here also each official excels himself to cancellation of his restricted power, but stretching himself to the powers of another, in order to be more expressive through others.

In contrast to the British Crown the independent India is operative through the elected President to the other elected officials as mentioned. There is thus a reciprocity among the different power-centres. Hence what is emphasized is a balanced operational process, of a centrally oriented Government, with a general sanction of the people representing through votes.

Moreover this Constitution, unlike the previous government under the British rule was a law enforcement on the Indian people. Now India's post-independent Governments is primarily meant for the Indian people. Here the 'Law' of the rulers has changed its shape as of a person serving others with a sense of self-sacrifice.

The Constitution of India, in its, so complexly varied structure is different from more or less uniform British and American Constitutions, as mentioned earlier.

Indian Constitution is not strictly formal, or an absolute command – but widely experiential, knowing that the confronted experiential process is never categorically presented, but may or may not be present, in an expected fixed pattern, but as an 'if-then' process, so one has to wait for the *event* to happen, and then act with an insight, to tackle the emerged situation.

It is not a Vedic interpretation of the *confronted universe* as it has emerged as a possibility, in the milieu of other infinite possibilities?

India demonstrates a transcendentally based confronted process, confronting almost all major religions of the world, with varied cultures stretching from the north to the south, and stretching again to some islands under Indian government, and from west to the east, including the pockets of many tribal people, with linguistic variations, to be taken into consideration, not as an absolute state but as a possible relative situation, with alternative possibilities.

Unlike the British Crown-based Constitution, and the President-based American Constitution, the Indian Government, by a contrast is headed by the President, with a difference. He confronts a vast population in explosion with varied alternative ideas and ideals, in contrast to an American President who as an absolute leader from the perspectives of American Constitution is operative with a Congress, represented by the Senate Members of the Democrats and Republicans as voicing the will of the people of the country, with choices for either the Democrats or the Republicans.

Often there is a possibility of an impeachment of a President or there is a move towards a Third Party, for a possible change. Now many Senators are reluctant to continue their long-drawn thinking hoping that some fresh blood will emerge to change the Constitution in a different direction, but they are also apprehensive of the capabilities of the new recruits.

It may be that the new comers may not be so capable as to be able to bring any change, or will continue the old ways with less capability of continuing the old tradition. All this follows from a postulation of unchanging Constitution.

But if from the very beginning people admit that the Constitution is people oriented, then the attitude would have been more pliable.

Indian Constitution is not simple, but it is an interrelated process with a complexity, as indicated above. But here India admits that the direct *will* of the *mass minds* emerges in new and special forms with historical process and circumstances.

As not a Constitutional specialist in a special way my objective is only to draw attention to the *general ideas*, surveying several types of *formally* oriented Constitutions, and also other types of cultural variety.

By a further stretch there is also a move to the Constitutions of the other major counties of the world – I am making an attempt to clarify some special features of Indian Constitution, by way of a contrast, as already indicated.

The speciality of the Indian Constitution in development through different terms of the elected members from the President and Prime Minister to the Chief Ministers of the different states and Governors is throughly dynamic in many-dimensional ways, depending on the different capabilities of these officials, in tackling the internal affairs of the country and the widely variant types of people, communities of various types of religions, cultures and customs, and the changing processes of the World-situations operative with various ideological schemes.

India's basic detached and transcendental attitude is expressed in political sphere of experiential process from the initial stage of Gandhi and Nehru, to constitutionally elected Prime Ministers, from Nehru to later Prime Ministers.

India's is a vast variant sphere of reference with the novelties of continuous transcendent modes of tackling the confronted empirical situation, with some kind of detachment. Not that 'detachment,' is not present in other types –*formal*, or *specifically culturally* oriented.

Indian Government, at least has a claim for running the 'democratic process of a Sovereign Socialist Secular Republic' by the free-will of the poorest of Indian peoples of various levels, trends and dispositions.

The individuals, stetched to the villages and remote tribal areas and islands – are considered as parts of Indian society. Insightful Indian people want to know the special types of their psychic dispositions or *bents*, so that no *harshness* is imposed on them.

India, to this century, bears a respectful attitude towards human beings, and also to animals, not in a anthropological-biological sense but in the depths of India's varied transcendental perspectives, from the myths to the sciences. Human heart and animal heart in depth flows along the same channel.

Only recently in America such feelings as animal ethics has emerged in medical sphere to stop the experiments on animals even when it has been found that biologically human organisms are not so similar to the animal organisms. Simultaneously 'with animal ethics,' other types of 'ethics' as 'medical ethics,' or 'business ethics,' and other ethics have already emerged and other types are likely to emerge.

My suggestion to stop the number of ethics, following Occam's law of seeking brevity would be to welcome a transcendent background and foreground for all aspects of experiential process as confronted. There is no ultimate, there is no command, nor an 'ultimate text or sacred scripture'—all are just possible options to be chosen with meditative insights.

So experiments with animals have to be conducted with restraint. And in the case of confronting any experience an attitude of restrain and caution in called for.

In Indian history, even to this date, the monkeys have a history of their protests, if something wrong to them is done they in a body assemble before a law court for justice. This flouts the view that the animals have no capacity to understand the human mind.

History of all countries, have a record of horses, dogs, elephants or cows. In someways they are not only conscious of the human mind, and but they can communicate their minds, and they are willingly responsive to human needs, and they serve or save their masters in critical situations, even sacrificing their own lives.

It is for these reasons – the dogs, monkeys, horses and elephants are utilized for various purposes. Dogs are watch dogs, police dogs and an another eye for the blind.

Both Karan Singh and Narayan Keshavan try to establish this character of India's democracy, symbolically represented through its replica of three visible Lions, shedding a lion as behind, at Sarnath, during the reign of Asoka, on an abacus, which has a 'Dharma Chakra' (Wheel of Cosmic-Divine Law) in the centre, a bull on the right and a horse on the left. This is a symbolical suggestion, to be supplemented by other symbols.

There is also the national tricoloured flag. The insightful persons admit its shortcomings and frailties, but hope for a progressive better future, as if to match hope with reality, in the path towards progressive Truth (Satyam eva Jayate – postulating only that the 'Truth' is ultimately triumphant).

This is an acceptance of a symbolical possibility through the acceptance of a *Truth* that is not ultimate in a fixed conceptual-empirical sense, but a possibility with a meditative insight.

India's confronted conceptual-empirical process is widely pragmatic, symbolical and experimental as a possible scheme – neither finally subjective nor finally objective – it transcends both these confronted conceptual and empirical processes, in ever-present self-shining possible ways of Sankara, a Vedantist of the eighth century, going beyond the Upanisadic-Buddhist process.

India's higher types of Intelligentsia, even of the varied different religions exhibit in India, for centuries insightful orders of experience with no ultimates. They are all tolerant attitudes, in various possible alternative ways, assimilating the best aspects of all major religions of the world.

It has a comprehensive philosophic-psychologic attitude, making accommodations for the development of all experiential processes from the myths to the sciences and technologies.

Nehru, the first Prime Minister, by his dedicated work, with the work-load of 17 hours a day took responsibility of the whole country, through India's great organization of the Congress Party to which he gave some concrete shape by the help of his father Motilal and by the insight of Gandhi, from a broad secular perspective.

He took note of religious varieties, and of the order of people, classes and communities, or parties as constituting the country from within, and also taking note of the different foreign countries, pursuing the best of various political ideologies.

By his policy of nonalignment, after Indian independence Nehru could maintain a speciality of India's foreign policy, by courageously facing the representatives of the Superpowers, with a due respect and recognition, without siding with any of the confronted powers.

Before Indian independence, he was operative through the Indian Congress, often differing from Motilal, Gandhi or other senior members in favour of the younger people like Subhas Chandra Bose and Jayprakash Narayan.

But Nehru differed from them, also, to maintain his individuality, and he became progressively conscious of the conflicting processes, of the European powers, and of the surrounding Asian powers, having varied interests of their respective dominant personalities and peoples, becoming critical of the fascist process, and as well against the British and French colonial possessions.

He often courageously expressed his views without caring for any Western Powers, thereby raising India's prestige to attract the attention of the United States, Russia and other countries of the world.

He did not side with any particular power or personality, forgetting his nonalignment as a political attitude, in tackling any international situation.

His foreign policy, through nonalignment was not neutralism, but a detached decision, with freedom, in any critical situation.

He was always alert not to displease the Superpowers, by showing any partiality for any, and with the Asian countries, he knew how he had to be operative with a Buddhist attitude, to dive deep into the fundamentals of the cultural heritage of the Asian minds in general.

But he was also aware of their change of attitudes after the British, French or Dutch rules of the different Asian countries, as mentioned in several of books on Nehru and of the politics that prevailed among the South and South-East Asian countries during his time.

Nehru was operative with a great alertness and skill, by taking recourse to continuous travels to those foreign countries and attending conferences.

But all these possibilities of his nature flowed spontaneously, revealing his compassionate heart in critical situations, as in times of family mishaps, or at the time of the unforeseeable events that happened to India, during the periods of riots before and after Indian independence and at the time of Gandhi's assassination.

Following some principal ideas of Rasheeduddin Khan's paper, on Nehru, entitled, 'The System-Builder' in *India News*, December 1988 – I found that Rasheeduddin depicted Nehru, from various perspectives. According to him: Nehru had a vision of transcendence, Indo-Islamic elegance, tribal passion, and Western liberalism all rolled into his composite personality.

With these many-sided qualities of Nehru, he thinks further that Jawaharlal Nehru ceaselessly strove to give shape to a vision and devised a system that should serve India long after he had left the scene.

He had a capacity to solve some problems among various religious, social and political order of people, he had to tackle as a president of the Indian Congress, or any other responsible capacity of the India's Congress Party, in the milieu of other differing parties or persons.

He was truly the renaissance man and the builder of modern India, in dual way, by finally resting on some kind of transcendental vision, admitting the human frailties, as everything is a possible selective process, without any claim of finality.

There are lives which do not end with death, but continue to live in the heart and consciousness of the people of India. Nehru was such a personality, representing India through its different periods. At the sametime, Nehru awakened the Indian minds to a modernizing process with the advanced ideas of science and technology, without which the status of the Indian masses, could not be levelled up.

Khan by his comprehensive survey establishes his point. In other words, he suggests that Nehru's political scheme was pragmatically conceptual-empirical, but progressively towards post-conceptual-empirical insightful modes as transcendent possibilities.

Hence, his attitude, in every sphere of experiential process was widely comprehensive and tolerant, and truly stretched to India's advanced Vedantic philosophical directions.

Agreeing with Khan this writer feels that Nehru had prepared the ground for India and its people, both before and after Indian independence, enabling its people to face other nationalities of the world, as equals, and with dignity, as conscious of civilization in progressive development, from before the Christian era, to the present political perspective.

India is one of the developing nations of the world with a sense of tolerable equality, and maturity, if not in wealth, at least in the demonstration of civilized attitudes of refined behavioural process.

This enlightened attitude is conspicuously present and represented in the life and activities of Nehru, from the days of the British rule of India, to the period of transition. He then confronted a tragic phase with the partition of India – into India and Pakistan.

Nehru transcended the so critical periods. His was an insightful attitude, helplessly placing himself to any critical situation, with some kind of hope for the better, believing in the ultimate goodness of man. In this attitude, he received a great shock from the Chinese invasion of India.

If one minutely studies the book, entitled *Nehru: A Political Biography* by Michael Brecher, an American politician, one will find faithful representation of the life of Nehru, from his early stage. His was a unique life-history, with detailed reference to Nehru's family life, education, marriage with Kamala, and their daughter Indira. He also refers to Indira's husband Firoze Gandhi and children. But Nehru – placed within this changing milieu – chose to continue his political career, following his father Motilal Nehru and Gandhi.

His was also an open protest against the British rule, apparently violating its laws, for which he suffered the inhuman treatment of the Indian police under British control and direction. His life in different jail-situations, with some facilities, or no facilities are pathetic tales, depicting his psychologically isolated existence – a punishment, hindering a smooth and normal relationship with his parents, wife, daughter, grand-children, and often complicated by the serious illness of his wife and finally terminated by his wife's death.

His was undoubtedly a life full of sorrows. But he could detach himself from some such psychological sufferings which often paralyse other individuals. But he could overcome them with a Buddhist detached spirit, by occupying himself in writing books and letters, which were published when he was out of jail.

In my *Dynamic Facets of Indian Thought*, I have utilized several of his publications, pointing out the superior order of his treatment of a subject – of an Indian sage, exhibiting a superb transcendental psychology and spirit. But all-through his career, he remained a man with great practical insight, in the normal socio-political milieu.

Nehru remained steady with his main objective of forcing through Indian independence, not just idealistically as may be generally understood, but with a depth of understanding. Many of the co-fighters of India of his time, for independence, of his party, and other opposing parties were critical of his attitude, but he always maintained his own individuality and composure.

His political and moral fights, challenging Gandhi for his dilatory tactics, brought him closer to Subhas Chandra Bose, of Bengal, and Jayprakash Narayanan of Bihar, irritating the seniors of the Gandhite groups of the Congress Party.

His was an indication of a broader understanding of socio-political process through scientific and technological process, and through progressive planning, which Gandhi was slow to accept.

All these international problems Nehru could tackle, maintaining his integrity and dignity, keeping himself aware of the European conflicting affairs with a great deal of insight.

His was also a direct study and observation of the Imperial British and French policy of colonialism, by his frequent travels to Europe. He was opposed to warring spirit of the Fascists, Marxists et al. He was not also supportive of the Communist policies of Russia and of other countries following, with some difference, some form of Communist ideas, as of the Chinese.

Nehru, both before and after independence, was always for democratic social process. All these processes, in the context of Nehru are well-depicted with wonderful skill by Brecker, with deep feeling and understanding.

Taking the full responsibility of India and its people, to later critical periods of India's history, since the beginning of the day of India's independence to Nehru' renewed Ministry for three terms to his death – India seems to proceed, following the path of Buddhism to start with.

After facing the shock from the Chinese, Nehru entertained a reflective attitude of a synthetic or integral Vedantic vision, with a sorrow and tolerance, and respect for many paths to truth. He admitted with a liberal spirit, India's confluence of Indo-Muslim culture, and the passion of the tribal and urban cultural process, as assets.

His was a vibrant nationalist spirit with a secular touch, and democratic socialism with the best elements of Western civilization, so that he could attract the attention of a literary giant like Bernard Shaw, and a scientific genius like Einstein of his time, and at the sametime, after Indian independence, the words of understanding in contrast to Gandhi's politics and activity.

Finally he could get an affectionate care of Gandhi as the foremost of his disciples, even after the tragic partition of India.

The scientists of India of Nehru's time, from C.V. Raman to Homi J. Bhabha and Shanti Swarup Bhatnagar, all responded to his call for advanced scientific outlook, by providing him with the possible schemes for India's progressive economic-technological prospect and development.

He had differences of opinion with the above scientists, but had also an agreement with them, to face critical situations for directing India to a broad transcendental orientation towards life. He had to wait for improving India's economic condition through scientific and technological advancement.

In response to their specialized scientific and technological suggestions,

Nehru could open up several research centres in different parts of India, and at the sametime, he could also feel the priority and importance of Defence of India, as a newly emerged independent country.

He judiciously invited P.M.S. Blackett, the British Nobel laureate, to submit a report about how scientific problems of defence could be addressed to India. Based on Blackett's Report, a Defence Science Organization was started in 1949 with Technical Units of services. This resulted in the formation of Defence Research and Development Organization in 1958. India's scientific and technological development was continued. This Defence programme was enlarged continuously through the successive ministries, so that the present Prime Minister, Narasimha Rao, could boldly say why he will not be inclined to sign NPT (Nuclear Non-Proliferation Treaty) or attend any five nation conference.

India's nuclear programme categorically is for peaceful ends only, or for keeping the Indian Ocean free from any foreign intervention (cf. *India News*, 1-15 March 1992).

After the chaotic situation of partition, the 500 Maharajas of the different Indian States, had the option to declare their independence, but Nehru's insightful Deputy Prime Minister Sardar Patel, with a tremendous speed compelled the Maharajas, to join the Indian Union, for the good of their future, offering them not much time, but preventing them also from any anti-patriotic attitude towards India.

For the common interest of the defence of India, and for the prosperity of India's millions – their support was indispensable. Most of them agreed with Patel, with only two exceptions – Maharaja of Junagarh and Nizam of Hyderabad. Patel dealt with Junagarh tactfully, but for getting control over Hyderabad he had to use the police force to compel Nizam to join the Indian Union, otherwise Nizam was ready for a conspiracy with Jinnah of Pakistan.

Without consulting any, to avoid delay, Patel allowed no time to Nizam, but brought Hyderabad under India's control. Nehru and Rajgopalachari at once approved of Patel's successful political move, which demonstrated a keen insight and quick decision.

India came to a stable condition, and Nehru, with Patel's skill and his success in the general election after the death of Patel soon after, stabilized India's condition. He could strengthen the foundations of parliamentary democracy, and this was a continuation of an over-all success, through later progressive five year plans in scientific and technological ways tackling many critical situations.

His simultaneous emphasis on the agricultural projects, made the country progressively independent in the food production for the rapidly increasing population of India.

Long after Indian independence, and after Patel's death, Nehru adopted the defensive policy of absorbing India's Goa, when the Portuguese Government invaded India, India successfully managed the Goanese problem, through its people, and through the able operations of the Indian Army, under the command of General Chaudhury, within a short time, without sustaining any casualty.

People of India, during Nehru's life-time as India's Prime Minister, were critical of his failures as not a good judge of men, or of his wrong policy of pleasing China, considering its Buddhistic heritage. But really nobody can be perfect, having a perfect foreknowledge of a person's mind or of a temper of a country, in the political sphere, specially when China was experimenting with some form of Communist ideas.

But undoubtedly Nehru had a leadership in initiating the development of the scientific and technological processes, and devising the process of a superb foreign policy of nonalignment with the Superpowers, and strengthening the inner stability of many of the developing powers, influencing them differently to foster the process of nonalignment, as a counterforce to balance the activities of the superpowers.

So one could see in Nehru, the vision of a gradual dissolution of the attachment for an obstinate ideology of communism, and the dissolution of its power, inspite of its potentiality for influencing the whole world through some kind of indoctrination.

Now that dream-world of the fixed ideology is towards a gradual demise, through Russia's initiation of a new socialist policy, through a scheme of *perestroika* one can hope for a progressive world of peace. It has now become evident that the world is for change, so alternative opinions of thoughtful men have a reasonable place. There is no use, fostering an obstinately assumed ideology. Democratic ideas have to be entertained with admissions of changes and reformulations.

So democratic policies are in change, both within and outside a country, by way of progressive contrasts, with some modification and peaceful gesture, by abandoning the tenet of no compromise with the opposing ideological frameworks of Communism, as in the past historical periods.

The world as a whole now has stepped into a process of constant change, in a different sense by a kind of detachment, as the world has now become closer to one another for scientific and technological advancements, and fast communicative process by tolerating its progressive gestures towards some changes in democratic ways, going beyond – from cold war, detente, and other intermediate processes – to progressive peaceful coexistence and finding out several preventive processes to eliminate wars from international situations.

The prospect for all countries lies in progressive talks about disarmament. The world is gradually towards self-development by a mutual exchange of ideas, and opening up possibilities for trade and commercial relationships. In the post-Gulf period of world-history, this is possibly the only alternative to experiment.

With this widely changing patterns of the world-process, in all aspects of experience, I subscribe myself to no fixed *ism*, or ideology, and this philosophic attitude I have developed in one of my books with the title, *Experience in Change and Prospect Pathways from War to Peace.*

While admitting the changing grounds from the *ism* to the *processes* – I am able to see the progressive character of India's democratic process, both in comprehensive philosophic ideas and also in realistic ways of insightful scientific and technological projects, so that all can be commonly interested in profitable business, with a liberal attitude. Common good or interest – through trade and commerce – in the long run is most fruitful.

So, India is in continuous change, due to changes of socio-political conditions within India, and also outside of India. The Indian prime ministers also have to be always on the alert, both with regard to internal and external affairs to step up India's international political status.

The prime ministers in turn exhibit their progressive capacity to be appropriately operative with the changing internal and external affairs. There are problems both from within a country and also from outside, due to change of policies and activities of a country, and of the international contending powers. So the prime ministers are educating themselves in advanced politics, as fast as possible, after their basic education at colleges and universities.

In this connection, I often remember the activities of Nehru, Patel, Gandhi and others, of the Congress and other parties, including the Muslim League under the leadership of Jinnah et al., both before and after independence of India.

All these personalities were well-educated and most of them were successful lawyers. But what has to be emphasized is progressive understanding of the transcendent possibilities with a detached observation of both the subjective ideas, and the objective situations, so that transcendence from both the observed orders becomes a further transcendent possibility.

In my early years I used to think that Partition could be avoided, but now I find it was inevitable, and it was good for a speedy transfer of power and transition to India's independence, otherwise any delay would have spoilt the case of India's independence.

To me, as I now feel, it was not ideologically oriented, it was transcendentally oriented, for, both Nehru and Lord Mountbatten were operative with detachments. These aspects of their character and attitude were most judiciously displayed by Brecher.

These personalities were dual benefactors of India's stepping into the process of independence, through some possible processes of 'dominion status,' under the Commonwealth of Nations under the British Crown, but finally this transitional process was abruptly ended by the stage of complete freedom of the Indian Republic.

It was necessary for India's defence and also for India's making of the

Constitutional process and progress, with Mountbatten's direction and prompt decision, with a quick approval of British Prime Minister Attlee. Nehru's independent role lay in his insightful acceptance and rejection, suiting Indian frame of reference, in consultation with Viceroy's Constitutional adviser – V.P. Menon, also Nehru's friend.

The entire process was often scrutinized by the sober judgment of Sardar Patel, who displayed a wonderful capacity in bringing 500 independent Maharajas or rulers under the aegis of Indian Union, as mentioned.

In Nehru one could see that during the critical period of the inhuman killings of men, women and children, of the both Hindus and Muslims, he had to decide for Partition instead of civil war.

India's unity was not so easily obtainable. He stopped the Hindus from killing the Muslims, in frenzy, in Delhi, by drawing attention to India's secular ideal, by piercing through the rioting mobs.

Often people of India or the foreigners misjudge Nehru as an atheist, but Nehru's attitude was all-through transcendental, indifferent both to theism and atheism, for, the transcendent is beyond the limitations of both.

So long as a man remains attached to subjective and objective ways, he remains confined to the intellectual forms. What is needed is an ascent to transcendent post-intellectual wisdom mode – beyond theism and atheism.

Religion, including sciences, needs continuous change and progress by a transcendence. This is evident from Nehru's activities during the period, before and after India's independence.

This superb attitude is reflected conspicuously, on the occasion after Gandhi's assassination, and at his funeral.

After looking at the deadbody of Gandhi, Nehru detached himself from his immediate shock, and went out to inform the crowd, in glowing terms, from the depths of his feelings; his expressions crossed the boundaries of time and space, when he said, 'Our beloved leader, Bapu, the father of the Nation, is no more.

Then with several modifications and hesitations he uttered – 'The light has gone out, then immediately correcting himself, he said that the light was not ordinary, it was a continuously luminous for many years, in India and in the world, when he was alive. In a moment, he again reminded us of the right path, and said taking this ancient country to freedom.' Nehru's expressions were flashes of quick changing emergent processes, and he glimpsed them all and immediately expressed them.

I am quoting with changes, to bring out in his statements the flickers of transcendental elements, which are clearly manifest, when Nehru in the Constituent Assembly, said with a depth of his sorrow and feeling for Gandhi that, 'as an individual and as a Head of the Government of India, he was ashamed of his failure to protect the greatest treasure, that we possessed.'

'He (Gandhi) was not of the common clay that all of us were made of . . .

A glory has departed and the sun that warmed and brightened our lives, has set and we shiver in the cold and dark.'

I had an experience of a 'painful shiver' with an inner vibration, when I heard this news in Sri Lanka.

Nehru's statements vibrate with a Christian-Buddhist Spirit, and are ended by saying that – 'We mourn him; and shall always mourn him because we are human, and cannot forget our beloved Master. But he would chide us if we merely remain in that mournful attitude... Let us be worthy of him.' This is undoubtedly a mystic statement which gripped him at that moment. Nehru was not a sceptic but a transcendentalist.

Nehru's final statement was in tune with the statement of the Buddha, which was stated to Ananda. Though Ananda remained sorrowing, Nehru came out of the grips of 'Sorrow' to declare the perennial Truth of the Buddha's 'Path', with a courage: To give up the sorrows and follow the 'path' – the 'perennial Truth.' Truth in progressive 'transcendence.'

All these linguistic flares and heart-penetrating ideas, of Nehru, through lucid English constitute a treasure for the world to share and enjoy.

In the true spirit of his masterly English, he should have emphasized the need of English as an instrument of communication for the different parts of India, and also of the world as a whole.

But he should have as well emphatically suggested that English should continue, not for a limited period, in India, but for ever – abandoning his pragmatic solution for a formula of linguistic process for the North and the South. From East to West, North to South languages flow into one another in mixtures and emerge into fresh languages.

He should not have given a chance at all for reorganization of India through linguistic preferences, for a wrong political initiative. His suggestion for change should have been through a convenient cultural emphasis, by the development of each regional language in relation to an 'intercultural oriented transcendent linguistic process.'

For stepping into the new order of historical process English in India had gained that status. Now, in tune with the prospective international milieu, of market economy one can freely accept English as a language of choice, separating it from any restricted politics. In this lifting process one can as well give a relative importance to the cultivation of one's own mother tongue or language, and also many languages of the world for both commerce and for knowing the cultural wealths of the different parts of the world. Trade and commerce flourish and take new directions with the operations of many languages in togetherness for day to day life.

Besides the political area there are other areas culturally more important. Life is based on many dimensional processes.

What Nehru said in the context of Gandhi's unnatural death, was nothing short of the expression of the Buddhism, that is sacredly preserved when the Buddha was demonstrating his meditative ascents and descents (*Parinirvana* as a meditative discipline for practice) to his disciple Ananda. But Ananda was overpowered by a sorrow, and so he became conscious of the possible demise of the Buddha, though the Buddha wanted him to understand his message with a steadiness and calmness – the inevitable Truth (the Death that happens to all).

Those who read the story of Indian independence as a mere narrative, do not understand either its historical process of independence, in its different stages as pointed out above, or its gradual 'formation' of the various stages of Constitution from the President downwards, only to preserve the 'Integrality' of the federal Provinces and Territories of Indian Republic, with concern, as a secular Union State.

The credit goes to the later British Government of Attlee and India's self-chosen first Governor General, Lord Mountbatten, during the transitional period and also at the first stage of the Indian Republic. Here again the roles of Nehru and Sardar Patel and other associates are important.

A consideration of these and other necessary things is related to India's progressive Constitutional development, for any Constitution is not a static phenomenon, but it takes progressive evolutionary shapes to transform itself suiting the minds of people at large of the country and of the evolving and transforming milieu of the World from varied aspects of experiential processes, from cultures to the sciences.

The later Prime Ministers, of India after Nehru became operative, in these multifarious directions with success and failure, to tackle India's internal and external problems, with insight and skill, as far as became possible. Hence the passage into the next chapter is a natural transition, to consider the roles of the successive Prime Ministers of India.

Having now a glimpse of the continuous progress in the spheres of science with great details taxing many minds with some impatience I can see a further prospect beyond Einstein's both Special and General Theory of Relativity, in prospective field-processes, transcending all restricted positions and possibilities in the sphere of science also.

The world will gradually learn to relate all aspects of experience from the myths to the sciences – as continuously penetrating into one another, to come out ever-fresh in all emergent situations – by a transcendence.

The mission of mankind as a whole is not to find an ultimate but to pursue a vibrating process with infinite possibilities. This ever-changing and developing perspective should be the chosen path for India to follow. These aspects of experiences will be progressively clarified to the stage of Conclusion of this work. This world in its humanist relationship is a mixture with emergence of various experiential possibilities in togetherness.

In the next chapter more of Nehru and other personalities of India will be given from Durga Das's journalist perspective.

If there is some repetition or even a kind of prolonging a process one has to tolerate it for, one will find, in Durga Das as an eminent journalist – a display of a power of treating the same subject with a freshness in focussing the entire process, where his role was indispensable. It was a change of history – through interconnections of great personalities to future novel possibilities, with a speed, insight and foresight.

REFERENCE

1. This chapter is evidently long as it has many objectives to fulfil through the studies of several important books for shaping the complex process of India's 'Constitutional Process' leading to an advanced democratic form of Government, apparently to be confronted as a secular process, but really one is enabled to ascend to a transcendental order of experiential process from the myths to politics. It is an ascent to a New Historical process with a New Politics.

   To traverse along this complex progressive path, in intercultural and international contexts from a widely comprehensive aspect I have chosen to develop the main ideas of the following books – with a penetrative insight as far as I could bring in with the energetic continuum of my organism, pursuing the disciplined path of Buddhist mode of life – where the learning lies in a total detachment from all restricted intellectual forms.

   These chosen books are: (i) *International Politics in the Atomic Age* by John H. Herz, (ii) *The Words of Gandhi* by Richard Attenborough, (iii) *The Government and Politics of India* by W.H. Morris-Jones, (iv) *Cases in Comparative Politics* edited by James B. Christoph, and (v) *Nehru: A Political Biography* by Michael Brecher. The relevant details of these books have been considered in relation to one another, with necessary insightfulness.

CHAPTER VII

India's Politics with Some Retrospective-Prospective Visions of a Changing and Advancing Constitutional Democracy

In this volume I have attempted to mould philosophical thinking not merely taking into consideration, the abstract conceptual-empirical modes of Western thinkers in progressive changes, as the philosophical milieu for presenting Indian philosophical deliberation by way of progressive contrasts but I have also indicated its further consequent possibility.[1]

I have further gradually placed the insightful transcendental modes of India, furnished from the time of the Vedas to the present day movements – as the confronted conceptual-empirical processes, with alternative possibilities, for discursive purposes.

Here I have remained cautions not to fall into the traps of any ultimate, in any order of experiential process. But I have treated them as possible expressions of human mentality.

My attitude all-through was extensively dual, with an insightful understanding of the advanced currents of Western and Indian thought. My deliberations were for application of this philosophical thinking to the different aspects of experiential processes of my interest.

With reference to the major Asian countries – China, Korea and Japan – my attitude was varied, in accordance with their differing cultures.[2]

The contrasting attitude as developed in advanced Indian thought, as it stands, can be comparable with the advanced line of Whitehead's process-philosophy in contrast to the early Western conceptual and empirical abstractions. It is a progressive passage towards post-intellectual wisdom modes – non-different from the Buddhist progressive concentrations.[3]

I have further drawn the attention of the readers to the chapter of Part I – in the context of shaping philosophical deliberations of some of the contemporary Indian philosophers, their judicious discretion to mention briefly their individual life-history, pertaining specially to their family-history and educational career in outline to a transcendent possibility of their choice.

Each philosopher's life-history in the land of his birth and his later movement within or outside the country, forms a background to understand his preference for some emergent philosophical attitudes and ideas, and his interpretations of philosophic deliberations.[4]

He may be conscious of the Western, and Asian thought – in contrast to Indian thought in general, but the question is whether he is free from the attachment towards the different cultural processes – as to express his creative philosophic attitude, with some insightful understanding and freedom.

The value of a philosophic thinking depends on his final decision, in the shaping of philosophic expressions as creative manifestations, with no premeditated boundaries.

This kind of deliberative freedom I have seen well demonstrated in Haldar, Humayun Kabir, Ranade, Sharif and Wadia – mentioned in the context of the Part I: on Philosophy. I have taken note of some well-known personalities, belonging to different religions: that developed in India, with liberal attitudes.

In such observations of mine, I have found out that philosophy without one's own life-history is a tree without its roots, hence not a real philosophy.

Taking advantage of that dynamic process as a background for my reflection – I shaped my philosophic, scientific and political deliberative consciousness, in interrelation and in complex emergence – suggesting that a person's travel to another country, and a fairly long stay in that foreign country or countries – changes his country-centric disposition.

This complex process, as arrived at is not a process of reduction or simplification pursuing one's formal mode of reflection, but it is a continuous confrontation or observation of changing experiential environment – indicating progressive complications, and a facing of confronted universe, from moment to moment, in continuous transition. All of the above representative philosophers were alive to their confronted environments throughout their life-times. I remain present as providing some evidence in my associations with them during the different sessions of the Indian philosophical conferences through several years, in India or outside India.

India's Independence was not just a phenomenon, dependent on an individual's opinion, but it was a confrontation of a complex emergence realized with a difficulty, where many persons with insight were operative, to give it a focal shape. It was a cooperative enterprise, where the participants were many – Indians and non-Indians.

It was an earned political frame of reference – with a great exertion through the roles of some powerful personalities, including the role of journalists and politicians – Indian and non-Indian, in cooperation – shaping, with a detached attitude, the progress of the advanced international events.

But for the journalists's scrutinizing outlook and extraction of facts from an individual diplomat's minds – nothing is fruitful as a journalist brings to birth events as of a child born from a mother's womb. Ajournalist's job is of a midwife in practice.

A journalist quickens the process of birth of an event, for his contracts with many politicians of different countries. An ordinary person has not got that privilege. Hence his operative facilities and schemes have a speciality. This quickening and focussing of historical events with a rapidity, and ingenuity of an active journalist is, in an aspect, a manufacture of events with a psychological display, where many persons are involved with relevant questions and answers.[5]

After India's independence from the British Raj, India became operative with a new frame of reference with a sense of equality with other foreign powers of the world, operative in extending its capacity to grasp the multifarious range of experience from an agricultural sphere to the satellite-oriented country of the present times as provided by India's intelligentsia with the handymen from the India's *masses*.

But this kind of *new game*, in the political sphere was not a possibility, when India was still under foreign control. The new-born child was not there. Independence of India implies the birth of a new child – that has to grow and develop by its inner capacity and with environmental exchange of ideas.

Given this opportunity of independence, India could have been placed in a chaotic situation – which is well-expressed in a book, with a title, *Problems of the Developing Nations: Readings and Case Studies*, by Lewis P. Fickett, Jr. (Thomas Y. Crowell Company, New York, 1966). Here there is a reference to six major developing countries – Algeria, Tunisia, India, Pakistan, Indonesia, Thailand – the colonies of the three major European Powers – French, British and Dutch.[6]

In this book the developing aspects have been presented in five perspectives – philosophical, sociological, economic, military and political. But India's inheritors avoided all possible problems with a deep and far-sighted vision—managing the present and controlling the future processes, with the best principles of a dynamic democratic process, in continuous change with a fluent Constitution as mentioned.

These intelligentsia – politicians, diplomats and journalists of India – with their specialist knowledge of science and technologies were operative with the peoples of this international space-age – hence, according to my observation India's population has always to be viewed from two dynamic bases of polar opposites: (i) Higher orders of India's intelligentsia (so called elites) constitute one order, and, in contrast, there is (ii) India's mass population of varied levels of social conditions, differing in understanding and capacity, with ordinary knowledge of life and aspirations to pursue progressive orders of experiential processes. There is a difference between the two, but there is no open clash between them. In India there is an admission of difference with no argument.

But these emergent personalities were for India, and not for the foreign government to be utilized for serving it.

In this connection, I will not always insist on a compulsory academic programme of education, but only I will grant man's natural aspiration for learning as a simple soul – a village savant, an intellectual giant, in a different sense, with a common sense. Such a person, may as well be worth noting, in contrast to an academically learned person of a city. Such a soul brings in a variety, and a new spirit.

Often, these types of personalities will ignore mere schooling as in the case of Sri Ramakrishna. He chose to pursue his ecstatic meditative experience, thinking that school education is going astray from the religious experience to a worldly experience, which from his perspective, was a lower order of experience. He contributed towards the emergence of an exalted religious consciousness which was spread to foreign countries with revolutionary effects.[7]

One other type may be like that of a Rabindranath Tagore, who thought that in schools a fixed curriculum is provided. The schools cannot go beyond it. Hence it is restricted. Whereas he is free to study everything without any imposed restraint. He will, therefore, choose to pursue his natural dispositions for learning varied things, going to schools is not only wasting one's time, but also a kind of derailing oneself from a right tract. This is another variety and spirit. It is like the forest bird with a freedom of movement to the infinite skies, in contrast to the caged bird with no freedom. The reference is here to the symbolically oriented Upanisadic myth.

This attitude, and expression of his mind—brought him a recognition through a nobel prize.[8] In this connection one must as well note his favourable elite family environment and his own innovative activities parlance and refined taste.

There may be still another type, or just a mysterious life of an M.N. Roy, learning more than any ordinary person through non-academic media – and becoming all of a sudden an adept in political spheres, or in music or even in mathematics. In the sphere of mathematics, he had a challenging capacity to stand on par with Sulaiman, high court Judge of Allahabad, or with Einstein – the world renowned physicist.

He held revolting ideas of communism, which was not upholding Europeanized form of Communism, but revolutionizing by another extraneous concept, which actually hits the capitalist citadel or social structure. He had a foresight of a Gorbachev in breaking ideological boundaries, paving the path towards a common world of interchange through a market economy, with a compassionate attitude. Karl Marx the originator of evolutionary process of going beyond the conflicts within a capitalist society – was not beyond his pre-conceived form.

India is a land of mystery – where there are roles of many alternative types in non-linear dimensions.

Roy's was a case of self-study, with a culminating originality. Hence one cannot neglect the mass-aspect of the population of a country. This is a rare variety with a new dynamic effulgence. India fosters this new variety of personality-process.[9] It is a process of going beyond (transcending) all restricted intellectual forms.

The communist ideas of Russia were revolts against the German basis in Marx and Engels, and beyond the Russian, there were other forms, as of the Chinese, Korean or of the Vietnamese – all were different types of revolutionary political ideas from their respective cultural perspectives – were understood by Roy as possible alternatives. Hence any fixed *ism* – prepares the grounds for varied revolting attitudes from different cultural bases. This as a concrete experiential process is not provided by a restricted abstract intellectual form.

Roy's view was that Communism will not be able to set aside capitalism without – a stretching of Communism – to Asian market economy which is a new and concrete experiential process with progress. This happened to be a suggestion of Roy, with Lenin, long ago. He was not in the path of Stalin, but he sided with Trotsky, and had to escape from Russia with great difficulty in igcognito. His was a new adventure with a risk.

The revolts were all for a search for a new avenue. To my mind these are various transcendent wings operative behind all confronted situations. I am thinking of these exceptional peoples operative differentially, but bringing about inconceivable social changes in the world by stepping out of an obstinately oriented frame of reference.

Hence to my mind events take place as a natural course, but often it is differently oriented by some extraordinary personalities and by some journalists who give birth to certain events by snatching them out from some politicians with powers by some insightful methods, employed by journalists in interaction. It is a kind of a new manufacture of a product not known before.

These journalists of different locations are different types of personalities that emerge, by a process of self-creation. So that to my mind events and personalities operative behind populations are not simple naturally oriented personalities with any fixed or ultimate character, they are trained through some exceptional personalities in some mysterious field. Hence I quite agree with the advanced Buddhist process that develops events and personalities in progressive post-conceptualist and post-empirical directions.

I have developed my deliberations on philosophy, physics and politics in these radical directions all through this volume. No subject is exclusive deliberation on the subject itself, but a constant stretching to other subjects, bringing about necessary changes from alternative perspectives.

I do not believe, therefore, that knowledge emerges in a regular direction, by taking recourse to schooling as normally prevalent during and after the British rule, replacing India's ancient traditional system from the early Vedic period, or a later tradition of Indo-Persian process, the complex fabric of: Hindu, Buddhist, Jainist and Islamic systems, during the Muslim rule of India. Then also during these periods *events* emerged and *personalities* were born, or they shaped themselves with varied insights, with some specialities.

But the historical currents bring into play not only the evolution and development of a country's own cultural processes – but also it confronts the varied cultural processes that enter into a country. My reference is here to India's history and the impacts of the foreigners in successive periods, ending with the British rule of India. In this context – India has a many-dimensional speciality. With India's independence, it is taking multifarious 'emergent turns with infinite possibilities.'

Educational process is a complex phenomenon, it takes various possible alternative turns as the human mind is multifariously divergent. India's population is a complex expression of these various experiential processes, besides the academic processes that are also operative, and run parallel to these processes. The world-process is not merely a reproduction but a creation of novelties.

So my interpretation of confronted events and personalities is not simple, but complex. It is for this reason I do not ever take events or personalities as simple naturalist expressions.[10] Hence the confronted events and personalities are grounds for novel emergences.

Each one of the processes has transcendental basis and a possibility, in alternative directions, unfathomable by man's formal intellectuality.

In my life-history, I have found the activities of all these and other exceptional personalities, emerging out of these and other alternative processes, down to India's complex historical processes, with a modern turn, from a hoary past – before and after India's independence.

I have mentioned about the past and the present, indicating their multifarious emergent ideas in relation and prospect, shaping the pages of my different volumes of *Dynamic Facets of Indian Thought* in some details, as my self-oriented symbolic meditations.

I am not referring to other adepts, in the spheres of music, dance, magic and multifarious other fields of experience and excellence, emerging out of many complex bases, as I do not possess such special capacities, but I enjoy observing or experiencing them, coming back frustrated often watching the varied types of extraordinariness of the specialized persons in their respective spheres of interest. But about these also I have referred to in the first volume of my *Dynamic Facets . . .*, and other books. During my recent travel through many principal cities and places I observed the changes of India's youth, their growing sense of service to country from military services – with a spirit for self-sacrifice for country's defence.

Viewed from these polar extremes I would always expect from a scholar-writer to refer specifically to what kind of 'people' he/she has in his/her mind, in any occasion – pertaining to Indian people:- (i) intelligentsia or to (ii) the masses.

Both are *transcendentally placed* – hence, knowledge emerges in both these groups or centres in mysterious ways, so watching the two corners of the experiential process we cannot hold the view that what a man thinks of is categorically knowable in one form only – there are multifarious aspects of knowability. Hence the sphere of knowledge is widely indefinite, and its emergence in any direction among *intelligentsia* or among the *mass* is not dependent on a categorical form, it is a statement in a form of *may be*.

Hence when I read a book by Durga Das *India from Curzon to Nehru and After* (with a foreword by Zakir Hussain Former President of India, John Day Company, New York, 1970),[11] I found Durga Das, as a journalist, associated with many news papers of India, and also operative with an extraordinary capacity, before Indian independence, its political processes in varied critical situations. He also was in the battlefield of the Second War as a journalist. He marked the discriminative attitude of the British and the Americans – but they utilized the superior and dedicated service of the Indian soldiers as participants in both of the two World Wars. They frankly acknowledged the superior roles of Indian soldiers, but could not develop a non-discriminative attitude towards the Indian soldiers even at the War-situations.

This is human nature. Knowledge, also, is not confronted in the same form. People with *power* shape in one way, and people without power shape it differently, with some helplessness and acceptance – in tune with the historical process and changing environment. This is a new kind of tolerance one has to cultivate in human society, with a patience for a future possibility for mankind.

But after the end of the Second World War and after the victory of the allied powers, when the British during Attlee's Ministry resolved to hand over the administrative power to India, the British became concerned to find out to which power – The Muslim League or the Congress – the British would have to transfer the power. It was during the Viceroyalty period of Wavell in 1945.

Both the Congress and the Leagues sensed the proximity of freedom, but the tortuous game of political chess was not over.

The British Commander-in-chief took the fateful decision of committing to trial, three officers of Subhas Bose's Indian National Army – a Hindu, a Muslim and a Sikh – before a military court at Red Fort, Delhi but it was a misfire, as Lord Irwin's proposal earlier for an All-White Simon Commission on reforms. The defence of the accused was largely conducted by Bhulabhai Desai.

Jinnah too, attempted to figure in the affair. He sent word to Shah Nawaz Khan, the Muslim on trial, that he would. defend him if Shah Nawaz had dissociated himself from other accused. Shah Nawaz flatly declined, remarking: "We have stood shoulder to shoulder in the struggle for freedom. Many comrades have died on the field of battle inspired by our leadership. We stand or fall together." What a 'conscience, what an exalted character'? An attempt to drive a communal wedge through Jinnah, among the members on trial, by the British

Commander went back frustrated.[12] It reinforced the process of Indian union.

Hence as already pointed out, the events do not happen naturally or as what one expects to be, but it emerges as a surprise and a new undreamt of event takes place. So history sometimes does not repeat.

I have indicated here that the mental processes move in indefinite multi-dimensional ways as the particles and waves of the Heisenberg's advanced post-Einsteinian physics. Mine is a suggestion for such a stretch in the sphere of politics to new inconceivable emergences.

After this failure, two months later, after Wavell's announcement of 19 September 1945, there came the strike in the Indian wing of the Royal Indian Air Force, and mutiny in the ranks of the Royal Indian Navy. The prestige of the Congress shot up as a result of INA trial, followed by the revolt in the Indian military. As the new kind of events emerged the Cabinet Mission was despatched by Attlee with a new insight. It landed in New Delhi towards the end of March 1946, with a plan for a 'Union of India,' with a provision to draw up a Constitution by a Constituent Assembly.

In this milieu of unpredictable dynamic milieu Durga Das brings out in his memoir all of these, emergent events, dealing with 'India's evolution from a colonial dependency to a democratic republic.

In this work, with five divisions, Durga Das tells the story of the British Raj under the Viceroyalty of Lord Curzon at its height, and its failure, with a gradual decline and dissolution of the Governmental processes through successive Viceroys.

He was talking of the different emergent events, as historical processes that changed India's environment. Then emerges the process of transfer of power with a partition of India into India and Pakistan.

This process through several commissions, was only a delaying tactics, so the events are to be shifted backward to the past, when it came to a breaking stage after the Simon Commission of 1928, when during the Viceroyalty of Lord Irwin, inspite of his pro-Indian gesture, he could not give any pro-Indian garb. Still then all its members of the Commission were British, but as the luck would have it, it happened to send Clement Attlee, who after Churchill became the Prime Minister of Britain. The events then had a new turn. This is a mystery in historical process.

It was he who clearly had discerned, the immediate need for the grant of India's independence and also felt with an insightful study that for the government of India, with its vastness and variety of Indian population, States and Territories, the British system of Parliamentary Democracy will not suit. India needs a complex form of Constitution. He thought since then that some kind of Presidential form of Government, with a modification to suit India would be a need. If different parties are admitted, I would like to suggest, they are to be progressively directed towards a feeling of one integral India.

Attlee, as I discerned saw complexity within events and also in the personalities, and the emergence of unavoidable situations, demanding immediate tackling with a favourable attitude towards the emergent Indians. In the previous chapter I have given a shape to the dynamic character of India's Constitutional process, as addressed to an advanced democratic process and possibility – which is not a commanding script, but an admission of a changing process, and it is people-oriented. So my suggestion was for a continuous education of the India's population for taking full advantage of the Constitution in evolution and development – in course of History. I am for a new dynamic living 'Politics.'[13] India, through its insightful personalities came to this stage, which Attlee foresaw.

Inspite of such ideas governing Attlee's mind, there were counter currents both in England and in India. So the resultant process of India's independence was an emergence with a great process of India's independence was an emergence with a great suffering for the personalities – both Indian and non-Indian – who became interested in India's politics. This birth of independence is a stepping into a new dimension of historical process and possibility.

What became a *reality* was only a possibility, before it emerged, or it became a resultant process, or an actuality passing through several throes and ordeals in which many personalities were involved. In all these operational fields, there were some important roles displayed by Durga Das as a journalist.

Indian democracy has been constituted with these different insights and elements – as we have already discerned – for the framing of the Indian Constitution with its project towards people's government suiting India's cultural background.

Durga Das, as a journalist in his contact with Burnham – proprietor of the Daily Telegraph – and Clement Attlee, a member of the commission – had some role, but later some actual process was set in process – with Attlee and his group, as British Prime Minister, who could implement the process of transference of Government to real Indian independence in 1947 – through Lord Mountbatten, India's first chosen Viceroy – an insightful friendly gesture in tune with India's transcendental process.

To my mind it was a kind of transcendental expression, operative through some able and willing personalities, but everything of their activities was not visible earlier, till the right time came in *historical process*, when the British failed to continue further its delaying tactics, with a display of diplomacy. History speaks itself in a new language.

The Government of New India, with Lord Mountbatten as the chosen first Governor General of India and Nehru as the first Prime Minister began the initial stages of the government of independent India. It was a friendly gesture for the British and the Indian people.

According to Durga Das, Nehru as Prime Minister reached his peak in the mid-fifties, when he bestrode the stage like a colossus. The half centuries dividing these two men of destiny (Curzon and Nehru) saw the emergence of Gandhi as the leader of the non-violent revolution, the exit of the British with the creation of the two independent and separate dominions of India and Pakistan, and the evolution of a new policy in the sub-continent. These are novel elements on record in India's history.

The panoramic glimpses that I got from the study of Durga Das's book, were not patent by a mere pursuing of some books or newspapers, but it was also a pursuing of a different kind of role that a keen journalist was operative with, bringing out from the minds of the politicians, operative with their different positions and capacities. My statements are feeding processes – emerging from my life as I directly experienced from its early periods.

I read from a journalist's concrete observational sphere, and minutely observed his role, and also the roles of the different politicians or people well-trained for political service in India who came to India during the British period and they faced in India since the time when Indian people in general became conscious of political freedom from the foreign rule – India's insightful personalities had to step into a realistic sphere of human experience.

This aspect of tackling any situation, cannot be obtained without the eagle-like visions of some journalists. For the training of a journalist mind – a teacher in this line of observation and experience – is an absolute necessity. This has been well-demonstrated by Durga Das.

Reading the details of this book, I also found that the roles of the Indian politicians who paved the path towards Indian independence were perhaps more important than the ministers from and after Nehru – carrying forward the special type of democratic form of government, as evolved from the Constitution of India.

In this connection, one has also to give credit to the Constitution-makers and also to those who successfully could follow or carry out the fundamental principles of the Constitutions in confronting the changing environmental processes – both internal and external.

This evolution and development of Indian Constitution was effected with the speed of the progress of advanced scientific and technological age. India's confronted environmental situations, in many aspects became manifest or remains to be manifest in course of time, with a changing foreign policy, and with more complexities and subtleties. I want to emphasize these dynamic currents as they were my direct experiences of my youthful periods facing most of these personalities.

Durga Das here acknowledges the able guidance of the insightful journalist, K.C. Roy, an editor controlling the reporters of the Associated Press of India (API), and other presses – with clashes and reconcilliations, or in critical situations, giving up his services, seeking a new avenue with an abandonment.

The pathway of an independent minded journalist is a hard way to follow, and it is equally difficult to train the minds of the young reporters, unless there is a kind of mutual understanding and a total understanding of a team spirit (rapport).

The journalists are trained in a hard way, studying with intelligence and insight, the leading dailies of London, United States, and the recognized journals of India from India's *Statesman* to *Hindu*, *Times of India*, or *Amrita Bazar Patrika*, run by some private organizations. The standard of these dailies and editorials contained in those newspapers were on par with the British and American counterparts, so that the editors had equivalent reputations and one was treated in the same spirit without any discriminative attitude.

Durga Das being expertly trained as a journalist and reporter had an access into the political parlance between the Viceroys and the Leaders of the Congress, Muslim League et al., and also, persons with any Commission to introduce some changes in the government of India, or bringing some proposals regarding Indian independence. These innovative changes in the British Rule of India, were primarily controlled by the India House at London. In India Office there was a person as Sir Bhupen Mitter with great interest for India's prospects towards independence.

The rule of India by the Viceroy and the staff under him, including the Civil Servants was a continuous chain so that it was a consolidated process, breaking of this fabric was difficult. It is comparable with a general's march into the core of the battle-field.

A subtle understanding of this fabric and its function is not easily understood – ventilated penetratively by some challenging suggestions of such a journalist as Durga Das.

This writer after studying Durga Das's book feels how ignorant he was before his minute study of this book. Now he also feels that the study of the general books on politics is an acquaintance of the mere abstractions, it is not really understanding the complex *political process*, in its realistic phases. A theoretical knowledge of politics, from books and journals is not a concrete field-study of the object. Here events do not happen merely but they are shaped and chiseled by some dynamic insightful personalities. These personalities must develop an interrelated and international consciounss.

Without a concrete pursuing of an active personality as Durga Das, a person, who was more or less my contemporary I could not be aware of the events that transpired then before and after Indian independence.

The ideas of a *dynamic politics* henceforth, to me are from this penetrative perspective of my life-process. From this perspective, again, I feel that India's renowned politically oriented personalities, and the journalists of the types of Durga Das, played their roles – to bring about the process of transfer of Government from the holders of the strings of the British Raj to India's strong personalities, who gave birth to the Indian independence. It was a gift, not as a gratis, but an earned gift. Now after independence, the election commissioner must be operative with progressive insight bringing continuous changes by properly educating Indian minds in contrast to the election-schemes of other democratic and non-democratic countries by way of progressive contrasts.

The journalists by their questions to the persons concerned extract from the core of the opponents' thinking some prompt answers. The politicians, otherwise, left to themselves will ordinarily resist refusing to divulge their inner secrets. The journalists play the role of midwife, in making a safe delivery of a child from a mother's womb. Their roles are also akin to those of skilled psychoanalysts, with many-sided experiential processes, conversant with India's varied history of the evolution of the history of the meditative processes of transcendental attitudes, confronting the perishing and emergent processes of the Changing Universe. It is knowing the Buddhist process in its inner currents.

Without going into the details of this book, I shall often utilize the contents of Durga Das's book, in shaping of India in its aspect of internal developments and external foreign policy – both before and after Nehru. This process and role of Nehru I described in earlier chapters of this book. I also referred to the activities of the prime minsters – from Lal Bahadur Shastri and Indira Gandhi, and to others in process, in a previous chapter. But all with reference to Durga Das's book. Hence the readers are advised to consult this book whenever necessary.

When, *knowledge* is so defined, as an emergent possibility, in alternative ways there is no assertion that is categorical or linear – the assertion, if any, pertaining to any subject or event becomes meaningful in a possible sense.

The element of *indefiniteness* in every argumental form is a continuous form in the passage from the philosophic to the scientific and technological spheres, or from the spheres of economics or politics, to other spheres, in the present day senses. Hence I could see a *process-character* in all these subjects, and the individual who deals with these subjects makes them dynamically related, raising each to a focal point.

More of this specific line of thinking for progressive well-being of the peoples of this World-environment will also be clarified in the later stages of this chapter.

In the present context of referring to the different Prime Ministers of India, after Nehru, one by one as already indicated – I had selected only some of their complex deliberations, and activities, in the context of shaping their political ideas, in the background of their family-history, and their careers, only briefly, by a sort of pick and choose.

They all passionately tried to improve upon the inner condition of India, in historical process, facing the internal situation and also the external political events, affecting the prospect of India.

This consideration was not from the perspectives of India, alone, but also from their early life and education, their travels and experiences in some other major locations. Hence, in each case, there was some kind of background and then their tackling with the responsibility of a prime minister. Their desires were sacred wishes but success was not in their hands. Life and one's career are only in confronted phenomena, full of uncertainties. One has to accept something as transcendental, and functioning is necessary, but has always to be construed as a 'dependendent process' over which an individual has no absolute control.

Similarly my life and my career or my *ideas* that I cherish in my mind, in interpreting these personalities and their activities are *dependent* on many factors, over which I have only a relative control. My thinking or its expressions were never by way of a categorical assertion. In my life-process I received a lot of impacts from my jobs in different places, outside India, if I consider from the perspectives of my own life-history. My aim in life therefore was keeping myself open for receiving more and more experience, in my pursuing process.

There is no such thing as *thinking*, as such, in a categorical sense. A man thinks, within a socio-cultural milieu specifically with a specialized attitudes or with some meditative attitudes admitting alternative possibilities. My choice was never categorically oriented, I tried always to approach facts from a meditative aspect, admitting of my life at different places, referring to my teaching profession – it was for the clarification of my changing visions regarding my changing knowledge of philosophy, physics (sciences and technologies) or politics as grounds for my reflections. They served me as dynamic facets or focal points for my reflective activities, providing me with a *knowledge* and also some strength simultaneously. But these statements are only alternative possibilities. My knowledge, if any, is relative, circumstantial, or pragmatic.

I believe in change of my inner experience, not in just regular ways, but in many-dimensional ways, my mental dispositions changed. But I do not think that this change was good or bad. The change was/is broadly *experiential* by some kind of stretch, but not dislodging myself from my inner integrality, or some kind of enlightened inner identity, I was always alert to preserve my transcendental insight as a means for my continuous self-development.

Hence I have developed a *vision*, for India's improvements of internal welfare and well-being, both from an ideal and real perspectives, always acknowledging, India's need for ventilating an atmosphere of a wider attitude towards economic and political development. India's attitude is not selfishly oriented, it is always for the welfare of mankind from an international and intercultural perspective. This statement is a confession of truth.

My consideration of the different activities of these Indian Prime Ministers had been from my perspectives, relying on some relevant materials I obtained, or I chose to utilize. These formed the items of new curriculum of my progressive teaching methods. I obtained this kind of attitude in the past from my early education at schools, colleges and universities of India and at other places of the world, where I was lucky to be located, in various teaching professions.

I served the other countries of the world, with a detached and sincere attitude, in conformity with my changing and growing wisdom modes in process. I have cultivated a spirit of *detachment* in my learning or in my teaching.

In this connection, I would like to mention that I collected some newspaper cuttings on India, from those different locations – Sri Lanka, Great Britain and the United States – of India's foreign news earlier and specially from 1964 onwards, to indicate how I had chosen to review my past experiences, on those complex bases, in the light of my changing experiences of the present world-situation, though I did not ever think that I would be so occupied later, in writing about India, specially choosing India's political philosophy from an international and intercontinental perspective.

At least my views as presented in this work will be construed as one of the alternatives. It is a lesson in process, but at the sametime a *receiving*, and then a *giving*.

In my choice of the newspapers – or in the study of new book in varied subjects I was not swayed by the currents set up by them but I was operative with a freedom, to 'receive,' with scrutiny and appreciation, to formulate my new attitudes, with the background of my growing inner experiences in development, so that my 'giving' does not become repetitive, but a process of 'giving' with some 'enrichment.' Therefore, I did not indulge in mere thinking, enjoying some abstractions, dissociated from my personal experiences.

If there was some monotony in my statements, I submit that it was the expression of some truth I was encountering in my experiential process, behind my reflective activities.

In this connection, I would like to mention, my historical changes of my life and career in different environments – stretching from India, to different parts of the world.

In the early periods of my academic life and my employment, and from the first appointment, at the Indian Institute of Philosophy at Amalner, and then to Patna – I experienced the spirit of a research scholar, and at Chapra, as a teacher of philosophy, I enjoyed a spirit of a 'responsible teacher.'

I taught there at Chapra mostly by myself, as Rajendra College was receiving a small government grant. So I taught almost all subjects coming under philosophy – psychology, logic, ethics and religion from West-East perspectives. With this background, when one of my early colleagues, a friend, studying at the Patna College, and then in London School of Economics, a student of Harold Lasky – out of his goodness and friendship, recommended me to Sir Ivor Jennings, Vice-Chancellor of the University of Ceylon – I responded to his request, even when the Japanese were bombing Ceylon.

Sir Ivor offered me an interview, and I was selected to serve as a lecturer at the University of Ceylon. This added to the intensity of my teaching in a foreign country, specializing in Buddhism, and Western thought. Going to the new place I had to study hard to get over my deficiency, confronting other scholars of the place. I wanted to know more from some of them, who were seniors to me or appeared to me as somewhat exceptional. There was an interaction.

There I had the privilege of obtaining some further opportunity of teaching and research experiences, along with some eminent Sri Lankans, employed at the University, and the Indian, German, British and American scholars who used to come as visiting lecturers or as guests at the University of Ceylon, during my employment.

In contrast to a person of India – whose cultural process begins from a reference to the Vedas and the Upanisads – the Ceylonese stress was on Buddhism, in contrast to the Upanisadic thought of the transcendent Self. The Ceylonese will be sceptical of a stable Self. I did not understand the intensity of this philosophic contrast between the transcendent Self and the contrasting outlook beyond it before my arrival at Ceylon then in 1944, before its independence. So I had to study some local literatures written by the scholars of this place. I had a number of Buddhist Monks in my class, and I had associations with the entire Pali Department with its learned professors and students.

In confronting the new situation I had to improvise my thought and language in a completely different way. I, therefore, became used to use a different language when I talked with a Ceylonese and with an Indian. I had to think out a way to retain the subtlety in the experiential process of the knowledge of the 'Self' as providing a *transcendent experience*, not of the *knowledge* as obtained from any sense-organ – but as a double-consciousness – or a consciousness upon consciousness not always dependent on the sense organs, or as a mere consciousness, arising out of the sense-organs.

The double-consciousness simply is not merely seeing as seen, through a sense-organ, but it is knowing about the possibility of seeing on account of seeing, through a sense-organ is possible.

A sense-organ, as one's eyes – does not provide merely one vision, but makes one aware of another vision – the continuous presence of that transcendent possibility, on account of which one has a capacity of vision. It is a vision of a vision – a double-consciousness – an ascent to a transcendent process, that is continuously felt as present – and from this process, there is a possibility for experiencing a continuous existence of confronted process. It is lifting oneself to a transcendent ground, as an ever-present field of continuous consciousness.

But the Buddha suggested something beyond these orders, saying that this 'I' with a claim of existentiality is a changing 'continuity'. It is existent, but really non-existent and more than those presented forms, transcending all intellectual forms by a dialectic of 'neither-nor' when no specific intellectual form has any place to exist. It is soaring in the region of an empty space, with an elevating experiential continuity. This transcendent continuity is not established by an intellectual form, it is established by transcending all intellectual forms, but at the same time it does not refuse to give it a place. But it is not restricted to a place, it is that which transcends all phenomenological existentiality. It is only a pathway for a further *possibility*. It does not set up an ultimate existentiality.

This extended knowledge I had to continuously feel. This is an additional philosophical experience which I did not know before, nor did I feel the need of such a reflection. It was due to change of place, and my continuous impact with the students and persons present in this island. This formed the new item of my teaching, and I continued this process further.

This kind of new cultural environment was a critical situation for me, which I did not get in Chapra, where I remained unchallenged with my restricted knowledge due to the local environment. I spoke a different language, a different kind of thinking in the exchange of my ideas, because there was no public demand, as people in general were used to think of the ultimate presence of a transcendent Atman-Brahman process. The new natural and social environment modified my mind, in a new direction switched beyond any Atman-Brahman complex.

Coming to Ceylon – a change of vision – had to be granted, and that had to be shaped from philosophy to politics.

Some changes were possible by my yearly attendance to Indian Philosophical or Science Congress as a delegate representing my college. I met at the Indian Philosophical Congress teachers and research scholars of Buddhism from various parts of the world – Europe, America and Asia. But Ceylonese variety of Buddhism was different, and could be cultivated in relation to other South Asian countries, close to Ceylon, where Ceylon had an important role. Somehow I had to be gradually aware of them, without letting others know that I was not aware of them. This type of knowledge is a *knowledge* by way of a *contrast*. It has a place in philosophic parlance, and in an advanced aspect. I have developed this trend continuously in the development of intercultural philosophical deliberations, as each provides a kind of contrasting outlook. Here I detached myself from the milieu of Atman-Brahman frame of reference, but I tried to fit myself with a transcendent possibility.

My going to Ceylon gave me an opportunity to think by a contrasting way, detaching myself from holding fast to the Atman-Brahman tradition of India. I became tolerant to hold an alternative idea, by a contrast with a further choice of other alternatives. To me the possibility of 'no-self' provided a further 'open space' for continuing further insightful possibilities.

Knowledge emerges from the study of the relevant books on a subject, but the insightful understanding of a subject emerges in contact with one's association with the learned environment, interested in one's subject of specialization, and of the specialists of other subjects, with whom one has to interact. Here in this case, in two different countries of two different types of people.

Besides one's studies and interests there is a political and social environment, in the milieu of which, one lives and interacts. So, politics, economics or sociology are not abstract subjects of study, they are continuous learning processes, within varied cultural milieux. This varied experience, may result out of one's staying at different placed for some longer period to be conversant with the basic cultural trends of the place – and as developed by its peoples.

With these varied observations, my sense of responsibility as a teacher, in Ceylon changed. With a conscious new disposition as a foreigner I also became gradually conscious of the role of India's diplomatic officials and their functions, both before and after independence.

This kind of concrete change of feeling towards philosophy and politics, in interrelation would not have arisen in my mind had I remained in India. It is a truth, which I cannot ignore. I felt within myself not merely a sort of mental change, but also an organismic change, being actually located in a different place in a foreign country.

I also could not stay confined to entertaining my knowledge of philosophy, alone, I had to try to understand something of other subjects, and very specially history, because of my colleagues in history, specializing in some areas in Indian history. But here also there is a difference in understanding a subject like history from theCeylonese perspective. This perspective differs from Indian perspective.

But the people in Ceylon were specialists in the history of Ceylon, and also of British history. I had also to be aware of subjects like economics and politics, as there were professors in those subjects too, from India, with some specialized knowledge of Indian and Western thought in relation, confronting local professors in their subjects. Similarly, I had some Indian professors in different scientific subjects with specialization, I had to probe into their 'subjects' a little, as twice I represented Indian Science Congress in 1956 and 1957.

While, in Ceylon, I actually organized an International Session of the Indian Philosophical Congress in 1954, with Prof. Malalasekhera as Joint Secretary. I was indeed lucky in getting a chance to represent two organizations of India – science and philosophy.

Radhakrishnan was then the Vice President of India. He gave full importance to that 'International Sessions of Indian Philosophical Congress,' saying that 'the Indian Philosophical Congress is meeting outside India, for the first time.' He expressed his sorrow saying that 'owing to the Council of States' being in session, he could not leave Delhi till the 25th December. Therefore it was impossible for him to attend the Conference. He hoped that the Conference will be a great success in the new environment.' (13 December 1954). I became elevated by his letter thinking that I was not merely a representative of two learned organizations but I had stepped into a status of a political representative too.

Coming to a new country, a person imbibes immediately a new perspective, and changes his expressions, confronting the people's mind in different a way, with some subtle changes, in one's basic consciousness, as inherited.

These responsibilities gave me an opportunity to learn various other subjects, from my interrelated perspectives. Here I did not think myself as a teacher but as also a representative of India. The Ceylonese also took me, as a foreigner, even when there was a cultural connection between the two countries.

We, my wife and my colleagues of the University of Ceylon from India were often invited by these foreign diplomats, and we, too, had a chance to observe them and their functions, from the Ceylonese newspapers, then we understood the problems that emerged between the Indian immigrant residents in Ceylon, present and past – and also the problems, social and political, arising from the British interests in Ceylon and the migration of the India's labouring classes, including the technical or educational specialists, who came to Ceylon, in different capacities.

There were certainly all sorts of experiences, which do not arise in the mind of one who remains confined to India, or at any particular place or province within India. Hence change of place, and life in those places, give new occasions or confrontations with a demand for new ideas. But here also, the preparedness to know different subjects, lies in an individuals' discretion, and one's inner willingness to enrich oneself with the flowing stream of knowledge. Again this knowing-process from a foreign country is not just *thinking* with a basic thought in terms of one's inherited learned pattern but an interrogation with contrasts.

Philosophical and political consciousness of different dimensions, emerges with the change of locations and with the interactions with different types of people of one's country and of the various types of foreigners confronted. So consciousness is not just *thinking* but learning to think in many cultural aspects by way of many-dimensional *contrasts*.

These are the reasons why I am mentioning about these personal matters. These are real concrete expressions for the changing philosophical consciousness, and my forced learning of many other subjects associated with philosophy from history and politics to the sciences. My different experiences helped me to remove my *ignorance* and my effort to know different subjects from a new cultural background, this helped me to gain some kind of confidence and strength in my knowing process.

Finally when all sorts of Indo-Ceylon problems arose, with some further complexities I began to feel about the two countries – India and Ceylon – with a sense of difference, but also with some affinities, from an inter-cultural perspective. Hence the new insights that emerged were not abstractly philosophical, political and cultural, but all tinged with my experience in concrete situations – as direct experiential processes. Man's sphere of interrogations – becomes multiplied, and a man becomes more intellectually alert but at the same time, learns to be detached from a specific intellectual or cultural perspective.

When the two countries became independent each started with a different kind of history. The Ceylon had to face as the Indians, the period before and after independence from the British rule. The problems that emerged were widely stretched from politics to economics, and to problems of human relations and human rights and obligations.

I had to live and teach in confronting these complex processes. Therefore my visions changed, philosophically, economically, politically and culturally. The principal contrast in my life here was a feeling of contrast as a foreigner. My teaching, therefore, was not merely just a reproduction of my learned experience from books, but it became differently oriented – according as I confronted the changing situations.

I had to develop a different and novel skill, not what I practised in teaching in India. Hence mentally I became different, cautious and diplomatic – and with a programme for knowing more, to maintain my prestige in tact.

I leave out here lots of problems, for the specialists to deliberate upon, with relevance and accuracy. These problems I did not understand, in real depths, before I came to Ceylon, taking up my teaching job. Had I remained in India, my visions of the world would have remained India-centric, or abstractly bookish and miserably stagnant.

Knowledge develops by progressive contrasting outlooks. Movements to foreign places provide such opportunities. But going to foreign places is a luck, or a chance event. A man in change has to confess what the parlance goes on in his mind. It is not just talking about one's self, but it is relating to concrete facts that one confronts. It is process of progressive subjective-objective interrogations of different dimensions and stretches. It is progressively creative and changing of confrontations.

It is life. It is not a view, opinion or pursuing an *ism* but a concrete confrontation and participation, with progressive challenges.

On Sabbatical leave from Ceylon twice – 1951-52 and 1958-59 – I was offered the privilege of passage-money to travel to England, as a foreigner, from Ceylon, with a contract service with my family. I was given the responsibility to do some further researches on my subjects of interest. Here some details about my life is given to indicate my *changing ideas*, to confront or tackle some emergent situations, hence these should not be considered as mere narratives, but as items for new confrontations and opportunities for making myself courageous and self-conscious. It is a sphere of a novel philosophic consciousness which takes note of a process of different kinds of consciousness—an emergence with a detachment from the past modes.

During those years, I attended the seminars and select classes by some eminent professors of the University of London, by taking an admission at the university as a post-doctoral scholar, under a supervisor. These direct experiences of mine, helped me to understand the spirit of Western thought directly, specially of the major British thinkers of this century.

Sometimes I used to have discussions with the professors and sometimes by holding some special seminars on some Western thinkers, whose primary interest was mostly concerned with Western thought, with no reference to Indian or Asian thought in general. This gave me the impression, that most of the Western thinkers, with a few exceptions, indulged in a kind of thinking, which, to my mind, was a kind of 'soliloquy' or even a 'partiality.'

To insist on this attitude of maintaining exclusively a Western, attitude, with a plea for specialization, is somewhat egocentric, without any element of a dialogue, in contrast to most of the Indian thinkers, whose visions are not arbitrarily one-sided.

The Indian scholars, when they teach either the Eastern, Indian or Western are also concerned with the advanced Western thought. They also relate it to intercultural processes, or to some similar or not so similar advanced trends in Indian thought, or sometimes Asian thought in general, as one could see in the deliberations of Indian thinkers like K.C. Bhattacharyya, S. Radhakrishnan, M. Hiriyanna, S.N. Dasgupta, D.M. Datta and others.

Most of the deliberations which I experienced in England gave me the impression that the British mind, in general, was confined, with an obstinacy, not to know other cultures, or not expressing their incapacity to learn other cultural processes, or just saying that as specialists they are not concerned with subjects in any intercultural mode of deliberations. This undoubtedly is an arbitrary decision. Specialization is a mode of deliberation, but it does not exclude comparison with other cultural processes by way of contrast. In that mode, more depth-processes can be arrived at.

In London, specially I learnt their teaching and research skills, but all restricted to Western thinking. I had the chance of attending the Seminars of A.J. Ayer, an eminent exponent of logical positivism, but several times I drew his attention to dynamic India's logic from Vedic-Upanisadic perspectives with a stretch to advanced contemporary Western trends, and also the opposing attitudes as expressed in Buddhist historical process of its non-positivist thinking of any sort, or the Jainist attitude from an ultimate disciplined perspective, not definitely knowable, so long as a man is not completely freed from the lower orders of knowledge.

When the Jainas consider the different levels, for example, I pointed out that they do not talk from the Western perspectives of 'abstract conceptual or empirical thinking,' nor do they think, as confined to the restricted mental or 'thinking level' – but whereas the Jaina attitude was a meditative control of sense-organs and of the operations of the mind, to rise beyond the operations of the sense-organs or of the mind is difficult.

Hence the Jainas speak of different kinds of logic, and levels of psychological thinking. Really there are different types of 'logics' or 'psychologies,' and for that matter, different 'types' of subjects within a 'subject.'

These types of thinkers – who accept or admit different varieties of experiential processes, within a subject, and from intercultural perspectives—talk differently. Coming to America I found my students and some of my colleagues showed interests to hear me. This indicates their openness of mind.

They also want to *transcend* the confronted spatio-temporal process, and the ordinary spheres of mind's disposition of becoming subject to jealousy, hatred or covetousness by a different level of understanding, and they want to distinguish this type by choosing a special word disciplined process, not ignoring the fact, that any knowing is a disciplined knowing. But Western psychology treats man as harbouring the instinctive bases all through life, without trying to detach themselves from any particular instinctive urges.

The Jainas claim to search for rising to trans-empirical and transmental spheres – they indicate that knowledge from the ordinary to transcendent levels lies in a rise to an indefinite sphere, as examining alternative pathways, over which a disciplined man, when freed from instinctive urges has a relative control, with progressive ascent towards Omniscience, which confers an individual infinite powers.

But Ayer will not deliberately pay any heed to that comparative intercultural attitude, his was a deliberate *suppression* of the study of intercultural processes by an arbitary method suiting his own mind, or for addressing himself to one mode of reflection, which he thought *logical* in his exclusive sense of the logical, examining only some statements which could be empirically verifiable, and excluding other statements which are either abstractly metaphysical or purely imaginary. In this process, he gave birth to a group of thinkers in his line, who became specialists in that line.

But inspite of this restricted attitude, he was liberal enough to watch my research-procedures in Whiteheadian contexts. He frankly admitted that he would not be very helpful, as he did not subscribe either to Jaina thought of India, or to Whiteheadian thinking of the West. He, therefore, confined himself to develop the exclusive logic he specialized. He will not have any hesitation for formally accepting me as a scholar working with him at the University of London, pertaining to my subject of interest.

Besides these academic experiences, I understood from my talks with general English people who had been once in India, and others who had never visited India, or only some members of their family went, to India, in various capacities, as in military or in civil service. Their minds changed in general after

Indian independence. So my experience, coming to England changed, both in academic and public spheres, in contrast to my early life, if I were only in India. This formed a new food for my mind to deliberate upon all through my teaching and research activities, so long as I was in Ceylon or later in America.

And with this change of my mind, I lived in England as a research scholar, in two different ways, in two different periods, earlier immediately after Indian independence, but on both occasions, after the disastrous Second World War, immediately or after some interval. I was in England during the years 1951-52, and again in 1958-59. The British academic people expressed in different ways, during two different periods. The ordinary British people also expressed their feelings in different ways, with love and hate towards Indian people, in general.

I stayed in Ceylon from 1944 to 1964. During this long period I experienced change of Ceylon, in many spheres, religiously, politically, and finally linguistically, when they became confined to their own problems, which became dominant above other problems, and we as immigrants, felt insecure in every way. Sir Ivor Jennings, as Vice-Chancellor of the University of Ceylon, had to chose a position in England, as A Master of the Cambridge University, as the Ceylonese educated people and the Professors became nastily inimical to him. We as foreigners like Sir Ivor felt the insecurity. But Sir Ivor had some obligations, as an individual, and became inclined to help me, as a sympathetic foreigner, to get a job in some place in Africa, as Nigeria, where the British had some power.

After this period, with change of psychic experiences I, inspite of a great love for my increasing student community, and Ceylonese friends had to pass through a period of change and progressive insecurity.

But I had the gook luck, as a member of the Indian Philosophical Congress, when I came in contact with the foreign delegates from America, and my luck favoured, coming to the United States, and this opened up the frontiers of my new careers of teaching and research. This would have been impossible if I were at my first teaching job at Chapra, in North Bihar.

Here in this new environment, at America – first at the University of New Mexico, Albuquerque, and later at the California State College (later University) Hayward, and later still, with a part-time job at the California Institute of (Asian and now Integral) Studies, San Francisco, my experiences, transformed in multi-dimensional ways. My main job remained with the State University at Hayward.

In these different institutions I had imbibed three kinds of experiential processes: At Albuquerque I came in contact with both American residents and American Indians, at Hayward, I had progressive ideas of teaching from a comparative Indian and Asian (Chinese and Japanese) perspectives, and at San Francisco, I had more opportunities to teach courses on comparative studies – Western, Indian and Asian and South-East Asian Buddhist thought in contrast to North-East Asian Buddhist thought. These progressive experiences were enhanced with research experiences through supervising theses for M.A. and Ph.D. students. These varied ideas and experiences are progressively expressed through my writings from year to year.

With my varied experience, I could communicate with my American students and colleagues more freely, presenting to them my different experiences at different places, by contrasting attitudes. I find the people of Canada are also similarly open like the Americans as evidenced from my experience in attending some conferences in Canada on invitation, or in the examination of the research theses from other American universities and colleges, or from Canada's colleges and learned societies.

So my knowledge of the different subjects changed with progressive depths. I am lucky to enjoy these experiences, which transformed me in many different ways, I cannot imagine. So, with my varied experience, my ways of 'knowing' and 'communicating' through 'teaching' differed.

Only by imagining myself I can think of my present, in contrast to my life at my first location, in Indian environment.

If I had remained in India all-through my life, I would have dogmatized within myself with a self-satisfying argument. These varied experiences in different locations in India, or in other countries – came mysteriously, in my life-period which I communicated to those who came close to my life-history, but I do not like to communicate this private feeling to any person prone to mere ordinary orders of scientific thinking.

They might consider my varied experiences as merely autobiographical, having no academic value, but I consider that as concrete experiences which can be shared with others interested in that sphere, hence, they have both psychological and scientific values. Hence the sense of academic value needs a modification. There is always a transition from a subject of study, with the person interested in that 'subject' with his changing experiences.

They have values to my students and to those who came close to my life. Somehow they were influenced in different ways, of my life-experience.

To me, therefore, *thinking* – in separation from one's life-experience – is *vapid,* and *abstractly ghostly* expressions. Some students, and my colleagues agree with my thoughts and feelings when I express my mind to them in the context of some of my discourses.

Moreover varied experiences progressively emerged by way of progressive contrasts. When a person reflects, if he has direct experience of many cultural areas – his reflective activities are *multiplied* by progressive contrasts. They become multi-dimensional, otherwise a *thinking* remains a single-track process – a monotonous soliloquy.

I am again referring back to the paper I read in *India News* and my progressive understanding of it, again on 28 April 1992, in multidimensional ways, in contrast to my earlier reading of that newspapers, after this long interval.

I am examining the theses of the various institutions from the years 1964 as I had the opportunity to learn many things from them, by way of variety of subjects and the kinds of their choices and interests. In every case I had to give my critical comments. These changed my attitudes and perspectives. The theses from different countries reflect some specialities of the country, institution, and individual's tackling of a subject.

My appointment as a senior Professor of Philosophy by President F. Harcleroad, of California State University (formerly College) at Hayward, to teach Western and Asian Philosophy, was done with great insight, and with a compassionate attitude, on his part.

With my ignorance of the rules and regulations that govern a foreign country, I am sure, my continued stay at the United States would not have been possible, without his help.

How and why President Harcleroad liked my pattern of teaching and thinking, I do not know, but his compassionate psychological disposition, as expressed to me, changed my mind and elevated it with some freedom, to reflect from an intercultural perspective.

Here also I found most of the teachers were very much confined to Western mode of thinking, as in England. But some of them were interested in Indian thought and as well in Chinese and Japanese thought. But a few of them developed the Asian thought from the Indian basis of Buddhist thought as it differently influenced the basic cultural processes of major Asian countries of Tibet, China, Korea and Japan. So their reference to these Asian countries appeared to me as 'abstract isolated expressions.'

But President Harcleroad was a different type of personality. Harcleroad's was an expression of satisfaction from the reports of my teaching, from my colleagues, and students, and also increasing number of student-admissions, in my classes.

President Harcleroad kindly granted me a tenure after a year of service, and helped me to obtain, the Permanent Residentship at the United States, which changed the prospects of my life. He went beyond his power, and modestly told me that he will try his best for me.

I could formulate my future with a new insight. I became inspired and devoted myself to improve my scholarship further, since September 1965.

This improvement in me was because of my inspiration from President Harcleroad. Inner spirit of a man becomes extended in a spiritual direction from one's association with another person of a kindred heart. Human relation is not merely intellectual, it is post-intellectual or a detached attitude, following the intellectual processes in all its possible forms, and finally transcending them all, to place oneself to the continuous orders of insightful experiences, following the pathways of progressive concentration (*Samadhi*) and waiting for insightful *flashes* that change the entire manner of expression – a progressive meditative plunge. Here the transition is from abstract speculation to one's heart to heart relation, where one understands and appreciates one's indebtedness to another, who stretches himself in a compassionate direction.

In 1968, Haridas Chaudhury, President of California Institute of Asian (now Integral) Studies invited me to teach with him some courses of Indian, Chinese and Japanese philosophy. He also gave me the responsibility of directing researches in philosophy and psychology, of the Graduate students of the Institute. This increased the scope of my teaching and research abilities. During the later days of his untimely death, he made me one of the Vice-Presidents of his self-directed 'Cultural Integration Fellowship,' with a vision towards the development of West-East parlance in continuity and depths, through discussions and appreciations. This Fellowship, while retaining its own character, became the basis for a research Institute with the title: California Institute of Asian Studies in 1968, and later still, the Institute changed its name from – 'Asian' to 'Integral' – perhaps to extend its 'Field of Research.'

I am referring to these personal histories to indicate that with the changing environment, and with lucky opportunities, provided by some large-hearted personalities from Prof. Bahm and Prof. Alexander to President Harcleroad and President Haridas Chaudhuri – I could teach and do research with considerable freedom, I could improve the minds of the students, with inspirations, more than following some texts strictly, adhering to think with a Western bias.

My teaching all-through was interculturally based, with contrasting outlooks and stretches. Knowledge needs to be stretched to various cultural areas, to find out whether a statement is egocentric or a culture-centric and interculturally based. In the final stage of knowledge, a man is more or less free from restrictive knowledge. While teaching or writing, a man has to free himself from his restrictive attitudes. As my depth-studies of Buddhism and of Whitehead progressed, I became convinced, that there is no entity in any ultimate sense in confronted events, and as well in confronting one's own personality, one's mind has to be addressed to this non-entitative processes progressively.

But to realize this process effectively, one has constantly to concentrate, and give expression to that concentration by a continuous study of the relevant materials by writing from an intercultural perspective. I did that in a regular disciplined way.

I could confront various types of students, really inquisitive to learn something new, and this prepared the ground for my teaching many subjects, other than philosophy and religion, to come to my classes for learning, from the different parts of the world, both at the Institute and at my residence at Hayward.

Moreover a subject is not a subject in an exclusive sense, it is one of the aspects of experiential processes, in an interconnected sense, as interrelated to other experiential processes.

From these stretched visions of my interpretations of various subjects, in interrelation, I have sufficient reasons to declare that my treatment of the subjects – philosophy, physics and politics – is focussed through the triadic avenues that I have chosen for this book – where I claim my individual role as a phase of my reflective activity. I gradually become *transformed* recognizing the contributions of all cultures, places and persons – forming a human family.

During this period, I met many practising psychoanalysts from San Francisco, and some engineers from the University at Berkeley, or other serving in various companies.

They became generally interested in my teaching of India's Vedanta, and internationally known Whitehead, and joined some of my classes at the California Institute of Integral Studies, at San Francisco.

They provided me with some opportunities to confront them, and this confrontation helped me to shape the bases of my previous knowledge of philosophy and psychology, from West-East perspective, to progressively different types of thinking as demanded by the students from their different cultural backgrounds or from their different specialized studies.

Along with my teaching and research activities I enjoyed publishing my progressive experiences through my books on Buddhism and Whitehead, with different titles.

Here I had a chance to view things of India, more dispassionately than before, to shape Indian thought in all its aspects and in interrelation with the European and American thought, by way of contrast, in progressive developments.

All these different changing ideas inclined me to study different sciences from my changing perspectives. I gradually moved towards a scientific way of viewing my philosophic and psychological attitudes. My choice for choosing to read the above article, was gradually becoming more and more meaningful, and practically oriented.

My earlier interests in philosophical and psychological spheres with progressive scientific interests, by a contrast, prepared a ground for understanding many other subjects in course of time. I also moved towards economics and politics from intercultural perspectives – and I am finding new perspectives in the world of economics from my studies of some of the works of John Kenneth Galbraith, beginning with his, *The Affluent Society*.

Here he brings out the social changes brought about by the increasing supply of material goods by the business centres and the recipient consumers. It transcends the earlier economic processes of demand and supply, men start craving for *more* beyond their needs, paving the path for progressive 'Affluent Society,' to which the roles of the different Presidents of America became important as they were interested in obtaining public votes, from the affluents in many alternative senses. The earlier notions of demand and supply, took different dimensional orders complicating the earlier comparatively simplicist mode of understanding a subject. Laws of Economics – change with the different types of demands not known or imagined before.

But with the emergent problems of unemployment and varied human motivations the controlling processes of the different Presidents did not succeed, to remove poverty and sufferings of some unfortunate men, women, children, old, diseased, and the handicapped.

They emerged, as a resultant process of an affluent society as uncontrollable processes, for economic processes are widely interrelated with the different psychologies of the consumers. Hence economics is not unrelated to other social sciences, business studies or some sciences in general. Exclusiveness of economics as a scientific study changes in relation to changing and transforming international society.

My students of all types, even doctors and engineers of varying ages from twenties to seventies, felt inspired when sitting in my class. They assumed that I knew all subjects, but that was not the truth, but to prove my competence to meet their expectations, I had to study enough to let students remain under the spell of an illusion, as if I had mastered the cultural ideas in development of Asian, European and American thoughts, and also the major scientific trends.

To reach certain standard of intellectual understanding, I was compelled to develop my interests in varied subjects,' including the advanced sciences, and politics and economics, choosing some courses to teach, from an innovative aspect, relevant to the changing times. I chose some kind of scientific subjects, and also some aspects of studies of international politics, through the studies of some specialized journals and books to meet the demands of my students. Life is challenging, and an insightful teaching is equally challenging and autobiographical – with a stretch to intercultural and international forums.

My newspaper collections stem out from this disposition and choice of mine.

While in Ceylon, I used to attend the Annual sessions of the Indian Philosophical Congress, and also some sessions of Indian Science Congress. I was courageous enough to represent Indian Science Congress twice in Ceylon in 1956 and 1957, when I was appointed a representative to represent Indian Science Congress at the Science Association of Ceylon.

Earlier as mentioned, I had a chance to organize and hold an international Conference, of the Indian Philosophical Congress in 1954, as Joint Secretary of the Indian Philosophical Congress, with Professor G.P. Malalasekhera, then a temporary Vice-Chancellor of the University of Ceylon. He later became an Ambassador to America and Great Britain.

During the sessions of Ceylon International Philosophical Congress, I became intimately known to some American and British delegates, besides some well-known Indian and Ceylonese delegates and friends.

My switching back from the present to the past is for indicating that the present is built on the past with psychic revivals. I continued my scholarly activities, while in America, participating in various international conferences, representing the university or the institute, to which I was associated, or as a member of some Associations of which I happened to be a member.

In America since 1964, I was attending some special international conferences, in different parts of America and Canada, as a member of the American Philosophical Association, and other Associations, by some invitations, as a panel member or a Chair person, of a section, or even as a General President of a Panel of the Golden Jubilee Sessions of the Indian Philosophical Congress held in New Delhi (1975-76) with the World Philosophical Conference.

These again were my learning fields, giving me opportunities of gaining progressive novel experiences, with novel contacts. My inner personality was changed, developed and stretched to various subjects, and to various cultural processes, so I could be fluent and free, in shifting from one subject to another, one culture to another, without restricting myself to a specialized subject or culture.

With this continuous interest I extended my knowledge of philosophic and scientific studies, in relation and prospects. With a challenge – I formulated some courses with various departments, as anthropology, biology, geography, history, language and music at California State University at Hayward – as participants, and I impressed my students and colleagues, participating in those with my extempore introduction, and conclusion, commenting on the subjects represented by my colleagues in general.

These were great adventures of my educational career, delivering lectures extempore. In all these modes of teaching, I kept in my mind, the inspiring gestures and ideas of Radhakrishnan and Nehru. These two personalities developed an art of touching the minds of various types, with various subject interests.

I could have avoided giving any personal reference, but that kind of choice would have made my writing just formal or abstract, without any life-inspiring concrete reference to facts and concrete confrontations. As I am breaking that so called usual or normal or formally oriented traditions, I have followed a different way – giving from time to time the reasons of my preferences, indicating how I am changing and how I am confronting the apparent situations with a challenging attitude, from a different perspective, which I consider is an attitude but not very traditional. These are my *departures*, and from this *dynamic background* the confronted facts are given various alternative shapes. Here the human mind is freed from the stereotyped formalist mode of reflection, with no adherence to a form, already expired. It reminds one to think with the French thinker Bergson, why he captioned the formalist intellectual mode as a post-mortem examination of facts. Here the confronted experience is prospective living process or a continuity. Reflection without the change of the grounds, is ploughing the same soil, when it loses all elements of productivity.

From these I have been able to formulate my visions of philosophical studies in extension to scientific studies, and I could develop some interest in politics, when my students came with some questions about India and its changing politics.

With all respects to specialists in the sphere of politics, I have got much of an international outlook, and comparative perspective, but I always wished to perfect my knowledge in intercultural spheres.

In my observations of the writings and talks, with many scholars of the different universities, I observed that they may study their subjects or they may be specialists on politics, but their considerations of respective Constitutions, were expressed from their exclusive formal attitudes, and not with a reference to other comparable Constitutions. Hence they deal with dead-bodies in Bergsonian phraseology.

When they talked of people, they viewed them from a static or classical perspective, and ventilated their exclusive personalist attitudes. But, in this Chapter, I have pointed out that the emergent personalities from the *masses* often give a new direction to mankind, which the formalistically oriented intelligentsia – fail to give.

India, therefore, focusses its dynamic attitude from two living innovative aspects, from all spheres of experiential processes from the myths to the sciences, including the sphere of politics.

The specialists, in the sphere of politics, often do not distinguish between American, British, or Indian Constitution and its people, of different types. Some insightful people in the sphere of politics, reflect some progressive democratic Constitutions in interrelations, and, people of different parts of the world, in progressive modernizing directions. But I have given a detailed view of my notion of events and also of personalities or people in general, from varied perspectives – deviating from any formalist perspective. A living personality is not a perishing process but alive with prospective suggestions. Hence dynamically insightful in Bergsonian language.

To me there is no such thing as *people* in an *abstract sense*, but *people* of a particular place with a distinct cultural heritage, so they are distinguishable, with demands of varied types. These personalities again, have an inherited psychology, and a changing or transforming process if they are ready to study other cultural areas.

From these perspectives, which are not always on the surface, but somewhat unobserved, yet understandable meditatively, with insight – I intend to remind the readers of past experiences of the political activities of some Prime Ministers after Nehru, dealing with their distinct roles and contributions, and shaping of India's *democracy*, as an 'on-going process,' to further complexities. So all these considerations were from the perspectives of the people of India, and from India's Constitutional background, with a *changing prospective outlook*.

In this survey I observed also the roles of some presidents, ambassadors and scientists from America, or other officials from different parts of the world to India. So here also the impacts were from India, with some contrasting backgrounds brought to India from other parts of the world. New types of reflective activities ensue, due to foreign contacts and in the contacts with some personalities who are interculturally oriented.

But before I go to develop my ideas about changed and transformed India after India's independence – I like to refer to a very illuminating paper – Nehru's Address to US Congress, in the House of Representatives and the Senate, Washington D.C., 13 October 1949, soon after India's independence. There will be some repetitions, but it will open some new perspectives for politics in change.

It refers to the time when Nehru as Prime Minister of India visited the United State, and after staying in America for eight days he was interpreting time, with his emotion and experience. It was a psychological expression in an international contact.

His short stay gave him a new glimpse of his own mind, with a direct vision of America 'as he confronted,' as also in contrast to India's recently acquired democracy. He was also referring to the American democracy, as relatively advanced in the affluent order of progress.

But in evaluating these two countries by way of recognizing their differences – he felt the presence of a basic cultural affinity.

India is charged with Gandhi's *moral* fibre, but America also has a moral basis in the framing of its Constitution. These kinds of reflections reveal the depth-aspects of one's thinking, and Nehru was such a personality, connecting his past experience with the confronted orientation distinguishable from the merely abstract formalist orientation, and has a different and direct appeal to the confronted personalities. Nehru excelled in that art.

He had a speciality and he could make himself a likeable personality for his indication of some form of affinity between the two countries, from an emotional perspective. Nehru's attitude in diplomatic matters was polite and appealing, and at the sametime an eye-opener to some prospective outlook in which all will be commonly interested.

India had peacefully inherited a parliamentary democracy as transmitted from the British, with no alternative choice but with a partition. The drawbacks of which he frankly admitted were not so patent before the decision was made, but then after partition, one confronts the inheritance, with its consequences, which could not be anticipated before.

His graphic 'realistic description' was emotionally displayed as below: 'It was the cutting up of a living structure, of everything – all our services – whether the Army or Civil Services, transport and railways, communications, telephones, telegraphs and the postal system, irrigation and canals He gave a pictorial representation of the entire situation in a nut-shell..

Many families, domestic households, were suddenly cut in two. An extraordinary situation arose overnight. It created tremendous new problems, among which were upheavals, deplorable happenings and killings and then vast migrations. All our energies that ought to have been devoted to constructive effort, to economic betterment, which we had planned years previously, suddenly had to be applied to tackle these new problems. What a graphic description tinged with realistic flares? He was not critical of any one, but was placing facts in realistic ways, with a placidity, for others to understand as the matters were.

'We had no time or leisure or resources left to deal with the other and more basic problems' – was Nehru's next statement. But immediately there was a shift to another statement: 'Nevertheless the world did not stop. And we did try to deal with the basic problems to some extent.' Here Nehru demonstrated his dynamic nature and realistic confrontation. He was never a static phenomenon merely contemplating with a psychological depression. His was a 'courageous confrontation.'

Hence, with a natural strain, he observed after this bitter experience, the sequence of Indian politics – he had to remain always conscious of India's two inheritances – (i) the inheritance from the British rule, and (ii) from the split condition of India. His was therefore, a basic suffering, with an impatient attitude of a speedy method to adopt, to recover from this shock, and heal this split.

He had to address himself to tackle India's economic and political problems, with a provision for scientific and technological – speed-up process – to solve India's multi-faceted problems, and with India's defence from all types of inroads or even invasions, facing no more idealistically but with the stringency of a realist. A living personality to split itself into multifarious insightful expression – Nehru had that capacity.

Nehru, in his representation of facts, did not want any credit, but simply he pointed out what he had to do – to face the facts and to improve upon the situation. One cannot, therefore, dissociate oneself from the role of a personality in the execution of political activities.

Politics is not dependent on an abstract analysis, and remaining stuck up with it – it is a mobile attitude, confronting a situation, and a direct operation with insight and execution of things for improving a situation. A psychic approach and representation is a need, which is absent in many of the specialists in political thinking.

In the previous chapters, I had gone into the details of the partition of India into India and Pakistan, its changing Constitutions to the emergent Constitutional processes of Independent India and the role of the First British Governor

General or Viceroy, Lord Mountbatten, operative with the depths of a friendly attitude with India's first Prime Minister, Nehru. India in that critical situation had an extraordinarily capable Deputy Prime Minister – Sardar Patel.

They had to be operative with two successive, able and insightful Presidents – Rajendra Prasad and Radhakrishnan. They helped India to step into the transitional processes – with their extraordinary capacities, through charismatic expressions of their personalities.

Here one can also see in Nehru, how a person, who has an insightful vision of the past can be operative by confronted observation, with a foresight of the future possibilities. Nehru was such a many-dimensional dynamic personality. He could face and settle problems with the extraordinary Presidents without clashing. He used to solve many critical problems with their advice and insights. Hence, in my representation of Indian politics – I have always a reference to the transcendental psychic approach to facts of these different personalities. Each one of them were responsibly operative from their different positions, within the Constitution of India. These personalities, being transcendentally oriented became operative in unison, without clash.

When this initial process was given a shape the history of India was placed in the higher order of a historical current, terminating with the deaths of these fathers of Indian Constitution-makers. Hence the departure from this first Platform, was to the new spheres of deliberations of the later Prime Ministers and Presidents of India.

Here one has to see whether or not by these new personalities, the spirit of sacrifice of their private interests was also a continuing process or not.

They confronted the ambassadors of foreign countries and others in this operational field. India at this period, took various possible shapes, through exchange of ideas with the foreign ambassadors, over which an individual mind has no control. In this new mode of pursuit, I could proceed in varied ways and modes of deliberations, even thinking without the sheltered thoughts of India as shaped by earlier political leaders, I could understand the operative capabilities of many new personalities, with a progressive departure from a mere idealist or restricted realist stand.

I, therefore, recollect, and also collect some chosen incidents of some importance, about India, from my observations of the activities of the first prime minister, followed by other prime ministers in succession, understanding their responsibilities in critical situations, not like the British Kings or Queens of British history, who were operative only symbolically. The Indian Presidents were actively operative, they were creative personalities, as other great personages of India beginning with Sardar Patel, deputy prime minister, operative with Nehru.

Here the reference is, again, to Patel's wonderful capacity in bringing most of the Maharajas of India, with a three days notice, by a command to join the

Indian Union. His further efficiency could be seen in controlling the movements of Junagadh and Hyderabad States. Junagadh came to terms, with a little hesitation, but his tackling of the Nizam of Hyderabad to the Indian Union, had to be manoeuvred, with a demonstration of a military might without any casuality, and with a speed, otherwise, a moment's delay or hesitation, would have left India in a chaotic condition as seen before Indian independence, when the pro-union thinkers of India, would have remained occupied with several fruitless presentations of formulas, creating innumerable problems. It is an operation with a transcendentally oriented power which never fails.

This hesitation would have given a chance to the antagonists to sharpen their weapons for making a coalition between the dis-satisfied members of the Muslim League, and the Maharajas of the different states. Perhaps India's Union would not have taken a due shape at all. This would have spoilt India's further prospects after Indian independence.

The other prime ministers after Nehru – Lal Bahadur Sastri, Indira Gandhi, Moraji Desai, with a return of Indira Gandhi and later on Rajiv Gandhi – became operative, with insight appropriate to their times and they all successfully solved the problems of many critical situations. This is a unique kind of exercise of one's transcendentally oriented power. It become operative sometimes from this exalted centre. This power is an insightful attitude with a detachment. One has to cultivate this process. But the current of Indian politics, with an understanding of a compassionate depth-feeling for Indira Gandhi's contribution proceeded in a wrong direction. India behaved adversely, ending her activities by a horrible murder, dislodging one's faith in one's bodyguards, and thereafter, coming to the stage of Rajiv Gandhi, another fatal end, by a some foreign intervention, over which India could not be sufficiently cautious.

The Indian mind, at first started well casting some hopes of a luminous future, but due to some unthinkable foreign interventions, this glorious history of India was terminated, by a harsh end of this young personality by a horrible assassination by a bomb blast. Hence history, like all other experiential processes – from philosophy to the sciences – is an unpredictable possibility, touched through and through with indefiniteness.

History is mistakenly considered as a 'destiny,' but to my understanding, history manifests a realistic possibility. Man is placed in cosmic milieu, with a helplessness, and 'events' happen as an 'emergent possibility' in different possible ways, sacrificing human values.

In this tragic event due to foreign interventions India had to helplessly accept what happened to it. This caluminous event took place in the pure and sacred land of India, after the stepping in of the Ministry of V.P. Singh, and later in the interim period of Prime Minister, Chandrasekhar, during the heat of the election campaign.

At that uncertain critical period, things took a sort of departure from its earlier attachment for idealism that governed human life and aspirations in general. India was compelled to learn to remain alert to all possible unforeseen interventions. India has now to remain alert towards scientific developments, lest some inventions fall in wrong mischievous hands – instead of the destructive nuclear bombs, what becomes operative is a 'human bomb'. This is a process beyond idealist or realist complex.

If I think of politics I feel like thinking whether politics will play its old game of some form of mischievousness or tend to be free from such processes? Obviously the choice for mischievousness will repeat the past historical process, but India's politicians, in the frame of India's Constitution, in superb non-violent directions will have to express their novel skills to tackle all difficult situations.

We live in this age of alternatives and progressive uncertainties, but the distinctiveness of the modern age lies in coming out of these critical situations with progressive efficiencies, which were not presented earlier in the history of mankind. A new chance is given to humanity, for a progressive self-improvement for making man's life better and prosperous with scientific and technological facilities. Man has to cultivate a progresive detached attitude, freed from all terrorist dispositions.

After the assassination of Rajiv Gandhi Chandrasekhar was replaced by Narasimha Rao, when again the Congress came into power through people's vote. India, under his insightful and patient guidance, is progressively learning to tackle critical problems, including the financial matters.

The new finance minister Manmohan Singh is reshaping the destiny of India, to a better direction, in international contexts of a future, where 'rupee-currency' is becoming progressively appreciated, for the progress in 'Market. Economy,' where the various political ideologies are sinking into a background, and making democracy as a permissible process, to fiddle within the hands of its mighty players. But be it known the future of 'events' cannot be restrictively known.

But I am with my own fiddle to suggest India's importance of its 'transcendentally oriented' philosophy, psychology, sciences and politics to announce their uniqueness in their respective spheres with relational possibilities, and with progressive detachments.

I find that all experiential orders and aspects of experience move towards continuous changes along progressive realistic directions through progressive scientific knowledge and technological skills, with the creation of a set of keen and sharp intellects and also able technical assistants.

This new emergent process redetermines the character and efficiency of Indian people, so that India's population is not just a population, but it is in a transforming process, with progressive acquisitions of abilities from the masses of the village centres, and the developing intellectuals of the urban centres. It is for the people of India, to choose the better between two alternatives, in any situation. The relation as suggested here, between the intelligentsia and the masses operative in togetherness has a uniqueness of its own, not found in the newly emergent countries of Africa, after the Independence from the French, British and other Masters, for, in those countries the conflict between the intelligentsia and the masses – and also the conflict between the senior intelligentsia and the recently educated intelligentsia remains. [Cf. *Political Leadership in Africa* (Post-independence Generational Conflict in Upper Volta, Senegal, Niger, Dahomey, and the Central African Republic) by Victor T. Le Vine, The Hoover Institution on War, Revolution and Peace, Stanford University, 1967.]

Both orders – *Intelligentsia* and the *Masses* – in India are fast proceeding towards technologies pertaining to this spaceage touching the projects of outer space.

By a contrast, I can go back a little to the past to observe how the Indian President R. Venkataraman, with a courage and foresight was operative expressing India's progressively strong foreign policy in a realist direction by his visit to Three African States – Tanzania, Zambia and Namibia. In this mission he was conferring with the two Presidents for holding a free and fair election in Namibia, in its process towards independence.

Here India's the-then President's attitutde was bold, cooperative, and compassionate, with inspirations, so that both Presidents of Tanzania and Zambia were appreciative of India's contribution towards the freedom from foreign rules, dismantling of Apartheid. Hence to my mind, *politics* is not an *abstract* subject, it has a human aspect, and human orientation, where many problems can be solved by some gestures good for humanity at large, and with a foresight.

India's President declared that inspite of resource constraints, India is committed to make a modest contribution to the economic development of Zambia. It was nothing but a gesture of cooperation among the developing countries, ushering in a new international economic order. This kind of good gesture was demonstrated by American Presidents during the Second World War, for expediting Indian independence by compelling Churchill to change his policy, when America had to enter the war after Japanese attack of the Pearl Harbour.

The present Prime Minister, P.V. Narasimha Rao, was operative earlier as the Indian Minister for External Affairs and as representing the nonalignment movement in Harare. There he suggested, on 18 May 1989 that no group of countries of one ideological persuasion or the other or both can dictate.

These courageous statements by India's officials indicate that their expressions were no more vacuous idealist expressions of the past, but were all oriented from the depths of realistic understanding of the concrete situation of a country within the international milieu. But this realist attitude is interpersed with probabilities

It was an advancement beyond Gandhi's activities, in the post-independent stage of India, with a power, in international situation, and with a manifestation of a *compassionate* attitude in progress from the Buddha to Gandhi and beyond, with a judicious use of scientific and technological powers, indicating that India has real strength in the aspect of nuclear science but it sticks to its non-violent attitude, with a slight modification. India has to change – with scientific and technological improvements, so that its 'gestures' in different situations, in political parlance, in international situations will sound progressively realistic and not just idealistic with mere appeals. Appeals are heard by foreign powers, when behind them there are some real scientific and technological strengths. I am pointing out these things. If I am operative in this direction I am touching realistic spheres of *deliberations* and not harping on mere *abstractions*. My vision is addressed to transcendental orientation in observing the confronted experiential processes. This realist suggestion however is not restricted or categorical.

This long period of over 44 years of Indian independence, after the havocs of Partition was followed by the invasion from China during the time of Nehru, followed by invasion of Pakistan during the Ministry of Lal Bahadur Shastri, and again there was the intrusion of the West Pakistan Army to conquer the revolting attitudes of East Pakistan, from a democratic perspective, but to this West Pakistan was not mindful. India here was again threatened by another invasion, during the period of Mrs. Indira Gandhi.

History does not speak by itself, it is often expressed by *contrary currents* but some insightful persons, have to *confront* with a courage and steadiness. India had to play its distinctive role here, and not let history, flow, in its *Naturalist Order*, which is not rightly directed.

Here some insightful reflective personalities are to be operative. Those personalities are always a need for India, but they must as well remain internationally conscious, to retain India's cultural process, in tact, with a different modernized form.

They are records of India's progressive awareness of harsh realist attitudes, with concrete steps. These are tales of India's panoramic political history, with a developing potentiality of confronting the critical situations with success, and also its competence, in winning victories from the foreign invasions and maintaining India's capability, and defence powers.

India gave life to a new independent nation, with democratic possibilities, which was threatened to be smothered by the tyranny of the economically and politically powerful nations of the world, where somehow West Pakistan was pampered, that imbalanced India's internal and external problems, both economically and politically. These have to be tackled with insight and farsight.

Indira Gandhi's political gesture and tact was judiciously managed with no diplomatic help from any foreign power, in some vital aspects of India's interest, but it was managed through India's superb army efficiency and the sober diplomacy of Indira Gandhi, which prevented any antagonistic foreign intervention.

Indira Gandhi demonstrated a great insight and diplomatical gesture and move. She dealt efficiently to rebuff the foreign intervention of Pakistan, specially when Pakistan marched with its mighty army to conquer East Pakistan, and indirectly threatened India, during that advancing process with its military.

These efficient political manoeuvres of Indian Government, during the Ministry of Mrs. Indira Gandhi are skeletons of India's historical events, recording India's success. My later observations were on some recorded narrations from newspapers of America and the *India News* in America.

I mentioned in general, the roles of different prime ministers of India, in some further aspects, wherever necessary carrying forward some advanced processes, with realist touches. My interpretation of 'realist' attitude has to be understood in the dynamic prospective sense as I am developing and not with a preconceived disposition.

The immediate successor of Nehru, after his death in 1964, was Lal Bahadur Shastri, after election. The reference here again was with a little repetition, for it was a kind of an invasion by Pakistan, being pampered by China also. Here the problem was with regard to another front of India. The problem was regarding Kashmir. The reference is here to India's success of Tashkent Declaration, when both the contending parties, became inclined to come to a settlement, the invading Pakistan, through the armies organized by Pakistan's President and Military Head, Ayub Khan – and India, represented by Prime Minister Indira Gandhi, on 15 February 1966. The reference is here to an earlier period, immediately after Nehru's death.

They came to settle the Peace Treaty. It was an event after the sudden death of Lal Bahadur Shastri, in search for a possible settlement. The mandate was for the settlement of disputes through peaceful means.

The Tashkent discussions centred around the basic question of the renunciation of force, following the suggestions of Shastri. His main point was, whether or not, the two contending countries wanted to live in peace and settle their disputes without resort to force. The Declaration was also in the presence of Soviet Union and its helpful representatives. It was a creation of a 'moral attitude' in an international situation, no longer a subjective idea cherished by an individual. It was an international parlance with a new appeal.

When President Ayub Khan of Pakistan raised the question of Kashmir as the basic issue, Prime Minister Shastri, in his talks with him made it clear that it was not possible for India to deviate from its position that Kashmir was an integral part of India, and India's sovereignty over Jammu and Kashmir, was not negotiable (cf. *Declaration* No.I).

This settlement of the dispute between the contending parties, was settled by the Charter of United Nations, and the *Tashkent Declaration* was nothing but a categorical reaffirmation.

Under the *Declaration*, Pakistan had not only agreed to withdraw, but also had to promise to take recourse to peace for its own future interests. In this connection, India had a note of warning to China for its support to Pakistan, by construing Highways in portions of Indian territory under the illegal occupation of Pakistan.

This move, on the part of the Communist Republic of China had spoilt Indo-Chinese relation, after its invasion of India in 1962. That road construction from Mor Khun in Northern Kashmir to Khunjerab Pass on the Kashmir Shinkiang border was militarily a sinister move directed against India, contrary to the Tashkent Declaration and professions of goodwill and friendship.

Here China was going to reap the fruits of this road more than Pakistan. If this warning is not heeded, India will be compelled to take recourse to stringent steps, to strengthen its position through international help and UN. This demonstrates India's moral courage, in the political sphere, to face a critical situation. But behind this moral gesture, there was also a concrete improvement of India's military strength demonstrated during the period of India's direct clash with Pakistan.

During Narasimha Rao's ministry, China is turning towards a progressive friendly attitude towards India. India, as changing with some confidence, in the political front, is boldly confronting foreign powers. It is no longer idealistically oriented, as in the times of Gandhi and Nehru – but now with a harsh realistic gesture, knowing fully that the political environment is mischievously oriented by the interested foreign powers, which disturb India's internal condition and external relation with the foreign powers, in multifarious ways, not neglecting the progress in the military sphere for self-defence.

India is now in its drive towards the improvement of its economic condition of a country, not merely by its insistence on moral and transcendental attitude, but with a realistic know-how of carrying forward the peaceful attitude all through, in tackling both internal situations and external relations, making its economic fronts, progressively efficient, with business prospects through the avenues of the market economy in international context.

Hence these moves of the two ministries of Indian government were demonstrations of India's progressive strength in maintaining India's independence, with a courage and confidence, both economically and politically (weekly *India News*, 25 February 1966). The expression, non-violence has to be understood, neither idealistically nor realistically in any restricted formal sense.

But this courage of India could come in with India's progress of knowledge and achievements in science and technology. So India is going ahead in nuclear directions, openly declaring that its interest is not in creating destructive weapons, but for peaceful socio-cultural improvements.

Here India, by an adherence to higher humanist prospects of the world, holds a prestigious position, sometimes hoping against hope. My consideration of the present Indo-China's improved relation is based on my study of present newspapers stretched to this year. It is not merely a pointing out to its transcendental attitude, but a stern realist attitude to match with, scientific, technological skills, and with a progressive awareness of the powers of nuclear knowledge.

India's progressive successes in confronting critical extraneous invasions, are attempts towards a progressive powerful India, and the creation of India, and its people, both from within and without, with progressive developments in concrete scientific directions. Who knows very soon India's renowned scientists – like Chandrasekhar and Sudarshan, with their growing interest in astrophysics, and interest in particle science – will proceed to 'Neutrino Mass Huntings', to declare with cogency and ingenuity that, it is difficult to measure the exact mass in neutrino particles in contrast to photons, by progressive experiential technologies – when its measuring rod is through zeros (cf. Are Neutrino Mass Hunters Pursuing a Chimera?, *Science*, 8 May 1992, American Association for the Advancement of Science, Particle Physics).

That India is tolerably a strong and an efficient country, is evidenced by quick action by sending troops by plane in the early hours of the morning, in helping President Gayoom of Maldives island, a place in the Indian Ocean. India, with its strong military rescued the island's President, from a switch over to a military rule.

Before the Gulf War, India was quick enough to bring back to India by planes, its distressed residents numbering over 100,000. Here India demonstrated its affluence and capabilities as of an advanced country like America.

It must be remembered that India is a country, which is operative following its *transcendental* ideal, which is always reinforced by the path of Gandhi's non-violence and non-cooperation with the powers which are operative with force and violence, in a further realistic historical process, in technological and scientific ways.

India, now maintains its power, with a strength, but not selfishly for itself, but with a due consideration of its neighbours.

It is a new enlightened democracy, not just stringed with mere conceptual or empirical process, in unkind abstract cords.

This kind of internal change of policy, with corresponding changes of foreign policy, with a challenge, stems out from the time of the Buddha, and was reinforced in India since the days of Emperor Asoka, philosophically and morally, and in the present times, with scientific-technological ways of military or economic aids according to the need of the situation, but the motive is only for self-defence.

Emperor Asoka, followed the pathways of the Buddha, of a non-intervention, befitting his time, with a pursuit towards a peaceful solution by insightful negotiations. It was further reinforced by the great Moghul Emperor Akbar, through a progressive exalted attitude by a comparative study of the religious processes then prevalent in India. Akbar understood the core of all religions of the world, from Zoroastrianism and Judeo-Christian and Islamic processes, in one aspect, and the Indian religions in development in India from the Vedas and Jainism and Buddhism and Vedanta to further developmental aspects.

In the post-Buddhist period, India's development was in another progressive aspect – in scientific and technological ways, but not with any evil purpose, but for progressive improvement of the world as a whole, with the speed of nuclear energetic processes.

Religious experience in development is a gesture of love of one great family, exercising with religious current of energetic spirit for its perfection, following the principle of non-violence and coexistence.

This was not an exhibition of weakness, but an expression of India's cogent confidence, experience and courage in facing a critical situation. Non-violent method is a primary stage before the violent modes of operations, which must be avoided as far as possible, from all walks of life from religions to business.

This could be seen in all further realistic ways after India's independence, not merely by an appeal to transcendental compassionate attitudes, but now matched with some advanced scientific-technological ways. India is following the ancient path of non-violence with an insightful use of scientific-technological programme of defence.

India's gesture is one of cooperation, without any selfish motive, with a foreign policy of nonalignment and with the allied powers and in consultation with UN.

This expanded horizon of operations has been demonstrated well by the paper 'Year of the United Nations,' Report of the Ministry of External Affairs for 1988-89, devoted to India, the United Nations and International Agencies.

I am here observing impartially, with a distant view of India, in helping the world-politics to come to a perfect order of human service, with progressive humanistic attitude, fighting against Colonianism, Apartheid, Recognition of Palestinian State, Disarmament and against vicious use of nuclear assets, and harnessing them to progressive strides towards international space collaborations.

In all these spheres of progressive technological and scientific researches – India, with its limited economic resources, and restricted access to foreign politics – is going forward, opening up visions towards non-violence, as the best policy for mankind to choose, in an enlightened democratic aspect.

This adventure, in some unique way, with some hope and courage is not possible by not knowing of scientific knowledge in all spheres of experiential fields from agriculture to oceanography, and from physics to astrophysics. But this prospect, as I have pointed out – is in various alternative directions, considering that the confronted universe, as presented – is never categorical, but is tinged with a process and possibility of may-be.

My knowledge of all these spheres of experiential fields is extremely limited, but I am trying to understand the subtle aspects of this great area of experiential process, and am finding a glimpse of insight in pursuing the contents of the above paper mentioned.

I am proud of the various operational centres of India, viz. Vikram Sarabhai Space Centre (VSSC), Shar Centre, in Sriharikota island on the east coast of Andhra Pradesh, Liquid Propulsion Systems Centre (LPSC), with its development wings at Bangalore and Trivandrum, the test facilities at Mahendragiri, and Space Application Centre (SAC), at Ahmedabad. The primary responsibilities of this Ahmedabad Centre are to conceptualize, plan and execute projects and research programme leading, to practical applications of space technology. This centre is also actively involved in the conception, definition, planning and socio-economic evaluation of space applications programmes. The developmental programmes are for broadcasting, with training programmes with the health communication workshop. The National Remote Sensing Agency at Hyderabad is an autonomous institution. It is associated with other centres, and is operative with the Indian Institute of Remote Sensing and Aerial Photo-interpretation Training Centre in the country. The objective everywhere is to develop scientific potentiality for defence increase of agriculture products, farming and electric powers et al.

These scientific and technological centres can be taken in contrast to India's earlier references to various Ashramas of the renowned sages, or Yogic and Tantric experts from the days of the Vedas to the present century, not only in India, but in many foreign countries as centres for India's aspirations of universal peace, that humanity should cultivate. But all these programmes for self-perfection presupposes Man's disposition towards a peaceful coexistence.

This continuous development of non-violent attitude, with an insightful understanding of the progressive scientific assets is no longer soaring with some restrictive ideologies, of a bygone democratic world-political situation. India, in its dynamic forward march, fosters a spirit of a practice of coexistence, of the different nationalities of the world, with a relative access into science and technology, primarily for self-defence, and for expressing or providing some help to neighbouring countries, in their crisis-situations. India is relatively a major power in contrast to many smaller countries of South Asia.

The basis for this mode of deliberation of a world-power of international frame of reference, could be found in United Nations, whose character and operational scheme and vision of a United World, was idealistically suggested by Nehru since 1961.

A continuous idea in development has been given by K.K. Joglekar in his paper, 'India and the United Nations,' in *Indian & Foreign Review*, 31 October 1986, by a kind of recapturing of the past by a well-coordinated sequential statements, where it has been shown, if this Organization was not the necessary incentive to exist, the world-situation would have been more chaotic.

So one should not judge this organization, from its positive contribution, but from its general authoritative position and potentiality as a preventive force, to nullify the operations of the emerging minor powers, after their independence from the foreign powers.

The world today is in the post-Gulf War period – to visualize its activity – when the United Nations in cooperation with its allies, could not only terminate the Gulf War, by the exhibition of a superior military power but also the world had a chance to observe the consequent processes of the dwindling forces of the World powers – which could not have been effective, without the help of some remnant unassumed power of UN.

Here however I have no desire to applaud the military superiority, of the United States or of the Allies – but I am watching only the continuous life-history of an organization through the odds of historical process, and its emergence as like the phoenix of the past – before and after the Gulf War. Contributions of all nations of the world should multiply with a speed, detachment and no hesitation.

After India's independence from the British rule, India, since the period of Nehru, with its political policy of nonalignment, could wait to see the vanishing of ideology that governed the superpowers. Now the superpowers have given up their obstinacy to uphold their respective ideologies, each, with progressive changes, of some sort – is ready to accommodate other processes to fit in within its modified frame, by proceeding towards a common ideal favouring 'market economy'. Here every world-power had to accept confronted situation as important and not negligible.

The major Powers of the world are now ready for curtailing arms race, to prevent falling into the clutches of an economic crash – resulting out of over-spending their reserved budget for maintaining their respective military status and strength. Too much expenditure to strengthen one's military amounts to economic draining and failure to meet the needs of the people of a country.

In contrast to that foreign policy of joining the group of powers, controlled by two superpowers – India's governing bodies chose its path of nonalignment and successfully retained its international status, with a dignity.

Nehru, shaped its process and clarified the core of it by saying that it is not a 'vacuous idealist policy,' but it is a policy with 'realistic consequences,' which the world of superpowers have now seen, though th[illegible] unable to visualize, in the earlier stages, in their passion for dis[illegible]ing their [illegible]nergy-continuum with emotional outbursts.

India's insightful operational scheme has some insightful significance in the long run. This has become obvious in this changing universe since then specially after the Gulf War and the vanishing of ideologies, with a dissolution of Russian Union.

India is proud of its Nobel laureate scientists like C.V. Raman and S. Chandrasekhar, as through their advanced researches in their respective areas have shown that physics crosses its boundaries between the *particles* and *waves*, and runs into transcendent processes of *fields*, when it goes beyond to a further transcendent sphere of the aesthetic currents of the universe as a whole. There is no gulf between physics, astrophysics and aesthetics – all these aspects of experience – are crossing their respective boundaries to enrich themselves by their mutual loving touches, losing their entitative demarcations.

The scientific discovery of the neutrinos as having no conceivable 'mass' is a ground for further sphere. It may affect as well Chandrasekhar's research in astrophysics, in the aspect of measuring the 'mass' quantity of a perishing star. Not being a scientist in those spheres, I do not have any practical suggestions to offer, but I can suggest there is a need of a balanced attitude in developing any aspect of experiential process.

In Chandrasekhar's volume of eight reflective and illiminative essays on the interrelationships between poetry, aesthetics, philosophy and science, in Truth and Beauty – there is a new vision of a transcendent philosophy , touching thc core of scientific processes, in a new democratic direction (*India News*, 1-15 December 1990). Here human prosperity has to be understood as providing a balanced attitude towards a development of all experiential processes in interrelation. A new scientific morality has emerged, that needs progressive development, with a detached attitude and education stretching to entire humanity.

Nehru, with a foresight, became operative, through several of international conferences from 1954, inviting some willing nations of the world. He gave a foundation for his policy of nonalignment, in terms of the five-fold tenets of conducts, to be accepted by the willing governments of the various parts of the world. The first conference was held on 29 April 1954. But this course of policy did not succeed finally, as became evident with the Chinese invasion of India in 1962. Now India is taking a new initiative, and India's renowned scientists are equipped for tackling those critical situations with appropriate defence mechanisms of science and technology. This is India's second move; with scientific force, to human attitude towards a new morality with a foresight and detachment.

In this international conference at Bandung, all the major powers of the world were informed with a request to follow the five-fold behavioural policy, as a scheme of restraint. Here a Buddhist terminology was utilized and hence this five-fold process was named, Pancha (five), -Sheel (conduct or policy).

These are (i) mutual respect for each other's territorial integrity and sovereignty, (ii) mutual non-aggression, (iii) mutual non-interference in each other's internal affairs, (iv) equality and mutual benefits, and (v) peaceful coexistence. These behavioural processes are requisities for perfecting political atmosphere in international contexts, and these processes are to be supplemented by progressive new scientific insights and possibilities. Man has to step into new science, with a detached spirit and attitude, no more fostering one's animal nature.

These were first mentioned in the preamble of the Sino-Indian treaty on Tibet which was signed on 29 April 1954.

These principles of political conduct were not acceptable to the practitioners of power politics. But these principles attracted the attention of some of the nations of the world. This was evidenced by the First Nonaligned Conference at Belgrade in September 1961. Twenty-five Afro-Asian and one Central European State, and three Latin American States came in as observers.

The world is a centre of psychic actions and reactions – with choices and refusals. Hence the historical processes take varied turns over which a human mind or a unit of human mind has no control. But one has to respect this progressive psychical attitude, where there is always a place for human effort, willingness, for a continuous self-improvement. When everything else have been learnt, why not some behavioural processes with insightful attitudes for the good of human beings as a whole

Summarizing the contents of Nehru's meaning of nonalignment, negatively – and as well positively – it might be said that it is not siding (aligning) with the great power-blocs of the world, and positively, a lining up of keeping away from the power blocs which are military alliances. They spoil the possibility of peace.

The nonaligned nations, by contrasts will be non-military, throwing weight in favour of peace. What can be more democratic than this new economic-political move, not transgressing the continuous possibility in the advanced directions of moral and religious experiences, not going against the real scientific and technological developments which are pro-human?

These pro-human processes are not to be spoilt through murder and terrorism eradicating innocent individuals, not letting them live only for a few number of years, that nature provides, when it is not benignly operative. (Cf. for some general ideas, of *Revolutionary Terrorism* – The FLN in Algeria 1954-1962, played its role for deliberating itself from the French rule. This book is authored by Martha Crenshaw Hutchinson Hoover Institution Press, Stanford University, Stanford, California, 1978.) But that was an alternative left for the Algerians to choose in a critical situation of their country. But there is also a general way also. My suggestion is for the general way – with a development of a new insightful morality of the progressively emergent scienfic world as at present.

An International Conference for developing nations was held at Cairo in

October 1964, with participating 47 nations, and 11 observers, from all over the world.

I am for a progressive world of possibilities with no adherence to any theory or experience, as final or ultimate. I am gradually becoming convinced of the defects in the conceptual construction of an ultimate in sphere of experiential process – for, man's vision in every field of experiential process is progressive, and hence one has to accept its possible process character, abandoning all categorical assertions, with scientific insight.

The Third Nonalignment Summit, held its sessions in Lusaka in September 1970. In this conference, 54 States participated as Members and 9 came as Observers.

Combining the idealism of Nehru, and the concrete realism of Lal Bahadur Shastri, the Third Prime Minister of India Mrs. Indira Gandhi, by a contrast, said, the following: "Today's world is a single entity. We are deeply convinced that by staying out of military pacts – the nonaligned countries can use their collective wisdom and influence to tip the balance of power in favour of peace and international cooperation." To this insightful idea, scientific-technological process and plan have to be inserted.

The world today is united in peril, not merely the peril from nuclear destruction, but the more insidious daily pollution of our environment. Have the matter is not sentimental but an objective dealing with a concrete situation, with a scientific-technological plan as becoming obvious in international post-Gulf war situation.

Now all great powers have come together, in gradual emergence, with a progressive consciousness of one another's motivational displays towards some concrete practical solutions of some problems, in an international conference at Madrid, after the Gulf war with major superpowers and the delegates from some Middle-East countries, in their sincere move holding a continuous programme of peace, by progressive and liberal policies.

These powers are for opening up avenues for trade, and for common markets for the good of the humanity as a whole, no longer experimenting with the increase of armaments, polluting the atmosphere, land and ocean, and spoiling the resourceful nature, for the satisfaction of the crazy impulses of some persons or nations by terrorist activities, killing innocent individuals – men, women, children, old and the diseased.

In this resultant arrival at a critical world-situation, through concrete experience or confrontation – of complex political situation – some reflective nations of the world are trying to entertain some kind of improved consciousness, not visualized before. This leads me to reflect on India's progressive development of a meditative psychology, with a transcendent attitude as of the Buddhists through the centuries, canceling all confronted entities even one's own personality as *ultimate*, or one's created idea in the pursuit of many experiential processes from the myths to the sciences – or philosophic or religious experience as ultimates.

In the sphere of scientific deliberations, neither the minute nuclear particles, nor the waves, are ultimate. In the philosophic sphere, too, no idea even the material or spiritual is ultimate, in the religious sphere too, no God is ultimate. All experiences have realities as confronted processes, but not as absolute entities. They are momentary grounds for crossing their boundaries, to realize the transcendent interrelational processes beyond them.

From all these insightful perspectives beyond all confronted experiential processes, reaching the sphere of the present day politics, one can see that the so-called superpowers, through the centuries had played their roles, but lost that process, after some time, may be after some centuries.

Now they have come to a stage to realize the progressive values of stepping into the stage of the insightful understanding of a meditative psychology, which is an understanding of transcending all confronted conceptual and empirical spheres – and a stepping into a prospective transcendent inter-experiential sphere where human unity and feeling for mankind as a whole has to be felt as primary. Hence the place for Gandhi's non-violence and non-interference, with a protest, – is a transcendental operational scheme in terms of non-cooperation, in ordinary human relationship, or even against mighty rulers, without any fear.

India's move no longer is towards any soft idealism or realism, but a confrontation of the world with realistic changes, awakening itself in multifarious protective ways from some dogmatic slumbers of the past. Philosophically and politically, a nation cannot stay in a vulnerable condition with a mere idealist attitude. This is the reason, why India has to pursue a process of progressive self-defence, through scientific and technological ways.

Hence the development of science and technology towards nuclear efficiency in peaceful aspects of progressive types is needed from India – whose attitudes are not dissected by momentary emotions. This has been proved through historical records since India's independence.

But as the human nature is so obstinately oriented still, within India, some form of violence is continuing, making normal life impossible, by the murder of innocent persons who are for helping the Indian people in general. Within India also there is violence, there is not much of effort to solve problem by peaceful exchange of ideas, from a compassionate process of understanding, appreciation and insight.

India has no excuse in choosing the violent methods, but should have the insight from its own history in the progressive direction of peace, and also watching the world-process, towards a great effort towards peace, as evident in the continuous sessions at Madrid, of the different powers of the Middle East countries, and the major powers of the world, after the success at the Gulf War.

The might has certainly important roles to play, but at the sametime both

individuals and the superpowers must know in the depths of their minds that it is always preferable to live in a world where peaceful atmosphere prevails, and where man should, with a compassion allow every individual to live the only short life that he or she has got the luck to live, in the form worth living.

But with such insightful understanding – India must remain aware of its need for improving its progressive insights in nuclear science and technology for self-defence from any foreign attacks and for improvement of inner social and economic life with provisions for progressive facilities and conveniences. My suggestion is always for subtle discoveries, without show, but remaining aware of the possibilities. For this process to develop, India should make a provision for educating people with an adequate educational ways from the stage of schools.

Without a world-wide peace, and tolerable affluence, there may be improvement of knowledge in every sphere of experiential process, but if there remains no beneficiaries to enjoy these prospects of life is futile, and civilization turns out to be meaningless.

A realistic sense of progressive ideas is being gradually understood between Russia and the other emerging States within Russian Union, differing from Russia's fast breaking of Communist ideas by the introduction of Gorbachev's *Perestroika*, and the process of breaking the walls between East and West Germany, with prospective ideas towards peaceful coexistence.

Keeping this idealist attitude of progressive change from restricted ideology, is to confront the realistic world with caution and skill; it is not trying to tackle with an expression of pure violence, or pure non-violence, but avoiding extremes of both these alternatives, with an insightful show of some power, in some critical situations; it is a foremost concern of mankind to continue.

I would like to proceed with some contributions of India's progress towards an improved process of constitutional changes, in progressive democratic ideas with progressive schemes and strengths of economic improvements, through the powers of its Finance Minister Manmohan Singh, and educating the masses, gradually towards an understanding, according to their capacities, with freedom and choice.

What India and other countries need is an insightful choice by occasional international conferences for choosing higher and higher forms of business capabilities, supplementing them by some vocational educations, with a feeling for one another's smooth social life-process.

The Constitution of any country, should be made with progressive democratic ideals, but also with some progressive realistic vision as of the prospective economic prospects, through select and insightful educational programmes and novel election schemes.

The prospective vision towards some kind of profitable business prospects, with a kindly tolerant attitude prepares the ground for humanistic mode of an international life, without the exhibition of military prowess.

In India and also in many of the South Asian countries, as Sri Lanka, (formerly Ceylon) specially, the phobia for a national language, for an emergent 'Independent' nation, – created the ground for another restricted ideology for those countries, and spoilt the prospect towards a realistic political future with a goal towards advanced business possibilities. So the development of any language has to be insightfully discerned.

The struggle between two linguistic masses in the little island of Sri Lanka was injudicious. But this was a wrong historical move. They could easily flow through both the languages, with a transcendent language-process for progress, viz. through a return to English, with some concession, with the best elements of 'Ceylonization.' Here my appeal will be to my students –Singhalese, Tamils and to my many Buddhist monk disciples, who are in Sri Lanka and in the United States.

I will like to stand for English as a good communicable language for commerce and scientific studies, and also for development of international cultural processes as a whole. The new avenue opened through such a progressive English would have indicated a greater future for mankind as a whole and for Sri Lanka. The efficient communication through English language, was perfectly realized by the Sri Lankans in general when I was in Sri Lanka (Ceylon).

English then had become their mother tongue, unlike the Indians who were also efficient in both written and spoken English, but never allowing English to enter into their homes.

It was a foolish gesture to retard the achieved historical progress of a country, through the prevalent language, whether Sinhalese or Tamil. Both could have been elevated with transcendent spirit of 'Ceylonizing English' indicating that English is a modifiable language, it has a different possibility, which can be effected through the Sri Lankan intelligentsia and masses in varied ways, as indicated in my reflections of English and its prospects through 'Indianization.' That would have shown real 'independence.' I am for English in its relevant sphere as an efficient operational process, but it should not form a place in one's life at home, one's inborn language must be used at home, for maintaining one's spontaneity and jest in life-process or even to maintain a contact with India's teeming masses.

Now also the same possibility is open for this little lovable island – with its brilliant future with its many-sided resources – discernible from its land, with surrounding ocean, and people inheriting all the best elements from the foreign powers that dominated its history from the past to the present. I am only dreaming of its huge business-prospects, in so advantageously situated below India, with its traditional maintenance of a non-violent policy.

Rather than fighting, there is another easy and a reasonable way. One can solve the language-problem, not killing totally a foreign language, which dominated for more than a century. That foreign language served the inhabitants, with the advanced communicable ideas of uniting the people with the transcendent spirit of independence. There are other languages there, they can be cultivated for improved business relationship, and for perfecting the research studies on various languages of the world. Ceylon is a country where historically were established many foreign languages, including many languages of South Asian or South-East Asian countries, by conquest or by trade with many countries.

It, with its internationally oriented process helped the progress towards efficient ideas of communication, with scientific and technological prospects. There is no harm to cultivate one another's mother tongue at home, where there are various individual members with different capacities, not necessarily learning everything, language or anything in special.

The independent South Asian countries, in emergence, can foster the development of their different centres or states, through their respective languages, at home, or within their groups, but becoming operative differently through a third intermediary language, as English, French, Dutch or Portuguese, for communicating their different cultural varieties by progressive improvements of their native languages, following some advanced techniques that brought about an evolution and development of English language, also, in the past. English as prevailed in India and South Asian countries, was not merely English, but fertilized with the linguistic elements of these countries in general.

Rather, a vast country like India, can change the fabric of English language, by the inspiring linguistic flares of the cultures of the different areas or states, giving a new life to English language.

During the British rule of India, English was progressively developed and fertilized by such geniuses of India, as Tagore, Swami Vivekananda, Sri Aurobindo, Gandhi, Nehru, Radhakrishnan, Tilak, and others, too many to mention, and of Sri Lanka, Senanayeke (father and son), Bandaranaike, Coomeraswami, Malalasekhera, Parnavitana, and others, too many to mention.

The linguistic fights in India and Sri Lanka, are meaningless – a sad history for these countries, crying for independence of the distant linguistic areas, splitting the 'Unity' of the country.

Like Sri Lanka, India has some pockets within which the French language was widely spoken when the French rulers dominated in some pockets as in Pondicherry or in Chandannagore in Bengal. In many subtle cultural situations, the expertise services of such people can be used.

Goa continued long to this century, with its Portuguese language, and Portuguese rule. Sri Lanka was also ruled by the different foreigners including the Dutch, so some foreign languages are country's wealth. More one learns the foreign languages the better it is for trade and travel, and for knowing the cultural wealths of the different parts of the world. Any language is infinitely fertilized by different languages which carry currents of perennial cultures through the centuries.

Why not mankind learn to move with the different parts of the world as of Asian countries – like China and Japan, and portions of European Russia and its satellites – moving for world-markets for progressive trade, science and technologies, and for computer sciences, with telescopic efficiencies of progressive visions, of the galactic universe, with a break-neck speed?

All can feel an 'at-home-ness' with 'English', knowing the need of English, as the best means of communication, with efficiency, and also with some more languages if possible, to learn. Man has to go with a changing consciousness, by patience and insight, and not jump with some arbitrary selection of a language to fiddle with or fight with.

Sometimes a diversion of attention to other important events is a need. There should be ventilating windows all over in all directions.

The essential spirit of this noble objective was evident in a speech by Senator Edward M. Kennedy, which appeared in the *Washington Monthly* of November 1971, under the caption: 'Tragedy in Bengal: The Role of American Foreign Policy'. The main trends developed in this paper, in someway touches the fundamental ideals of India's principle of non-violence, both within and without a country. It is a kind of taking away of one's mind, from one's own people, to somewhere to a foreign land. It fosters a new vision with a humanist attitude.

It was a critical situation for India, with the intrusion of the distressed refugees from East Pakistan (formerly East Bengal), caused by the Army of West Pakistan to suppress political opposition in East Pakistan.

These statements may appear repetitive in some aspects, but its reiteration has some fresh values. The reference is to the brutal procedure of the advancing army of West Pakistan, to former East Pakistan (now independent Bangladesh) massacring innocent men, women and children, and then blandly announcing that everything was under control, with appeals for the refugees to India to return. It was not a diplomacy but a foolishness or cruelty.

By a thorough investigation into this critical situation of East Pakistan and India's West Bengal, Ed. Kennedy refers to this massive human tragedy, and suggests that what should be the reasonable role of America in this situation. Here he also refers to the horrible tragedies created by flood and coastal cyclone of the summer of 1970, decimating 400,000 people, and now again in 1971, several thousands killed by a *man-made disaster* by the armies of West Pakistan. His was an indication of a double tragedy that has fallen on some unfortunate people – which a powerful nation like the United States can be of some help.

Inspite of this kindly gesture of Ed. Kennedy, neither the then President nor the Secretary of State of America made a single public statement, condemning the West Pakistan Government's policy of violence and repression. It was a serious threat to *peace* caused by the massive inflow of Bengalee refugees into India. An appeal is here for a common sense, to act intelligently, with an experience of the moment. India had to take a positive step, defending itself from a foreign attack, and giving expression to its 'new democracy' in an international context.

It was also receiving the helpless refugees from East Pakistan – a large number of humanity, exceeding the population of some European countries, accounting for the 'why' of India's growing population.

Sometimes some general policy, gives an opportunity to a self-seeking country to misuse the privileges obtained from a donor superpower, and then it neglects the rightful claims of a weaker country or a counterpart, by exerting a forced process with the money and equipment. West Pakistan behaved that way, but America paid no heed to that. Here is uneven foreign policy and some kind of fallacies in international politics, caused by the superpowers, in controlling the minor powers, with some gifts.

Kennedy was courageous enough to point out that West Pakistan was wrong in its brutal policy of supression of the people of East Pakistan (former East Bengal), and it was America alone that could stop this by cutting off military and economic aids.

There is no meaning in mindlessly continuing the arms flow, as that will go against the fundamental tenets of American foreign policy. He also pointed out that neglect on America's part had resulted in the 'emergence' of a demand for a new state of Bangladesh. America must now support that process, for its demand, through its leaders of Awami League, was very reasonable, that it 'does not need any help, guns or money, but only wants that America remains neutral, without helping West Pakistan.

The appeal evidently was to awaken Americans, to a consciousness of its traditional role of a leading 'humanitarian nation of the world-community.'

Kennedy's final appeal to his Government was to give a warning to the military regime of West Pakistan. He was for 'rescuing America's foreign policy from a course, that had been disastrous both to America's best traditions and interests in South Asia.' Hence often, for the improvement of political vision some inter-political action is a need, and hence the need of such pointers, with a repetition. Human mind is a 'rough ocean' – often needing the blow of a benign current. Kennedy's was a blow of that sort.

But more than this mere gesture towards a right attitude, was also a further demand. Indira Gandhi took a courageous initiative, not at first taking a direct action, but waiting to take this final decision, patiently, till India's political integrity was threatened, and India, then, had to jump into quick action, amounting to war.

I will next draw the readers' attention to Prime Minister Indira Gandhi's courageous submission of India's dangerous plight that West Pakistan had wrought to India, by its invasion of East Pakistan, by forcing the refugee problem.

Indira Gandhi expressed her protest to President Nixon, when she went to America on invitation. Her complaint was by drawing attention to a realistic representation of the condition of East Pakistan. She pointed out that it was not one of a 'civil war' in the usual sense, but it was a genocidal punishment of civilians for having voted democratically. Her's was a superior order of political gesture, in the sphere of international politics.

It was a strange and cynical way of getting rid of one's opponents and of deliberately using helpless millions as a weapon against a neighbouring nation. Her main intention was to draw attention to the refugee problem, which had overburdened the economic condition of India. India's action against Pakistan was a novel political diplomacy – a scheme of a non-violent awakening of a civilized mind.

Her's was a direct challenge, with a novel progressive political consciousness, and applying politics with a moral fiber – and indirectly pointing out of a wrong foreign policy of America, inflicted upon India.

Refugees entering into West Bengal – as India calculated – daily were more than the population of Austria and Belgium. India had to accommodate all, with great difficulty with problems to emerge.

Her's was not the display of any political game or diplomacy, but drawing attention to America, and Nixon specially, to an international obligation, to safeguard America's objective of the preservation of peace and order in international situation. Her appeal was entirely on a humanitarian basis, as so well indicated by Ed. Kennedy.

Here this writer again draws attention of the readers, to understand that the mere analysis of *abstract ideas* in any sphere of experiential process – including politics, – is a vacuous gesture, with no humanistic expression. Hence, as a personality Indira Gandhi became operative to change the political atmosphere of the time, by pointing out that politics also should provide a place for morality.

She articulated her feelings with a profound expression of the depths of her mind thus: "I have come here looking for a deeper understanding of the situation in our part of the world, in search of some wise impulse which, as history tells us, has sometimes worked to save humanity from despair" (*India News,* Washington, D.C., 12 November 1971). Hers was a demonstration of a transcendent psychic power, with an appeal to touch the psychic core which is the basis for human existence and for a peaceful coexistence.

After this statement she with a decency said "I look forward to our discussions. I have no doubt that they will lead to strengthening of friendship and understanding between our two nations, and to a lighting of our path as we work together for peace in Asia and the world." This was dramatic, and a gentlewomanly gesture. This is a subtle gesture of perfecting 'international political process.'

I gave here my full credit to Indira Gandhi by her noble expressions as constituting a wise impulse, curtailing the language deftly to meet the-then situation.

Political activities by such personalities turns politics to a human science of international importance.

I am trying to shape politics in this *dynamic frame of reference*, breaking its citadels when it deals merely with some abstract terms, understood only by some select specialists on the subject, soaring high in an altitude which no body can reach. Mankind should know that *politics*, like all sciences is oriented towards the benefit of man. There is no 'German Wall' to separate politics from other experiential processes.

Indira Gandhi's answer to many questions was not only about the refugee problems on account of West Pakistan's invasion, but it was also, a demand for an answer to many other questions, regarding her talks with Nixon. Her further talks were India's treaty with Russia, whether that would affect India's nonalignment status and India's relations with China. When Nixon could establish some contact with China, whether that would settle the problems between East and West Pakistan after invasion. And finally the problem of India will remain, if major powers as Russia and America are allowed to dominate the Indian Ocean, then, what should be India's foreign policy? Politics is never an abstract subject unrelated to other aspects of life, with a true human spirit.

These were the political scenes, which then prevailed. Indira Gandhi was a pointer to these, to determine her further line of judicious action in the sphere of politics.

All these questions and answers of politics, then, during her Address at National Press Club, in Washington, D.C., 5 November 1971, were demonstrations of her insightful understanding of changing politics. Her's was also a discovery of a new direction to pursue.

Indira Gandhi, all through, maintained India's foreign policy of nonalignment, at an elevated level, to generate an international consciousness in this direction, and she also maintained India's true democratic attitude, as a nationalist attitude in tact. It was indeed a continuation of Nehru's political attitude, with a personalist touch, that makes politics a human science.

Mrs. Gandhi was full of acknowledgement for American economic help in the developmental process of India, but she always maintained India's freedom of speech, inspite of the American help – India progressively chose to improve her condition as to be progressively self-dependent, and become less dependent on foreign help – not merely from America, but also from other developed powers of the world. Only rising to this process will help India rise to the progressive path towards real independence.

She cherished this domestic policy of a secular attitude as essentially democratic spirit, which Panikkar, the historian of India, mentioned earlier. She undoubtedly gave birth to a cogent *new politics* which India dynamically improved.

Then she drew attention to the solution of the problem between West and East Pakistan, pointing out that India, will depend on the expression of the democratic aspiration of the people of East Pakistan or Bangladesh, without any imposition, for help, in which its own leaders, had a leading part.

India had to come in clash with West Pakistan only when India was invaded, and India had to rescue herself from the foreign attack. Here India's *dynamic politics* was given a new spirit for further activity.

This helped the emergence of a new independent country, without any imposition of its will, inspite of her demonstration of India's power. Indira Gandhi preserved India's dignity, and helped the development of an honest unselfish development of a democratic spirit in politics. Her's was evidently an attempt to discover a *new politics*. She initiated an advanced *democratic outlook*.

Without a reliance on a transcendent insight, this type of *new politics*, can never be contemplated. Indira Gandhi reached that dignified status.

In the concluding chapter a general summary of this work will be focussed, to give the work, its widely spread coordination, and its many-sided aspirations – to pave a ground for world-wide international ideal – through a *rejuvenated politics*.

REFERENCES

1. *Dynamic Facets of Indian Thought*, in four Volumes, by Anil K. Sarkar, the author of the present book, in general.
2. *Changing Phases of Buddhist Thought*, third edition, by Anil K. Sarkar.
3. *Whitehead's Four Principles From West-East Perspectives*, by Anil K. Sarkar & also *Buddhism and Whitehead's Process Philosophy* by the same author.
4. Reference is here to Part I, Ch.III of this Volume.
5. In this book philosophic thinking has been developed all through in the direction of a post-conceptual aspect to directly observed confronted empirical situation as it is, with all its nakedness, so that immediately after Indian independence – the constitutionally elected personalities from the – President and Vice-President to the Prime Minister and other officials responsible for the Indian Constitution – thought of improving India from a scientific and technological perspective. It first tackled to develop India's 'agricultural' condition to make India completely self-sufficient, in food, or adequate milk-supply and ox-breeding et al., and next India headed towards its

progressive processes of 'self-defence' to tackle the international situation.

6. India's political idea were directed towards India's internal welfare through improvement of agriculture, and modernization-process through scientific and technological methods, modernizing also other processes, making the country progressively self-sufficient in its process of 'self-defence', with a policy of nonalignment in foreign affairs.

7-10. Hence after Indian independence – the Indian politicians always pursued a dual attitude to improve the social life first, and to meet these needs – a provision for relevant education and employment opportunities were made, then the attention was given to the political sphere with employment opportunities for administration of the country from within and then looking into the foreign affairs.

On account of this balanced attitude India was not faced with the problems, creating a gulf between the 'Intelligentsia' and the 'Masses'.

In the context of India – between the two orders of intelligentsia the 'senior' and the emerging 'juniors' – there was no conflict, as in many of the American countries, which after 'independence' from the French, British or Dutch or other types of the European masters did fall into the traps of the generation gaps. But in India this did not occur, nor in India, there was any open conflict between males and females, as the Indian Constitution from the beginning gave the same rights to the males and females, as this process comes as having a natural human right. Presently India is training the females to take the responsibility of the aviation process of a pilot, which was thought as man's job only. Needless to mention we had India's Prime Minister as Indira Gandhi, and Ambassador as Mrs. Vijayalaksmi Pandit, and Governor as Mrs. Sarojini Naidu.

India as a secular State had a Muslim as President in Zakir Hussain, and Ambassador of the United States, like Abid Hussain. Needless to mention, India takes into consideration the merits of all deserving persons in every sphere for the progressive improvements of Indian Constitution to build up a fully democratic process – with transcendental objectives – confronting the empirical world conceptually and empirically oriented but not as 'absolutes.'

From India's cultural perspective – there was no gap between the Intelligentsia and the masses – as India always believed that from the masses, there is always possibility for finding an extraordinary person, in various aspects of experiential processes from philosophy and religion to the sciences and aesthetic orders of experiential processes. Man emerges with special qualities, though, there is a presence of the ordinary educative processes as provided in the schooling system of some form.

This kind of sudden possibility was always granted in India's cultural frame of reference from the Vedas to the present varied socio-cultural movements.

There remained some class-discrimination, which was not present when the first principles were formulated but it emerged with the different behavioural processes of the masses in general. The Indian Constitution, with an awareness, in its framing the Constitution – ruled out the discriminative attitude. But something contrary to this 'emerged' as prevalent in the different parts of the world in every situation, over which no one has an absolute control. They are all emergences in confronted naturalist process, but they need to be transcended.

The speciality of India's respect for the mass remained, throughout the centuries to this day, for the mass emerges in many novel process, which do not emerge always by apparent system of educational processes. There is always a place for uncertain emergent possibilities.

This kind of emergence comes from the masses or from the people in general, who develop themselves from their inner insights.

Excluding many such emergent personalities of different periods of Indian thought one can refer to the emergence of Ramakrishna as a 'religiously enlightened personality' more than the speciality trained intellectuals after the spread of the British form of education in India. From the study of the life-history of Ramakrishana, one knows how he could master the difficult processes of the highest orders of Tantric principles from Vairabhi Brahmani an itinerant female adept in that area of experiential process, and how also Ramakrishna could grasp within a moment the transcendent religious experience from another itinerant Vedantist Totapuri, which freed Ramakrishna immediately from his attachment for 'theism', his life-long love for the Goddess Kali, but through his genius he could make the religious process as a continuous movement from the lower to the higher. This realization in him, prepared the ground for crossing the boundaries among the different religions of the world, realizing the core of other great Western religions as of Christianity or Islam. He developed a process of non-difference among the religious experiences and also between the religious and other experiences.

Rabindranath Tagore consciously flouted the process of receiving any academic education from the schools or colleges, but pursued his own method of receiving and mastering the knowing-process in various spheres of learning, with a speed and accuracy, which prepared the ground for international recognition, through a nobel prize.

Similarly, M.N. Roy showed himself an extraordinary personality without any specific pursuit of education, but he emerged as a revolutionary in the political sphere, trying with the Communist ideas but finding its limitations, by making a sort of ground for determining its future possibilities: from the progressive aspects of deviation, from its basis in the Marx-Engel's frame of reference, with a departure, in Russia, and in other countries, from the Russian basis, with a further march to other countries. This different form of Communism in emergence from the different 'cultural background' of the different countries is very significant, for, that cancels the ultimateness of this process. From a different perspective Gorbachev could see its future in a different sphere, other than the political, in an international business-process of a market economy. M.N. Roy, foresaw the same process, by a shift from the elites to the masses, in the economic and business aspects of the poor peasants of the Indian villages.

From his concrete experience of the different forms of Communism Roy finally returned to India, his mother land, to give it a further touch and shape suiting the Indian cultural process, which always takes care of the teeming millions, not in the affluent order of human process. Roy's was initially an experience with Russia's great personalities – Lenin, Stalin and Trotsky and others – and by an escape from Russia, he could go to many countries of Europe, and to America and Mexico, and then finally he returned to India, with different experiences, to be finally operative with Jayaprakash Narayan and Gandhi, who also later inclined to this direction.

11. Reference is here from a journalist perspective of Durga Das, the author of *India: From Curzon to Nehru and After.* Here Durga Das refers to the gradual transference of Parliamentary form of Government from the British to the Indians with the initial stage of 'dilatory process' to an active and willing process of a real transfer when the circumstances changed, and when it was felt that for mutual good any further delay will lead to future chaos, and loss of good feeling between the two countries.

CHAPTER VIII

An Over-all Conclusion

This fairly long work with a title: *Triadic Avenues of India's Cultural Prospects – philosophy, physics and politics* – needs some general ideas for the readers to correlate the ideas of its different parts with different Chapters.

The book begins with its triadic bases, and runs into its three wings of philosophy, physics and politics, with different objectives to shape the entire work, not as a mere subjective process of an author but with a demonstration of an experience, that these two aspects of reflections – subjective and objective – are of distinct orders, and they are clearly distinguishable, not merely as relative currents but also as confronted processes – in the background of a transcendent process – which is continuously operative as a self-shining process as enlightening the pathways for the confronted subjective and objective processes.

They also provide for both in change and total perishing, making an open space for, with an emergent possibility, transcending both. This work, if viewed insightfully – is a platform for providing an ever-fresh emergent process to arise and then vanish, opening up the open space for further visions.

This *emergent possibility* is no longer just the former processes but renewed with a complete change, to begin a fresh cycle of life, as the Phoenix of old – a mythical bird living for considerable lengthy life, and later totally burning itself to a confronted fire, to gain a fresh lease of life.

Hence, in the Preface, as forming the part of the bases of this work, it has been shown, that Indian thought, as only subjectively oriented is wrong. Rather it is more than the *subjective*, as also, more than the *objective* when confronted in correlation – with a *transcendent possibility*, through *neither-nor* dialectic scheme, so that always there is an open field for a novel possible emergence.

Not to choose either the subjective or the objective, or in their combination, as interlinked – but they are treated as only as confronted processes and possibilities, with no fixed past heritage – to be abandoned constantly. Every fresh emergence of the confronted present, is a possible renewed instance of a confronted present, like the life and continuity of the mythical bird – Phoenix, as referred to.

In the 'Introduction' of the book it has been shown that India in its adventure towards the objective sphere wants to examine each of the personalities, as deeply conscious, no longer indulging in mere abstract thinking.

Each person's activity is a 'self-creative activity', where one transforms oneself with a progressive insight, and understands that the concrete prospect of Indian people, since independence, as a whole lies in the scientific-technological direction. As 'a confronted scientific and technological process in India' – it has a history of its own, with a speciality, based on a 'transcendental insight' with a possibility. Hence further advancement of the present century– is not a treading into an unknown path, it is a continuity and extension of the process, with modern touches.

So, both the *intelligentsia* and the *people at large* are operative in unison for the development of their economic prospects, with a speed. With this common objective, India is stepping into the sphere of a new democratic politics. This new politics is secular in spirit, and is progressively transforming from an intercultural perspective, where the prospect of India is not a selfishly oriented process – but a process with an interest for all human beings, of different cultures, with linguistic and other variations.

But knowing that human nature is operative in varied instinctive ways or in the ways of the enlightened intelligentsia, the Indian mind is always operative with an awareness in restricted or dogmatic ways of some such intelligentsia – so it makes necessary provisions for safeguarding their respective interests, not only by legal or other necessary ways, but also by progressive ways of educational schemes, for both the orders of people – intelligentsia and masses.

A country has as well to make provisions for self-defence, from any foreign intervention or invasion. In the modern world of science and technology, the intelligentsia must try to reach the intellectual level of the advanced countries, and prepare the capable masses with technological skills and rise to the subtleties of nuclear heights with the international consciousness for peace. The international attitude is a volatile process full of unpredictable possibilities. It is a domain where all types of intellectual and insightful persons dominate, changing the prospects of the world as a whole. So a really independent-country has to equip itself scientifically and technologically to that defensive order.

India is progressively proceeding in this advanced aspect with a reasonable speed, and with a safeguard of its interests. All these have been indicated in this book in sufficient details in different chapters and in relevant ways, and with a placidity.

These assumptions have no doubt some idealistic features, but India, in its political sphere, tries to rise beyond these to an elevated consciousness, with a realistic scrutiny and security.

India intends to provide these advanced triadic attitudes through its three wings ventilating the globe with a transformed atmosphere – to eliminate the contrary currents from this world. It is reluctant to repeat the past history with its drawbacks, in every sphere of life. India tries to get beyond its past historical errors.

So, it has a project of a re-study of the cultural processes of all countries, from an intercultural perspective, with progress, in the currents of process-philosophy of the Buddhist disciplined pathways throughout the centuries to the present century.

Some Indian and non-Indian scholars, including this writer, have also found out a kindred spirit expressed by A.N. Whitehead, both philosophically and scientifically. His understanding of mathematics, is a recognition of India's discovery of *zero*, outmoding the clumsy ways of mathematical counting and its representation of Greco-Roman civilization.

India has given a speed to the world through its mathematical symbol of *zero* (0) by releasing the tension of the mind, and replacing mere abstract thinking by lively confronted perception or seeing, with a clarity and insight, and with a possibility, eliminating the dogmatically and categorically oriented processes as fossils of the decayed past, so that mankind can progress with the speed of a horse, in a universe, that has been freed from abstract ideas, in infinite directions, with alternative possibilities. In the present world of science and technology the confronted cosmology has to be viewed with a galactic speed where science is operative with its telescopes of progressively wider visions with X-rays and gamma rays et al.

These new pathways as scientific-technological ways, are operative through satellites and Voyagers, through the help of advanced types of telescopes and through other gadgets, and stretching the human vision to infinity, in all experiential spheres – as symbolically indicated by zero (0) – confronting the positive integers (numbers from 1 to 9) – in its both ends – as to elevate the *confronted finites* to *infinite spheres*. (Vide the current weekly issue of Science, American Association for the Advancement of Science, Washington, D.C.) India's objective is to raise its country to this symbolic-mathematic heights.

Whitehead's also is an understanding of the confronted physical universe, as a continuous relatedness, not in a linear sense only, but also, in its multi-dimensional orders as a process and possibility – from a non-sensuous or non-intellectual order (indefinite 'relatedness') – to a transcendent possibility progressively. These spheres are to an emergent (definite) order of eternal objects and in their relations in propositional processes. It is also a further transcendence – to an order of conscious definiteness or intellectuality, and also beyond intellectual restrictedness with a potentiality for further emergent orders. These continuous transcendences are beyond any restrictive and restless sphere to the sphere of the infinite.

This transcendent process is towards a comprehensive divine order of epochal occasions, transcending all preceding orders, to abandon finally all formal orders to pass into creative or formless possibilities.

Here one can clearly see that the positivist logic and psychology that occupied A.J. Ayer's mind is considered as arbitrary selection for an exclusive analysis of a special type of sentence which has been arbitrarily chosen to fit in with a project to eliminate metaphysical statements. This book totally denies this halting and restrictive attitude—as a ghostly emergence in the domain of philosophic thinking.

The advanced modern science of physics and astrophysics is observing the galactic Universe – outmoding the so-called special and general theories of relativity, and any fixed projects of measuring the distance between the earth and the stars, or between the different galaxies, are some advances no doubt, but who knows anything finally about the future of the galaxies, if there is such an emergence as of a series of possible supernovic effulgences? The world-process is towards an unrestricted possibility.

In this present day advanced notions of no constancy view of cosmology the traditional mode of reflective attitude has no place. We are soaring in a Universe of 'if-then' process and possibility, in every sphere of experiential process – symbolically with infinite speed.

In this general development of reflective activity, as Whitehead, discerns, there is always a movement beyond Ayer and others of his line, and also beyond any scientist with a proneness for any specific theory.

So, rightly Whitehead, also, goes beyond the scientific restrictive relativist modes of reflection of Einstein, from the special to general theory of relativity, with a back and forth possibility. But the real rescue from this relativist and restrictive universe lies in a transcendent sphere – beyond either of these restrictive spheres.

Here the two currents – Buddhist and Whiteheadian – meet in a transcendent sphere with a reciprocal understanding of their basic differences, and final meetings in prospective spheres of infinite possibilities, denouncing the confronted world as something categorical – positive or restricted.

It is a passage from a transcendent formless philosophy to a similar and not similar scientific physics of formlessness – prescribing a transcendent political consciousness beyond all restrictive ideologies towards a progressive secular democracy in an international order.

Here one can take advantage of Bernhard Mollenhauer's interpretation of Whitehead as a pursuant of an *Eternal Quest for a Divine Flame* in his paper: 'Whitehead and the Journey of the Flame,' mentioned earlier in Introduction to this book.

The different Parts of this book are further concrete details, updating India's cultural processes in intercultural contexts. They point out that all confronted orders of experiential-processes from philosophy and religion to science and technology are possible processes – emerging and perishing – with no ultimateness but are just there.

They are dependent on a transcendent ever-shining background and foreground of any presented experiential-process, which is the middle.

The transcendent order as meditatively experienced by a series of disciplined persons – from the Vedas to the present is beyond the reach of sensuous and intellectual processes. Hence it is *not Seen*, but is the *ever-shining subject*, making the *Seen* possible.

The realistic march in advanced scientific-technological ways has now come to a stage of decision of priorities whether an affluent country will spend for the satellite stations in some 'outer space' – to determine certain scientific principles, as the feeling of weightlessness, as a human experience, and spending enormous sums of money for obtaining such a knowledge – or whether man should remain interested in solving other problems of the terrestrial world – in the improvement of human relations within one's country, or among humanity as a whole.

Both these alternatives have values to be attended to with insight and judiciousness, in relativity, with prospective possibilities, never as ultimates. There is no ultimate in any order of confronted experiential field. The compulsive attitude has always to be mellowed by insightful alternative possibilities, to eliminate conflicts and tensions in human socio-political processes.

To my mind, without going into further complications and inquisitiveness – I suggest, if there is a feeling for a transcendent experience to realize, I would like to repeat the initial reflective activity of Vedic people. They did not launch any theory of the specific beginning of the universe, at a certain time, or by way of a causal principle, or by a Creator.

They stood with profound silence visualizing that there is no 'Power' as such, in any entitative ultimate sense, but only a "Perennial Process and Possibility" in a symbolical way, of a *beginningless universe*, in its dual aspects – of evolution and involution – interpreting the evolved or the apparent confronted universe as 'a manifest order,' as transcendentally based expressing symbolically some kind of involution-with a pragmatic suggestion.

Hence the evolved or involved processes of the Universe or Cosmology viewed from the advanced aspects of both philosophy and science are to be understood from a transcendent basis or an ever-furnishing ground for open-space for deliberation, for eliminating the restrictiveness of the human mind, and to prepare itself with an enriched consciousness to confront something new.

The transcendent possibility – is always an unread page, somewhat akin to a non-dual involved transcendent continuity – process and possibility – not a 'repetition' but a rise and fall of the processes of progressive enlightenment.

The supposition is of an integral reality – not grasped intellectually in a linear intellectual way – but in a post-intellectual cyclic and meditative possibility.

Here the ground for the confronted visions never remains a static phenomenon, but continuously perishing and emerging processes – not in one space but in multi-dimensional spaces. Hence the project as suggested, is not considered from a restricted conceptuality but in infinitely oriented many-dimensional possibilities.

The entertainment of creation or causal theory may look scientific and satisfying, but now with advanced scientific advancement and meditative deliberations, of the advanced scientists – Indian and non-Indian – enlightened personalities of the world as a whole – excluding none – intelligentsia or the masses – that all are in some senses, in the same tract with new advanced insights.

Just to repeat: Einstein passes into a prospective 'field-theory' – beyond his two theories of relativity – and Heisenberg similarly passes into a prospective transcendent processes and possibility beyond his shelters of *particles* and the *waves*.

Hubble and others, as trying to measure the distance between the galaxies are pointing to their 'failures,' rather than hitting at any 'absolute certainties.'

But still the philosophic and scientific adventures and discoveries were not in vain – they can be fostered for ever, with prospective insights – to renew the visions to confront the infinite stars with wonder, as the exalted situations with supernovic possibilities – discarding restricted intellectual visions, even of the black holes, without remaining confined to the confronted presented universe with no *ultimateness*.

The function of this present book stands for a triadic process and possibility – from philosophy to the sciences and both to politics – to furnish ever-fresh visions to pursue and remain wondering of its precious achievements without canceling the transcendent self-shining background and foreground for experiential possibilities. The Light is ahead of the 'Light' of the light-house located in 'the middle' as a 'searchlight' indicating a forward march, from a fixed location. There is an ever-present 'Tachyon' beyond all lights, with a new speed.

The consciousness of *zero* – as symbolically standing for infinity both, at the beginning of the positive numbers (from 1 to 9), and at the end extends the horizon of human vision beyond any theories. The confronted theories, in the middle, throws light at its extreme background and foreground of its process. The 'Middle' does not exist as such, but it looks before and after – as a search light, admitting human limitations, yet an 'altar' for some sort of help or guidance.

My suggestion will be in this advanced stage of human history to cultivate the further avenues of progressive knowledge in any field of experience from philosophy and religion to science and technology as grounds for psychic visions that a politician needs for making the Universe truly humanistic in a secular sense, getting rid of any *ism* or *ideology*.

This *referential possibility* can be effectively continued with a further clarity and insight if freed from controversial theories or ideologies.

What is wanted of all persons, craving for knowledge, is not a controversy, but a pursuit with a meditative placidity and freedom, even in the sphere of advanced sciences.

Like philosophy facing '0' (*zero*) or the unfathomable infinity – on both sides of the positive numbers from 1 to 9 – the advanced cosmologists and particle physicists are facing *Neutrino Mass* as a 'treacherous testing ground' for physics and cosmology: 'If the 'Neutrino' has any 'mass' at all, then there is physics, but if there is 'massless Neutrino' then physics becomes 'non-existent or ethereal.' In that case it can pass through the entire earth 'like photons through a sheet of glass.' This is 'landing in no man's land,' in the words of a University Professor of Chicago – a cosmologist, like Michael Turner.

It is physics at the limit, agrees Haxton: 'One has to have some sympathy for the poor experimentalists,' he repeats. 'These guys are all out there, trying to push the realms that have never been explored before.'

It is a research in no man's land. Add to this is theoretical predisposition to find a 'non-zero mass.' This pursuit has become a reality through researches in particle physics, since 1973.

Researches thereafter by 1980, have come to a tentative conclusion, through researches along the path of 'electrons,' where the argument is that 'if neutrino has a mass,' then, it is a tiny residue of energy corresponding to that mass. But that remnant process is of the 'neutrino rest mass.'

Physicist Fred Reines of the University of California at Irvin announced that neutrinos can change 'identity' as they travel, oscillating from electron neutrinos to one of two other neutrino species – 'Muon' and 'Tau Neutrinos.' Such oscillations were de facto evidence for a 'non-zero mass.' with an 'in-definiteness.'

Closing up this long-drawn research activity I would like to conclude that 'It is very difficult to measure, zero.' The world as a whole is mysterious, if not paradoxical.

But from the perspective of Indian philosophy from its beginning to end—if the confronted presentation is towards 'Zero' – it is not towards an entity or something positive. It is a project towards the unfathomable – an indefinite that haunts all confronted something (cf. 'Are Neutrino Mass Hunters Pursuing a Chimera?' *Science*, 8 May 1992, pp.731-733, American Association for the Advancement of Science).

If from philosophy to sciences, the adventure is towards nothingness (zero) – why not politics as well pursue the same path, and sympathetically, confront human beings, so long as they are visible, for a peaceful coexistence, without chasing them out as entities, leading the world-process to conflict, chaos or war?

Let placidity and non-violence prevail for mankind to persist for a while, without trying to extinguish them abruptly, with an uncontrollable passion, with a show of radical merciless Intellectuality.

Man's abstract intellectuality creates innumerable phantasmagoric visions – for its inborn restlessness – and it enjoys killing them all, but fails to learn the transcendent insightful vision – that finally if there is no static phantomor mass to kill, then why there is a rush for *'killing it*? We are all *homeless*, confronting no 'ultimates'.

If mankind cultivates a *reasonable patience* – there is a great prospect ahead – in a compassionate direction. Only *Path* is but no static traveller or shelter, nor a matter nor man. A traveller is from zero to zero – the apparition of mass in physics has been found to end up in masslesss neutrino. The sphere of astrophysics waits for losing itself continuously in unpredictable supernovic efflorescences.

Index